Shaping Future 6G Networks

Shaping Future 6G Networks

Needs, Impacts, and Technologies

Edited by

Emmanuel Bertin
Orange Innovation/Institut Mines-Telecom

Noel Crespi
IMT, Telecom SudParis, Institut Polytechnique de Paris

Thomas Magedanz
Technische Universität Berlin/Fraunhofer FOKUS

This edition first published 2022

Registered Offices
John Wiley & Sons, Inc., 111 River Street, Hoboken, NJ 07030, USA
John Wiley & Sons Ltd., The Atrium, Southern Gate, Chichester, West Sussex, PO19 8SQ, UK

Editorial Office
The Atrium, Southern Gate, Chichester, West Sussex, PO19 8SQ, UK

For details of our global editorial offices, customer services, and more information about Wiley products visit us at www.wiley.com.

Wiley also publishes its books in a variety of electronic formats and by print-on-demand. Some content that appears in standard print versions of this book may not be available in other formats.

Library of Congress Cataloging-in-Publication Data Applied for

ISBN: 9781119765516

Cover Design: Wiley
Cover Image: © sutadimages/Shutterstock

Set in 9.5/12.5pt STIXTwoText by Straive, Pondicherry, India

C9781119765516_281021

Contents

Editor Biographies *xiii*
List of Contributors *xv*
Foreword Henning Schulzrinne *xix*
Foreword Peter Stuckmann *xxi*
Foreword Akihiro Nakao *xxiii*
Acronyms *xxv*

1 **Toward 6G – Collecting the Research Visions** *1*
Emmanuel Bertin, Thomas Magedanz, and Noel Crespi
1.1 Time to Start Shaping 6G *1*
1.2 Early Directions for Shaping 6G *2*
1.2.1 Future Services *2*
1.2.2 Moving from 5G to 6G *2*
1.2.3 Renewed Value Chain and Collaborations *3*
1.3 Book Outline and Main Topics *4*
1.3.1 Use Cases and Requirements for 6G *4*
1.3.2 Standardization Processes for 6G *4*
1.3.3 Energy Consumption and Social Acceptance *4*
1.3.4 New Technologies for Radio Access *5*
1.3.5 New Technologies for Network Infrastructure *5*
1.3.6 New Perspectives for Network Architectures *6*
1.3.7 New Technologies for Network Management and Operation *7*
1.3.8 Post-Shannon Perspectives *8*

2 **6G Drivers for B2B Market: E2E Services and Use Cases** *9*
Marco Giordani, Michele Polese, Andres Laya, Emmanuel Bertin, and Michele Zorzi
2.1 Introduction *9*
2.2 Relevance of the B2B market for 6G *10*
2.3 Use Cases for the B2B Market *11*
2.3.1 Industry and Manufacturing *11*
2.3.2 Teleportation *13*
2.3.3 Digital Twin *15*

2.3.4 Smart Transportation *15*
2.3.5 Public Safety *16*
2.3.6 Health and Well-being *17*
2.3.7 Smart-X IoT *19*
2.3.8 Financial World *20*
2.4 Conclusions *22*

3 6G: The Path Toward Standardization *23*
Guy Redmill and Emmanuel Bertin
3.1 Introduction *23*
3.2 Standardization: A Long-Term View *24*
3.3 IMTs Have Driven Multiple Approaches to Previous Mobile Generations *25*
3.4 Stakeholder Ecosystem Fragmentation and Explosion *26*
3.5 Shifting Sands: Will Politics Influence Future Standardization Activities? *28*
3.6 Standards, the Supply Chain, and the Emergence of Open Models *30*
3.7 New Operating Models *32*
3.8 Research – What Is the Industry Saying? *33*
3.9 Can We Define and Deliver a New Generation of Standards by 2030? *34*
3.10 Conclusion *34*

4 Greening 6G: New Horizons *39*
Zhisheng Niu, Sheng Zhou, and Noel Crespi
4.1 Introduction *39*
4.2 Energy Spreadsheet of 6G Network and Its Energy Model *40*
4.2.1 Radio Access Network Energy Consumption Model *40*
4.2.2 Edge Computing and Learning: Energy Consumption Models and Their Impacts *41*
4.2.2.1 Energy Consumption Models in Edge Computing *41*
4.2.2.2 Energy Consumption Models in Edge Learning *41*
4.3 Greening 6G Radio Access Networks *42*
4.3.1 Energy-Efficient Network Planning *42*
4.3.1.1 BS Deployment Densification with Directional Transmissions *42*
4.3.1.2 Network with Reconfigurable Intelligent Surfaces (RISs) *43*
4.3.2 Energy-Efficient Radio Resource Management *44*
4.3.2.1 Model-free *44*
4.3.2.2 Less Computation Complexity *44*
4.3.3 Energy-Efficient Service Provisioning with NFV and SFC *46*
4.3.3.1 VNF Consolidation *47*
4.3.3.2 Exploiting Renewable Energy *47*
4.4 Greening Artificial Intelligence (AI) in 6G Network *47*
4.4.1 Energy-Efficient Edge Training *48*
4.4.2 Distributed Edge Co-inference and the Energy Trade-off *49*
4.5 Conclusions *50*

5 "Your 6G or Your Life": How Can Another G Be Sustainable? 55
Isabelle Dabadie, Marc Vautier, and Emmanuel Bertin
5.1 Introduction 55
5.2 A World in Crisis 56
5.2.1 Ecological Crisis 56
5.2.2 Energy Crises 57
5.2.3 Technological Innovation and Rebound Effect: A Dead End? 57
5.3 A Dilemma for Service Operators 59
5.3.1 Incentives to Reduce Consumption: Shooting Ourselves in the Foot? 59
5.3.2 Incentives to Reduce Overconsumption: Practical Solutions 60
5.3.3 Opportunities. . . and Risks 61
5.4 A Necessary Paradigm Shift 62
5.4.1 The Status Quo Is Risky, Too 62
5.4.2 Creating Value with 6G in the New Paradigm 63
5.4.3 Empowering Consumers to Achieve the "2T CO_2/Year/Person" Objective 64
5.5 Summary and Prospects 64
5.5.1 Two Drivers, Three Levels of Action 64
5.5.2 Which Regulation for Future Use of Technologies? 65
5.5.3 Hopes and Prospects for a Sustainable 6G 65

6 Catching the 6G Wave by Using Metamaterials: A Reconfigurable Intelligent Surface Paradigm 69
Marco Di Renzo and Alexis I. Aravanis
6.1 Smart Radio Environments Empowered by Reconfigurable Intelligent Surfaces 69
6.1.1 Reconfigurable Intelligent Surfaces 70
6.2 Types of RISs, Advantages, and Limitations 72
6.2.1 Advantages and Limitations 74
6.3 Experimental Activities 78
6.3.1 Large Arrays of Inexpensive Antennas 78
6.3.1.1 RFocus 78
6.3.1.2 The ScatterMIMO Prototype 79
6.3.2 Metasurface Approaches 80
6.4 RIS Research Areas and Challenges in the 6G Ecosystem 82

7 Potential of THz Broadband Systems for Joint Communication, Radar, and Sensing Applications in 6G 89
Robert Müller and Markus Landmann

8 Non-Terrestrial Networks in 6G 101
Thomas Heyn, Alexander Hofmann, Sahana Raghunandan, and Leszek Raschkowski
8.1 Introduction 101
8.2 Non-Terrestrial Networks in 5G 101
8.3 Innovations in Telecom Satellites 103

8.4 Extended Non-Terrestrial Networks in 6G *105*
8.4.1 Motivation *105*
8.4.2 Heterogeneous and Dynamic Networks in 6G *107*
8.5 Research Challenges Toward 6G-NTN *107*
8.5.1 Heterogeneous Non-Terrestrial 6G Networks *109*
8.5.2 Required RAN Architecture in 6G to Support NTN *109*
8.5.3 Coexistence and Spectrum Sharing *110*
8.5.3.1 Regulatory Aspects *111*
8.5.3.2 Techniques for Coexistence *111*
8.5.4 Energy-Efficient Waveforms *112*
8.5.5 Scalable RF Carrier Bandwidth *113*
8.6 Conclusion *114*

9 Rethinking the IP Framework *117*
David Zhe Luo and Noel Crespi
9.1 Introduction *117*
9.2 Emerging Applications and Network Requirements *118*
9.3 State of the Art *120*
9.4 Next-Generation Internet Protocol Framework: Features and Capabilities *122*
9.4.1 High-Precision and Deterministic Services *122*
9.4.2 Semantic and Flexible Addressing *124*
9.4.3 ManyNets Support *125*
9.4.4 Intrinsic Security and Privacy *126*
9.4.5 High Throughput *126*
9.4.6 User-Defined Network Operations *127*
9.5 Flexible Addressing System Example *127*
9.6 Conclusion *129*

10 Computing in the Network: The Core-Edge Continuum in 6G Network *133*
Marie-José Montpetit and Noel Crespi
10.1 Introduction *133*
10.2 A Few Stops on the Road to Programmable Networks *134*
10.2.1 Active Networks *134*
10.2.2 Information-centric Networking *135*
10.2.3 Compute-first Networking *135*
10.2.4 Software-defined Networking *136*
10.3 Beyond Softwarization and Clouderization: The Computerization of Networks *137*
10.3.1 A New End-to-End Paradigm *137*
10.3.2 Computing in the Network Basic Concepts *138*
10.3.3 Related Impacts *140*
10.3.3.1 The Need for Resource Discovery *140*
10.3.3.2 Power Savings for Eco-conscious Networking *141*
10.3.3.3 Transport is Still Needed! *141*
10.3.3.4 How About Security? *141*

10.4 Computing Everywhere: The Core-Edge Continuum 143
10.4.1 A Common Data Layer 143
10.4.2 The New Programmable Data Plane 145
10.4.3 Novel Architectures Using Computing in the Network 147
10.4.3.1 The Newest and Boldest: Quantum Networking 148
10.4.3.2 Creating the Tactile and the Automated Internet: FlexNGIA 148
10.5 Making it Real: Use Cases 149
10.5.1 Computing in the Data Center 150
10.5.1.1 Data and Flow Aggregation 150
10.5.1.2 Key-value Storage and In-network Caching 151
10.5.1.3 Consensus 151
10.5.2 Next-generation IoT and Intelligence Everywhere 152
10.5.2.1 The Internet of Intelligent Things 152
10.5.2.2 Industrial Automation: From Factories to Farms 153
10.5.3 Computing Support for Networked Multimedia 154
10.5.3.1 Video Analytics 154
10.5.3.2 Extended Reality and Multimedia 154
10.5.4 Melding AI and Computing for Measuring and Managing the Network 155
10.5.4.1 Telemetry 155
10.5.4.2 AI/ML for Network Management 156
10.5.5 Network Coding 157
10.6 Conclusion: 6G, the Network, and Computing 158

11 An Approach to Automated Multi-domain Service Production for Future 6G Networks *167*
Mohamed Boucadair, Christian Jacquenet, and Emmanuel Bertin
11.1 Introduction 167
11.1.1 Background 167
11.1.2 The Need for Multi-domain 6G Networks 168
11.1.3 Challenges of Multi-domain Service Production and Operation 169
11.2 Framework and Assumptions 170
11.2.1 Terminology 170
11.2.2 Assumptions 171
11.2.2.1 SDN-enabled Domains 171
11.2.2.2 On-service Orchestrators 172
11.2.2.3 Any Kind of Multi-domain Service, Whatever the Vertical 172
11.2.3 Roles 173
11.2.4 Possible Multi-domain Service Delivery Frameworks 174
11.2.4.1 A Set of Bilateral Agreements 174
11.2.4.2 A Set of Bilateral Agreements by Means of a Marketplace 174
11.2.4.3 A Set of Bilateral Agreements by Means of a Broker 175
11.3 Automating the Delivery of Multi-domain Services 175
11.3.1 General Considerations 175
11.3.2 Discovering Partnering Domains and Communicating with Partnering SDN Controllers 176

11.3.3 Multi-domain Service Subscription Framework *178*
11.3.4 Multi-domain Service Delivery Procedure *179*
11.4 An Example: Dynamic Enforcement of Differentiated, Multi-domain Service Traffic Forwarding Policies by Means of Service Function Chaining *181*
11.4.1 SFC Control Plane *181*
11.4.2 Consistency of Operation *182*
11.4.3 Design Considerations *182*
11.5 Research Challenges *183*
11.5.1 Security of Operations *184*
11.5.2 Consistency of Decisions *184*
11.5.3 Consistency of Data *184*
11.5.4 Performance and Scalability *185*
11.6 Conclusion *185*

12 6G Access and Edge Computing – ICDT Deep Convergence *187*
Chih-Lin I, Jinri Huang, and Noel Crespi
12.1 Introduction *187*
12.2 True ICT Convergence: RAN Evolution to 5G *187*
12.2.1 C-RAN: Centralized, Cooperative, Cloud, and Clean *190*
12.2.1.1 NGFI: From Backhaul to xHaul *191*
12.2.1.2 From Cloud to Fog *194*
12.2.2 A Turbocharged Edge: MEC *195*
12.2.3 Virtualization and Cloud Computing *197*
12.3 Deep ICDT Convergence Toward 6G *198*
12.3.1 Open and Smart: Two Major Trends Since 5G *198*
12.3.1.1 RAN Intelligence – Enabled with Wireless Big Data *199*
12.3.1.2 OpenRAN *202*
12.3.1.3 Scope of RAN Intelligence Use Cases *205*
12.3.2 An OpenRAN Architecture with Native AI: RAN Intelligent Controller (RIC) *208*
12.3.2.1 NRT-RIC Functions *209*
12.3.2.2 nRT-RIC Functions *211*
12.3.3 Key Challenges and Potential Solutions *212*
12.3.3.1 Customized Data Collection and Control *212*
12.3.3.2 Radio Resource Management and Air Interface Protocol Processing Decoupling *213*
12.3.3.3 Open API for xApp *214*
12.4 Ecosystem Progress from 5G to 6G *214*
12.4.1 O-RAN Alliance *214*
12.4.2 Telecom Infrastructure Project *215*
12.4.3 GSMA Open Networking Initiative *216*
12.4.4 Open-source Communities *216*
12.5 Conclusion *217*

13 **"One Layer to Rule Them All": Data Layer-oriented 6G Networks** *221*
Marius Corici and Thomas Magedanz
13.1 Perspective *221*
13.2 Motivation *222*
13.3 Requirements *223*
13.4 Benefits/Opportunities *225*
13.5 Data Layer High-level Functionality *227*
13.6 Instead of Conclusions *231*

14 **Long-term Perspectives: Machine Learning for Future Wireless Networks** *235*
Sławomir Stańczak, Alexander Keller, Renato L.G. Cavalcante, Nikolaus Binder, and Soma Velayutham
14.1 Introduction *235*
14.2 Why Machine Learning in Communication? *236*
14.2.1 Machine Learning in a Nutshell *237*
14.2.1.1 Kernel-based Learning with Projections *237*
14.2.1.2 Deep Learning *238*
14.2.1.3 Reinforcement Learning *241*
14.2.2 Choosing the Right Tool for the Job *242*
14.3 Machine Learning in Future Wireless Networks *243*
14.3.1 Robust Traffic Prediction for Energy-saving Optimization *244*
14.3.2 Fingerprinting-based Localization *244*
14.3.3 Joint Power and Beam Optimization *245*
14.3.4 Collaborative Compressive Classification *245*
14.3.5 Designing Neural Architectures for Sparse Estimation *247*
14.3.6 Online Loss Map Reconstruction *248*
14.3.7 Learning Non-Orthogonal Multiple Access and Beamforming *248*
14.3.8 Simulating Radiative Transfer *250*
14.4 The Soul of 6G will be Machine Learning *251*
14.5 Conclusion *252*

15 **Managing the Unmanageable: How to Control Open and Distributed 6G Networks** *255*
Imen Grida Ben Yahia, Zwi Altman, Joanna Balcerzak, Yosra Ben Slimen, and Emmanuel Bertin
15.1 Introduction *255*
15.2 Managing Open and Distributed Radio Access Networks *256*
15.2.1 Radio Access Network *256*
15.2.2 Innovation in the Standardization Arena *258*
15.2.2.1 RAN *258*
15.3 Core Network and End-to-End Network Management *260*
15.3.1 Network Architecture and Management *260*

15.3.2 Changes in Architecture and Network Management from Standardization Perspective *262*
15.3.3 Quality of Service and Experience *263*
15.3.4 Standardization Effort in Data Analytics *264*
15.4 Trends in Machine Learning Suitable to Network Data and 6G *265*
15.4.1 Federated Learning *265*
15.4.2 Auto-Labeling Techniques and Network Actuations *266*
15.5 Conclusions *268*

16 6G and the Post-Shannon Theory *271*
Juan A. Cabrera, Holger Boche, Christian Deppe, Rafael F. Schaefer, Christian Scheunert, and Frank H. P. Fitzek
16.1 Introduction *271*
16.2 Message Identification for Post-Shannon Communication *273*
16.2.1 Explicit Construction of RI Codes *277*
16.2.2 Secrecy for Free *279*
16.2.3 Message Identification Without Randomness *280*
16.3 Resources Considered Useless Become Relevant *281*
16.3.1 Common Randomness for Nonsecure Communication *281*
16.3.2 Feedback in Identification and the Additivity of Bundled Channels *282*
16.4 Physical Layer Service Integration *283*
16.4.1 Motivation and Requirements *283*
16.4.2 Detectability of Denial-of-Service Attacks *284*
16.4.3 Further Limits for Computer-Aided Approaches *288*
16.5 Other Implementations of Post-Shannon Communication *288*
16.5.1 Post-Shannon in Multi-Code CDMA *288*
16.5.2 Waveform Coding in MIMO Systems *289*
16.6 Conclusions: A Call to Academia and Standardization Bodies *290*

Index *295*

Editor Biographies

Emmanuel Bertin (PhD) is a Senior Expert at Orange Innovation, France and an Adjunct Professor at Institut Polytechnique de Paris, France. His activities are focused on 5G and 6G, NFV and service engineering, with more than 100 published researched articles. He received a Ph.D. and an Habilitation in computer science from Sorbonne University. He is a senior member of the IEEE.

Noel Crespi (PhD) holds Masters degrees from the Universities of Orsay (Paris 11) and Kent (UK), a diplome d'ingénieur from Telecom Paris, and a Ph.D and an Habilitation from Sorbonne University. From 1993 he worked at CLIP, Bouygues Telecom and then at Orange Labs in 1995. He took leading roles in the creation of new services with the successful conception and launch of Orange prepaid service, and in standardization (from rapporteurship of the IN standard to the coordination of all mobile standards activities for Orange). In 1999, he joined Nortel Networks as telephony program manager, architecting core network products for the EMEA region. He joined Institut Mines-Telecom, Telecom SudParis in 2002 and is currently Professor and Program Director at Institut Polytechnique de Paris, leading the Data Intelligence and Communication Engineering Lab. He coordinates the standardization activities for Institut Mines-Telecom at ITU-T and ETSI. He is also an adjunct professor at KAIST (South Korea), a guest researcher at the University of Goettingen (Germany) and an affiliate professor at Concordia University (Canada). He is the scientific director of ILLUMINE, a French-Korean laboratory. His current research interests are in Softwarization, Artificial Intelligence and Internet of Things.

For more details: http://noelcrespi.wp.tem-tsp.eu/

Thomas Magedanz (PhD) has been Professor at the Technische Universität Berlin, Germany, leading the chair for next generation networks (www.av.tu-berlin.de) since 2004. In addition, since 2003 he has been Director of the Business Unit Software-based Networks (NGNI) at the Fraunhofer Institute for Open Communication Systems FOKUS (https://www.fokus.fraunhofer.de/usr/magedanz) in Berlin.

For 30 years Prof. Magedanz has been a globally recognized ICT expert, working in the convergence field of telecommunications, Internet and information technologies understanding both the technology domains and the international market demands. He often acts as an independent technology consultant for international ICT companies. In the course of his applied research and development activities he created many internationally recognized prototype implementations of global telecommunications standards that provide the foundations for the efficient development of various open technology testbeds around the globe. His interest is in software-based 5G networks for different verticals, with a strong focus on public and non-public campus networks. The Fraunhofer 5G Playground (www.5G-Playground.org) represents, in this regard, the world's most advanced Open 5G testbed which is based on the Open5GCore software toolkit (www.open5Gcore.org), representing the first reference implementation of the 3GPP 5G standalone architecture, which is currently also used by many customers for testing against different RAN equipment in different use cases. For three years, he has actively supported the buildup of emerging 5G campus networks based on the Open5GCore considering emerging campus networks as the prime spot for 5G innovation.

His current research is targeting the 5G evolution to 6G, including Core-RAN integration (including O(pen)RAN integration), Satellite/Non-terrestrial Networks and 5G/6G integration, as well as AI/ML based 5G/6G network control and management.

For more details and a longer version look here:

http://www.av.tu-berlin.de/menue/team/prof_dr_thomas_magedanz/

List of Contributors

Zwi Altman
Orange Innovation
Châtillon
France

Alexis I. Aravanis
CNRS, CentraleSupelec, L2S
University of Paris-Saclay
Gif-sur- Yvette
France

Joanna Balcerzak
Orange Innovation
Châtillon
France

Yosra Ben Slimen
Orange Innovation
Châtillon
France

Emmanuel Bertin
Orange Innovation
Caen
France

Nikolaus Binder
NVIDIA
Berlin
Germany

Holger Boche
Institute of Theoretical Information Technology
Technische Universität München
Munich
Germany

Mohamed Boucadair
Orange Innovation
Cesson-Sévigné
France

Juan A. Cabrera
Deutsche Telekom Chair of Communication Networks
Technische Universität Dresden
Dresden
Germany

Renato L.G. Cavalcante
TU Berlin/Heinrich Hertz Institute
Berlin
Germany

Marius Corici
Fraunhofer FOKUS
Berlin
Germany

Noel Crespi
IMT, Telecom SudParis
Institut Polytechnique de Paris
Paris
France

Isabelle Dabadie
Laboratoire de recherche en sciences de gestion Panthéon-Assas (LARGEPA)
Université Paris 2 Panthéon-Assas
Paris
France

Christian Deppe
Chair of Communication Engineering
Technische Universität München
Munich
Germany

Marco Di Renzo
CNRS, CentraleSupelec, L2S
University of Paris-Saclay
Gif-sur- Yvette
France

Frank H. P. Fitzek
Deutsche Telekom Chair of
Communication Networks. Technische
Universität Dresden
Dresden
Germany

Marco Giordani
Department of Information Engineering
University of Padova
Padova
Italy

Imen Grida Ben Yahia
Orange Innovation
Châtillon
France

Thomas Heyn
Head of Mobile Communications Group
Broadband and Broadcast Department
Fraunhofer IIS
Erlangen
Germany

Alexander Hofmann
Department RF SatCom Systems
Fraunhofer Institute for Integrated Circuits
Erlangen
Germany

Jinri Huang
China Mobile Research Institute
Beijing
China

Chih-Lin I
China Mobile Research Institute
Beijing
China

Christian Jacquenet
Orange Innovation
Cesson-Sévigné
France

Alexander Keller
NVIDIA
Berlin
Germany

Markus Landmann
Electronic Measurements and Signal
Processing (EMS) Department
Fraunhofer Institute for Integrated Circuits
IIS. Ilmenau
Germany

Andres Laya
Ericsson Research
Stockholm
Sweden

Thomas Magedanz
Fraunhofer FOKUS
Berlin
Germany

Marie-José Montpetit
Concordia University
Montreal
Canada

Robert Müller
Electronic Measurements and Signal
Processing (EMS) Department
Fraunhofer Institute for Integrated Circuits
IIS. Ilmenau
Germany

Akihiro Nakao
The University of Tokyo
Tokyo
Japan

Zhisheng Niu
Tsinghua University
Beijing
China

Michele Polese
Institute for the Wireless Internet of Things
Northeastern University
Boston
MA
USA

Sahana Raghunandan
Department RF SatCom Systems
Fraunhofer Institute for Integrated Circuits
Erlangen
Germany

Leszek Raschkowski
Wireless Communications and Networks Department
Fraunhofer Heinrich Hertz Institute HHI
Berlin
Germany

Guy Redmill
Redmill Communications Ltd
London
UK

Rafael F. Schaefer
Chair of Communications Engineering and Security
University of Siegen
Siegen
Germany

Christian Scheunert
Chair of Communication Theory
Technische Universität Dresden
Dresden
Germany

Henning Schulzrinne
Columbia Univeristy
New York
USA

Sławomir Stańczak
TU Berlin/Heinrich Hertz Institute
Berlin
Germany

Peter Stuckmann
European Commision
Brussels
Belgium

Soma Velayutham
NVIDIA
Santa Clara
CA
USA

Marc Vautier
Orange Innovation
Cesson-Sévigné
France

David Zhe Lou
Huawei Technologies Düsseldorf GmbH
Munich
Germany

Sheng Zhou
Tsinghua University
Beijing
China

Michele Zorzi
Department of Information Engineering,
University of Padova
Padova
Italy

Forewords

Henning Schulzrinne, Columbia University, USA

The first few iterations of cellular networks, 1G through 3G, were largely telephone networks with mobility added on, including the choice of addressing through telephone numbers, signaling through SS7, and emphasis on interoperable voice services. 4G and 5G started the transition to an Internet-driven architecture, with remnants of the old architecture still clearly visible. But beyond the protocol choices, all existing generations were largely driven by the assumption that networks are operated by a relatively small number of carriers, typically with at least a nationwide service footprint, reliant on licensed spectrum and an assumption of mutual trust. 5G has started to focus more attention on using the same radio technology for both industrial and consumer networks, but the large-carrier mindset still pervades the design, with a tightly-coupled set of protocols and entities. This tightly-coupled model provides some advantages; it bundles a consistent set of features and technologies designed and packaged to work together, relying on a strict user management and authentication framework. However, this model comes also with drawbacks, such as the lack of flexibility to adapt to new technologies or use-cases, and having to rely on three or at most four carriers in most countries.

Since 3G, branding mobile network generations have had both a technical and a consumer marketing role. The generations provided checkpoints for equipment vendors, and made advances in technology that's otherwise largely invisible to consumers relevant and marketable. 5G is probably the first iteration where a transition in technology standards became a matter of national pride and an indicator of national or regional competitiveness, with promises of increases in consumer and societal welfare that may be hard to deliver. However, as the digital divide during COVID-19 illustrated, universal access to affordable broadband, typically at home, mattered more than higher 5G speeds in the downtown business districts and digital transformation is not assured by having nationwide 5G. Thus, technologists and policy makers working on post-5G efforts should be careful in calibrating expectations, given that wireless network technology may not be the most significant hurdle that prevent addressing key societal challenges.

It seems likely that we will see a much larger variety of operational scenarios in the next decade, from traditional vertical-integrated carriers to disaggregated carriers and to private or federated enterprise networks. Any future network architecture needs to be sufficiently modular so that it can scale down to unmanaged home networks and scale up to networks where participants have limited trust in each other. This suggests a much more flexible and much simpler authentication and roaming model than we have had in previous network generations. Here, 6G can probably learn from another wireless technology where "generations" have played less of a role – ubiquitous Wi-Fi.

Developments for IoT during the 5G standardization and deployment phase may also hold lessons that encourage predictive modesty for 6G. Rather than being the universal network that connects billions and billions of IoT devices to create "smart" buildings and cities, cheap home Wi-Fi and new low-cost technologies like LoRa, leveraging unlicensed spectrum, have come to dominate, with carrier IoT offerings falling short of expectations – indeed, retaining boring and obsolete 2G often seems to draw more interest than new 5G ultralow latency capabilities.

Previous generations of cellular networks offered their per-user speed as the headline advantage, but 5G is already showing the limitations of that approach, as few mobile applications are likely to be built that will rely on 1 Gb/s or above speeds. Thus, the key metrics will not be per-user throughput or latency, but cost per base station month, governing deployment cost in low-density areas, and cost per bit delivered, i.e., primarily operational costs. Environmental metrics such as energy consumption or electromagnetic fields (EMF) must also be considered. For many years, capital equipment has only accounted for about 15% of revenues of most carriers, i.e., the vast majority of expenses are operational. This argues for a simple, self-managed, and robust network, with as many commodity components and protocols as possible and as much re-use of available fiber access networks as possible, rather than infinite configurability or elaborate QoS mechanisms. The largest opportunities for improved operational efficiency and reduced complexity are in the control plane, not the data plane, relying for that on machine learning and automation technologies as detailed in this book. However, since 6G will serve as infrastructure, with concomitant reliability expectations, robustness, predictability and explainability of any use of machine learning will be more important than squeezing out the last percentage points of efficiency.

Despite all the changes in technology, the common thread across mobile technology generations has been a dramatic reduction in the consumer unit cost of mobile data, with new applications enabled simply because they became affordable. Thus, 6G will likely only offer a significant value proposition beyond a marketing tag line if it is engineered to minimize operational complexity, maximizes operational automation and ensures high availability. The Wi-Fi experience can offer lessons and might even offer an opportunity for convergence, where 6G radio access is just another PHY, with a common upper-layer stack optimized for a heterogeneous service provider environment that allows a wide variety of industry, academic and government users to rapidly and cheaply create new applications and an even wider variety of entities to offer access to network services. Deciding what to omit from 6G and leave it to other parts of the networking eco system will be as important as deciding what to include.

Research, particularly academic research, should be driven by the urgent needs of society, not just supplying patent-protected "moats" against competition, whether between companies or nations. 6G offers a unique opportunity to the research community to identify the best engineering approaches that enable universal, affordable, secure and reliable networks. This book provides an initial and valuable exploration of these questions.

Henning Schulzrinne
Columbia University, USA

Peter Stuckmann, Head of Unit, Future Connectivity Systems, European Commission

Recent years and in particular the COVID-19 crisis have shown us the importance of resilient and high-speed communications infrastructure. Trust and acceptance in connectivity infrastructure has grown as global societies have discovered its added value and the possibilities for remote working, but also for citizens' daily lives. Business has understood the critical importance of high-speed networks and technologies in maintaining operations and processes. The crisis illustrates both the potential that 5G networks have to provide the connectivity basis for the digital and green recovery in the short to mid-term, and the need to build technology capacities for the following generation – 6G – in the long term.

5G technology and standards will evolve in the next few years in several phases, just as deployment advances. Operators worldwide have launched commercial 5G networks in major cities. This early deployment will build on 4G networks and will aim primarily at enhancing mobile broadband services for consumers and businesses. Huge investments need to be unlocked for the more comprehensive deployment covering all urban areas and major transport paths by 2025. 5G technology is expected to evolve towards new 'standalone' 5G core networks enabling industrial applications such as Connected and Automated Mobility (CAM) and industry 4.0. These will be a first step towards digitising and greening our entire economy. The growth potential in economic activity enabled by 5G and later 6G networks and services has been estimated to be in the order of €3 trillion by 2030[1]. For such critical services, we need to ensure that 5G networks will be sufficiently secure.

R&I initiatives on 6G technologies are now starting in leading regions world-wide, with the first products and infrastructures expected for the end of this decade. 6G systems are expected to offer a new step change in performance from Gigabit towards Terabit capacities and sub-millisecond response times, to enable new critical applications such as real-time automation or extended reality ("Internet of Senses") collecting and providing the sensor data for nothing less than a digital twin of the physical world.

Moreover, new smart network technologies and architectures will be needed to enhance drastically the energy efficiency of connectivity platforms despite major traffic growth and keep electromagnetic fields (EMF) under safety limits. They will form the technology base for a human-centric Next-Generation Internet (NGI) and address Sustainable Development Goals (SDGs) such as accessibility and affordability of technology.

All parts of the world are starting to be heavily engaged in 6G developments. There will be opportunities and challenges concerning new business models and players through software networks with architectures such as Open-RAN[2] and the convergence with new technologies in the area of cloud and edge computing, AI, as well as components and devices beyond smartphones.

Firstly, success in 6G will depend on the extent regions will succeed in building a solid 5G infrastructure, on which 6G technology experiments and, later, 6G deployments can build. In this context, building 5G ecosystems will be of key importance, also because industry R&I investments tend to relocate where markets are more advanced.

1 McKinsey Global Institute, 2/2020, Connected World – An evolution in connectivity beyond the 5G revolution

2 More open and interoperable interfaces in Radio Access Networks (RAN)

Secondly, 6G will require taking a broader value chain approach, ranging from connectivity to components and devices beyond smartphones with the massive development of the Internet of Things (IoT) and connected objects like cars or robots. They also exist on the service side, with edge computing integrated in connectivity platforms and cloud computing enabling advanced service provisioning, e.g. for big data and AI.

One important success factor to create and seize such opportunities is to be a standard setter in 6G and the related technology fields. Both future users and suppliers need to shape key technology standards in the field of radio communications, but also in next-generation network architecture to ensure the delivery of advanced service features, e.g. through the effective use of software technologies and open interfaces, while meeting energy-efficiency requirements.

Spectrum resources are another key factor that will determine success in 6G. Whereas bands currently allocated for mobile communications will be reused for 6G, new frequency bands will be identified and harmonised. Industry and governments need to identify the opportunities related to spectrum that can be suitable for 6G and be made available with the potential to be harmonised at global level. 6G technology will also have the potential to make a further step towards a multi-purpose service platform replacing legacy radio services for dedicated applications. This could help the progress in defragmenting the radio spectrum and drastically enhance spectrum efficiency that will in turn free up new bands for 6G or other purposes.

Such outcomes in global standardisation and spectrum harmonisation need to be prepared by proactive and effective international cooperation at government and industry-level. This includes regular dialogues with leading regions and possible focused joint initiatives in R&I, standardisation or regulation.

I am looking forward to the creativity and ambition of the global research and innovation community to shape the new generation of communication technology throughout this decade.

Let's kick this off!

Peter Stuckmann
Head of Unit, Future Connectivity Systems, European Commission

Akihiro Nakao, The University of Tokyo, Japan

Mobile network systems have evolved from communication infrastructures to critical and indispensable social infrastructures over the generations. The 5th generation mobile network system (5G) has been getting deployed commercially since 2019 and is bringing new innovations, both in terms of technology and business models. New models of 5G private network deployments are indeed emerging, and the connectivity landscape appears to be more and more split between various players and domains. Beyond 5G networks are expected to be deployed around 2025 onward, and studies on standardization of 6G have already begun.

6G networks and services are expected to play a central role as the backbone of our future societies by tightly integrating virtual and physical spaces. Japanese governmental agencies have forged the term Society 5.0 to designate this future society that Japan should aspire to be. Following the hunting society (Society 1.0), agricultural society (Society 2.0), industrial society (Society 3.0), and information society (Society 4.0), Society 5.0 should achieve a high degree of convergence between cyberspace (virtual space) and physical space (real space). In this future Society 5.0, huge amounts of information from sensors in physical space are accumulated in cyberspace and analyzed by artificial intelligence (AI) to provide intuitive and near-real-time feedback to humans in physical space. This vision first drawn by science fiction authors in the early 1980s is about to become a reality. "Cyberspace... Data abstracted from the banks of every computer in the human system. Unthinkable complexity." wrote William Gibson (who coined the term of cyberspace) in his 1984 novel *Neuromancer*.

The recent COVID-19 misfortune might appear as a new step toward this Society 5.0, as we have re-recognized the need for enhancing and upgrading information communication infrastructure to ensure the continuity of our social activities, as well as the growing blurring between virtual and real relationships. On this road, it is essential not only to promote research and development of technology but also to consider the global environmental impacts (such as carbon neutral and green recovery), the social inclusiveness so that no one will be left behind, and the ethics and social acceptability of these forthcoming technologies.

This wish for a future better and enhanced society shall be and remain the underlying foundation for designing future 6G networks. It should bond all the stakeholders engaged in research and development of next-generation cyber infrastructure, 6G mobile network systems, to globally unite forces to define new requirements, use cases, and fundamental theories and technologies that must be realized for the next decade. These researches are also a way to progress for accomplishing the 2030 Agenda for Sustainable Development adopted by the United Nations in 2015, where one of the sustainable development goals is about building resilient infrastructure, promote inclusive and sustainable industrialization, and foster innovation.

Although it is just the very beginning of our journey for developing 6G mobile networking, we can assume that the next-generation cyber infrastructure will bring us communications features very close to human capability, such as ultralow latency, ultra-high capacity, ultra-large number of connected devices, ultralow power communication, stringent security and privacy, autonomy enabled by machine learning and AI, and ultra-coverage and extensibility including non-terrestrial networks, underwater communication, etc.

This journey will not only be driven by the telecom industry. Many countries have allocated frequency white space to private 5G usage and made open to non-telecommunication companies so that they can operate their own customized 5G networks. We believe that this "democratization" (i.e. making something accessible to anyone) of 5G networks will open a door to new innovations coming from the civil society as well as from industrial players. 6G will thus be the opportunity to conciliate various types of innovations: grassroots innovations coming from local players with new use cases and ad hoc solutions, radio and core layer innovations coming from Telco players, and also real-time software innovations coming from Internet player. Besides the regular migration path from 5G to 6G promoted by telecommunication operators and vendors, there is another evolution avenue possible, from private 5G to private 6G and then to public 6G because a lot more stakeholders may participate in the game of developing custom solutions tailored for their real use cases that may be eventually distilled and adopted as viable 6G technologies to be standardized.

Along with the editors, I hope that this book serves as a navigating compass in our endeavor for developing 6G infrastructure for the next decade, by providing the insights from internationally known distinguished experts.

Akihiro Nakao
The University of Tokyo, Japan

Acronyms

Abbreviation	Explanation
3GPP2	3rd Generation Partnership Project 2
5G	5th Generation
5GAA	5G Automotive Association
5GC	5G Core
5G-NTN	5G Non-Terrestrial Network
6G	6th Generation
AD	Anomaly Detection
AFL	Agnostic Federated Learning
AI	Artificial Intelligence
AIaaS	AI-as-a-Service
API	Application Programming Interface
APS	Angular Power Spectrum
APSM	Adaptive Projected Subgradient Method
ARCEP	Autorité de Régulation des Communications Électroniques et des Postes
ARIB	Association of Radio Industries and Businesses
AS	Autonomous System
ASIC	Application-Specific Integrated Circuit
ATIS	Alliance for Telecommunications Industry Solutions
B2B	Business-to-Business
B2C	Business-to-Consumer
B5G	Beyond 5G
BBUs	Baseband Units
BGP	Border Gateway Protocol
BN	Boundary Nodes
BOM	Business, Operation, and Management

(Continued)

(Continued)

Abbreviation	Explanation
BS	Base Station
BSS	Business Support System
BW	Bandwidth
CAPEX	Capital Expenses
CBRS	Citizen Broadband Radio System
CCNx	Content-Centric Networking
CCSA	China Communications Standards Association
CDMA	Code Division Multiple Access
CeTI	Centre for Tactile Internet with Human-in-the-Loop
CFN	Computer-First Networking
C-ITS	Cooperative Intelligent Transport System
CN	Core Network
COINRG	Computing in the Network Research Group
COTS	Commercial Off The Shelf
CP	Control Platform
CPM	Collective Perception Message
CPNP	Connectivity Profile Negotiation Protocol
CPU	Central Processing Unit
CR	Common Randomness
C-RAN	Cloud-RAN
CSAE	China Society of Automotive Engineers
CS	Channel Sounder
CSI	Channel State Information
CSI	Channel Side Information
CU	Centralized Unit
CUPS	Control User Plane Separation
D/A	Digital to Analogue
DCAE	Data Collection, Analytics, and Events
DC	Data Center
DDoS	Distributed Denial of Service
DetNet	Deterministic Networking
DFG	Deutsche Forschungsgemeinschaft
DI	Deterministic Identification
DINRG	Decentralized Internet Infrastructure
DMC	Discrete Memoryless Channel
DNN	Deep Neural Network

(Continued)

Abbreviation	Explanation
DoS	Denial of Service
DPI	Deep Packet Inspection
DRL	Deep Reinforcement Learning
DSCP	Differentiated Services Code Point
DU	Distributed Unit
EE	Energy Efficiency
EI	Enrichment Information
eLSA	Evolved License Shared Access
EM	Electromagnetic
eMBB	Enhanced Mobile Broadband
EROI	Energy Return on Energy Injected
ETSI	European Telecommunications Standards Institute
FD	Full Duplex
FDM	Frequency Division Multiplexing
FH	Fronthaul
FL	Federated Learning
FLOP	Float Point Operation
FPGA	Field-Programmable Gate Array
FR2	Frequency Range 2
FSS	Fixed Satellite Service
FSS	Frequency Selective Surface
GDPR	General Data Protection Regulation
GEO	Geostationary Earth Orbit
gNB	Next-Generation NodeB
GPP	General-Purpose Platform
GPU	Graphics Processing Unit
GSM	Global System for Mobile Communication
GSMA	GSM Association
GTP	GPRS Tunnel Protocol
HAP	High-Altitude Platform
HD	Half Duplex
HDFS	Hadoop Distributed File System
HDS	High Impedance Surface
HFT	High-Frequency Trading
HMD	Head-Mounted Device
HRV	High-Risk Vendor

(Continued)

(Continued)

Abbreviation	Explanation
HTS	High Throughput Satellite
ICDT	Information, Communication, and Data Technology
ICT	Information and Communication Technology
IDF	Identification with Feedback
IMT	International Mobile Telecommunications
IoE	Internet of Everything
IoT	Internet of Things
IP	Intellectual Property
IPFS	Interplanetary File System
IRTF	Internet Research Task Force
ISL	Inter-Satellite Link
ISTN	Integrated Space and Terrestrial Network
IT	Information Technology
ITU	International Telecommunication Union
JCSS	Joint Communication Sensor Systems
KPI	Key Performance Indicator
K-V	Key Values
LEO	Low Earth Orbit
LF	Linux Foundation
LISP	Locator/ID Separation Protocol
LPWAN	Low-Power Wide-Area Network
LTE	Long-Term Evolution
M2M	Machine-to-Machine
MAC	Media Access Control
MAMOKO	Molecular Communication
MC-CDMA	Multi-Code CDMA
MCTS	Monte Carlo Tree Search
MDA	Mandate-Driven Architecture
MDAS	Management Data Analytics Service
MEC	Mobile Edge Computing
MEO	Medium Earth Orbit
mHealth	Mobile Health
MIMO	Multi-Input Multi-Output
MIoT	Massive IoT
ML	Machine Learning
M-MIMO	Massive MIMO

(Continued)

Abbreviation	Explanation
MMSE	Minimum Mean Square Error
mMTC	Massive Machine Type Communications
MSS	Mobile Satellite Service
NAS	Non-Access Stratum
NAT	Network Address Translator
NCSC	National Cyber Security Centre
NDN	Named Data Networking
NF	Network Function
NFV	Network Function Virtualization
NGP	Next-Generation Protocols
NIA	Network Index Address
NIC	Network Interface Controller
NIN	Non-IP Networking
NN	Neural Network
NOMA	Non-Orthogonal Multiple Access
NRI	Non-Randomized Identification
NRT	Non-Real-Time
NRT-RIC	Non-Real-Time RAN Intelligent Controller
NSF	National Science Foundation
NWDA	Network Data Analytic
NWDAF	Network Data Analytics Function
OAM	Operation and Management
OBO	Output Back-off
OBP	On-Board Processor
OFDMA	Orthogonal Frequency Division Multiple Access
OPEX	Operational Expenditure
O-RAN	Open Radio Access Network
OSC	O-RAN Software Community
OT	Operation Technology
OTF	Open Testing Framework
OTIC	Open Testing and Integration Center
OTN	Optical Transport Network
P4	Programming Protocol-Independent Packet Processors
PAPR	Peak-to-Average Power Ratio
PCE	Path Communication Element
PCF	Policy Control Function

(Continued)

(Continued)

Abbreviation	Explanation
PE	Provider Edge
PFNM	Probabilistic Federated Neural Matching
PHY	Physical
PISA	Protocol-Independent Switch Architecture
PLC	Programmable Logic Control
PN	Pseudo-Noise
PS	Public Safety
PSCE	Public Safety Communication Europe Forum
QoE	Quality of Experience
QoS	Quality of Service
RAN	Radio Access Network
RANDA	Radio Access Network Big Data Analysis Network Architecture
rApp	Radio Application
RCA	Root Cause Analysis
ReLU	Rectified Linear Unit
RI	Randomized Identification
RIC	Radio Intelligent Controller
RIC	RAN Intelligent Controller
RIS	Reconfigurable Intelligent Surface
RKHS	Reproducing Kernel Hilbert Spaces
RLNC	Random Linear Network Coding
RRC	Radio Resource Control
RT	Real-Time
SA	System Aspect
SBA	Service-Based Architecture
SBI	Service-Based Interface
SDL	Shared Data Layer
SDN	Software-Defined Networking
SDO	Standard Development Organization
SDR	Software-Defined Radio
SE	Spectrum Efficiency
SFC	Service Function Chaining
SFP	Service Function Path
SGD	Stochastic Gradient Descent
SLA	Service-Level Agreement
SLA	Service Layer Agreement

(Continued)

Abbreviation	Explanation
SNR	Signal-to-Noise Ratio
SOM	Service Order Management
SRv6	Segment Routing Based on IPv6
TCO	Total Cost of Ownership
TDM	Time Division Multiplex
TDMA	Time Division Multiple Access
TIP	Telecom Infrastructure Project
TSDSI	Telecommunications Standards Development Society
TSN	Time-Sensitive Networking
TTA	Telecommunication Technology Association
TTC	Telecommunication Technology Committee
TTI	Transmission Time Interval
UAV	Unmanned Aerial Vehicle
UDN	Ultradense Network
UMTS	Universal Mobile Telecommunication System
UN IPCC	United Nations Intergovernmental Panel on Climate Change
UP	User Plane
UPF	User Plane Function
URLLC	Ultrareliable Low-Latency Communication
V2X	Vehicle to Everything
VHTS	Very High Throughput Satellites
VM	Virtual Machines
VNA	Vector Network Analyzers
VNF	Virtual Network Function
VPN	Virtual Private Network
VRU	Vulnerable Road User
VSAT	Very Small Aperture Terminals
WBD	Wireless Big Data
WWW	World Wide Web
YOY	Year over Year
ZB	Zettabytes
ZSM	Zero-Touch Network and Service Management

1

Toward 6G – Collecting the Research Visions

Emmanuel Bertin[1], *Thomas Magedanz*[2], *and Noel Crespi*[3]

[1] *Orange Innovation, France*
[2] *Fraunhofer FOKUS, Berlin, Germany*
[3] *IMT, Telecom SudParis, Institut Polytechnique de Paris, Paris, France*

1.1 Time to Start Shaping 6G

During the past 30 years, the successive generations of mobile communication networks have enabled major steps toward a more digital world. Each generation has featured comprehensive cellular network architecture, including radio access technology, access and core network routing, and a set of associated services (such as authentication and access control, mobility management, data transfer, or voice and messaging services). The 2nd generation (2G) brought the first fully digital mobility solution, giving birth to the mobile phone as a portable personal device and to the rise of text messaging. The 3rd and 4th generations (3G and 4G) introduced the use of multimedia services in mobility and enabled the advent of the iPhone and all the digital industry and services relying on smartphones (e.g. mobile Internet, applications, and marketplaces). The 5th generation (5G) should accompany the emergence of a nest of communicating objects, along with new devices enabling augmented reality, for both the consumer and the enterprise market. The path is already drawn for the deployment of 5G non-standalone (5G NSA) networks starting from 2019 (where only the radio part of 5G is deployed as a new access network) and then of standalone 5G networks (5G SA) starting from 2023 (where a new 5G core network is also deployed).

It may seem strange to start shaping the 6th generation (6G) of mobile communication networks while 5G is just starting to be deployed around the globe – given we are still witnessing a big gap between the high expectations surrounding the capabilities of new 5G networks and the functional limitations of initial 5G products and solutions. Moreover, 5G is quite different from previous mobile network generations in regard to its technological innovations, complexity, and targeted broad spectrum of applications, ranging from energy-efficient massive Internet of Things (IoT) and massive broadband multimedia to low-latency communication. In addition, every new network generation (including 5G) must

Shaping Future 6G Networks: Needs, Impacts, and Technologies, First Edition.
Edited by Emmanuel Bertin, Noel Crespi, and Thomas Magedanz.

strike a compromise between backward compatibility, disruption, innovation, and ability to enable completely new applications. This complexity takes time for the telecom industry to fully master.

In this context, should we focus on building 6G or first draw the lessons from 5G deployments and use cases? Every new network generation deserves around 10 years of research. The first generation of digital cellular network (2G) was commercially launched in 1991, followed by 3G in 2001, 4G in 2009, and 5G in 2019. Thus, now is the time to shape 6G, with a target launch in 2028–2030. Research on 6G effectively started around the globe in 2020.

1.2 Early Directions for Shaping 6G

1.2.1 Future Services

So, what will 6G look like? Will there be a killer application? This book discusses some future possible use cases, such as teleporting and digital twin; smart and autonomous transportation; digital services in cities, farming, and warehousing targeting environmental monitoring, traffic control, and management automation; or a fully digital commerce and payments experience, featuring resolution digital signage with facial recognition in retail, and augmented reality/virtual reality (AR/VR)-enabled e-commerce. Some of those use cases were also discussed for 5G. Before we can clearly assess the use cases, though, we have to see what emerges in the next few years, as 5G evolves and gains acceptance in different vertical markets.

Do we need a predefined mind-blowing application driver before shaping 6G architecture? That was not the case for previous generations, and uses like text messaging (for 2G) and smartphone-based mobile Internet (for 3G and 4G) emerged without strong support from the telco industry. So 6G should probably be seen more as the infrastructure on which innovative actors will build new digital services. Modularity, flexibility, and openness are key requirements.

1.2.2 Moving from 5G to 6G

5G is already a software-based end-to-end communication system, allowing the addition of new access and backhaul networks as well as new control and management functionalities and virtual network functions (VNFs). So, should the industry start building a new generation instead of perfecting the existing one? It is likely that similar to previous even network generations (i.e. 2G and 4G), which perfected preceding network generations, 6G will finally deliver what was promised years ago for 5G. Many research topics currently performed in the context of 5G evolution will also pave the way toward 6G. Therefore, most researchers may consider the need for 5G evolution as the driving force toward 6G. In fact, at the end of the decade, which represents the typical life span of a mobile generation to deliver innovations, 5G may have become an open extensible and customizable communications platform, representing a toolbox to build public as well as private mobile communication networks for any kind of vertical application domain.

While it might be realistic to assume that 6G will be an evolution of 5G, there are also voices who propose that 6G should be much more disruptive and revolutionary, due to the exploitation of new enabling technologies. New concepts like the post-Shannon theory and the use of emerging quantum computing technologies are just two examples of this line of thinking. To provide a scientific look beyond the rim, we address one of these topics at the end of the book (Chapter 16).

Moreover, while defining a new generation of cellular system every 10 years has a lot of advantages, as it enables deployment of a consistent set of features and technologies where all elements have been designed and packaged to work together, this model comes with a major drawback: the various components and technologies are tightly linked. It is therefore difficult to redesign one piece of the puzzle without touching the others. In a world being eaten by software, this may appear a bit old-fashioned, as modern software engineering relies on decomposing systems into loosely coupled entities. So, 6G could also be the opportunity to extend 5G into an even more modular framework where various parties can more easily add different components, keeping in mind the necessary trade-off among openness, reliability, and security, in order to achieve a highly trustworthy architecture. This would imply breaking or at least weakening the link between the radio access part and the core network part in the definition of this new generation, inspired perhaps by the idea of other wireless technologies, relying on unlicensed spectrums, such as Wi-Fi and LoRaWAN. Finally, 6G is also an opportunity to continue decreasing operational costs, using artificial intelligence (AI) and machine learning (ML) to extend automated network planning, deployment, and operation; the ultimate target being to enable real self-organizing networks.

1.2.3 Renewed Value Chain and Collaborations

Besides technologies and services, the business models of mobile communication networks are also evolving and will continue to evolve rapidly in the forthcoming years. Due to the ongoing fixed-mobile network convergence and Information and Communication Technology (ICT) convergence, future communications will be tightly integrated in enterprise applications. The global rise of 5G campus networks should be considered just the start toward 5G enterprise networking and the emergence of new business models and ecosystems. This also raises questions on the role of international standards and rise of open software stacks paving the way toward a new telecommunications ecosystem, in which virtualized network functions from different developers and providers can be dynamically orchestrated and integrated in a secure, reliable, and energy-efficient manner. The work on OpenRAN and the involvement of new players (e.g. Facebook Magma) can be considered a foretaste of these changes in the value chain of the entire mobile industry.

In mobile technologies, as in many other areas, geopolitical factors might mean a more fragmented future for the world. In their desire for digital sovereignty, different governments push national academic and industry researchers to generate as many intellectual property rights as possible while shaping 6G. Prefaced by insightful tech leaders from America, Asia, and Europe, with authors from all around the world, this book is an attempt to promote the collaborative approach used to enable academic and industry players with different interests to work together to shape a common future.

With this book, we aim to provide students, researchers, senior executives, managers, and technical leaders with a snapshot of current international thinking on the major 6G research aspects. Similar to 5G, 6G also represents an aggregation of different technology innovations into an overall complex system architecture. We do not have the ambition to catch every technology trend, but we believe that we are quite comprehensive with our present collection of expert views in 2021.

1.3 Book Outline and Main Topics

1.3.1 Use Cases and Requirements for 6G (Chapter 2)

The first point we address in the following chapters concerns the prospective services and use cases that could require a new generation of mobile communication networks. For defining 5G, enterprise needs – rather than the consumer market – have been the main driver. By the way, the rise of private 5G is a good illustration of the growing importance of the business-to-business (B2B) market for mobile networks. We believe these B2B needs will also be the primary driver for the evolution of 5G and the definition of 6G. Innovation in mobile networking will be pushed more and more by companies for their own needs, either by using carrier networks (e.g. with slicing solutions) or through innovative private 5G deployments. In Chapter 2, the authors introduce a collection of potential 6G services for the B2B market, in order to understand potential 6G drivers and the associated requirements. Authors consider services in eight different application domains: digital transformation of manufacturing, teleporting with holography, digital twin, smart transportation, public safety, health and well-being, smart IoT for life quality improvement, and transformation of the financial sector. The authors then derive the key networking requirements induced by these services.

1.3.2 Standardization Processes for 6G (Chapter 3)

The second question we investigate is how and by whom can 6G be defined. Previous generations have been framed under the leadership of the telco industry grouped in standardization bodies (e.g. 3GPP). However, new bodies are emerging, for example, with the OpenRAN alliance to improve openness in radio access networks of next generation wireless systems. De facto standards are more and more driven by providing software implementation within open-source communities, rather than by submitting written contributions to international standardization bodies. New actors are also emerging alongside the telco industry (e.g. Facebook with Magma), which is being reduced more and more to a few suppliers, and industry verticals are more frequently pushing their own needs and solutions (e.g. 5G Alliance for Connected Industries and Automation [5G-ACIA]). In Chapter 3, authors investigate this evolving role of standards for 6G. They also discuss the impact of the shift started in 5G from a standardization based on functional entities to a standardization based on Application Programming Interface (API). Finally, they raise the question of economic as well as political pressures on industry players that might lead to a fragmented ecosystem.

1.3.3 Energy Consumption and Social Acceptance (Chapters 4 and 5)

Lower energy consumption was already an important design criterion in the course of 5G research and development. But with climate change progressing, this requirement is becoming even more important for the design of 6G. Thus, another question to address is the environmental sustainability of 6G. We discuss this topic in two different chapters, showing two complementary viewpoints. In Chapter 4, authors look at technical solutions to provide more sustainable cellular networks, relying on an intensive use of AI mechanisms. First, they identify the main factors of energy consumption in mobile networks. They then provide a holistic approach for defining a more sustainable 6G based on AI training executed at the edge of the network. Authors of Chapter 5 argue that reducing network energy consumption per byte transferred is not a sufficient path when the bandwidth consumption is continuously increasing (rebound effect). The question of sustainability is therefore not only a technical question but also a social and societal issue. Here, marketing will be an important point, along with consumers' rising awareness of the impact of information technologies on global warming. Beside the eco-design of networks and services, the question of changing the way we consume network and service offers should be addressed, with a possible trend toward more digital sobriety.

1.3.4 New Technologies for Radio Access (Chapters 6–8)

Every new mobile generation features a new radio technology, which typically pushes for higher frequencies and, thanks to new coding and signal processing algorithms, enables much higher data rates and communication capacities to mobile users. In city centers, the deployment of new antenna systems for ever smaller cell sizes operating in ever higher frequencies is facing limitations with signal distribution through walls and windows, representing big challenges. Chapter 6 provides an overview of the development of new reflective materials to enhance coverage of urban areas, introducing the technology and the challenges of reconfigurable intelligent surfaces (RIS) for smart radio environments.

In 6G, terahertz (THz) communications represent this next big radio access network innovation. Chapter 6 describes the new technical capabilities and the research challenges to be mastered to exploit these capabilities. In particular, THz base stations will enable the seamless integration of sensing, localization, and communications, making new types of applications possible. At the same time, they put immense requirements on the core network to utilize these new capabilities.

While THz access networks feature small cell sizes and are likely used for indoor use cases, satellite networks have already gained momentum in the context of 5G evolution for outdoor coverage in rural environments, the maritime environment, and the sky. Besides satellites at different orbit levels, drones and high-altitude platforms have also emerged as so-called non-terrestrial networks in the recent past. The authors of Chapter 8 address opportunities and technical challenges to master the upcoming 6G network architectures, including the need for new mechanisms for dynamic access and backhaul network integration, as well as challenges in roaming and handovers in between moving cells and networks.

1.3.5 New Technologies for Network Infrastructure (Chapters 9 and 10)

Beside the radio technologies, a new mobile network generation is also an opportunity to reframe the underlying mechanisms of access and core networks. Chapters 9 and 10 discuss new requirements to address by the network infrastructure (especially on the routing layer), as well as the various solutions to progress on a more optimized implementation. First, one of the major drivers and the foundation for network convergence, the Internet Protocol (IP) also represents a major legacy and limitation factor in communications. The design of 6G is also an opportunity to question this foundational layer 3 protocol, as has already been done for other layers (e.g. Quick UDP Internet Connections [QUIC] for layer 4), in order to solve some intrinsic limits, such as addressing scheme or security vulnerabilities. In Chapter 9, authors investigate options to renovate basic low-level communication protocols to gain more efficiencies and flexibilities beyond the limits of the IP framework. Starting from some iconic application domains (e.g. holographic communication, Industry 4.0), they introduce future requirements for this new framework: high-precision and deterministic services with the right quality of service (QoS) level, user-defined network operations embedded in the packet delivery layer, semantic and flexible addressing, optimal support of various types of access technologies with a high throughput, and intrinsic security and privacy. These requirements should be used as a starting point to build innovative 6G protocol layers for both human-oriented and machine-oriented communications.

Then, as stated before, 5G has turned the mobile communications network into a pure software-based system, exploiting the innovations of software-defined networking and network function virtualization. In addition to cloud computing, which inspired new centralized service architectures, edge computing evolved as the new architectural principle to enable network function distribution for low-latency communications and efficiency in network data processing. Chapter 10 proposes to go one step beyond in considering the network more as a distributed computer board than as a provider of pipes and forwarding mechanisms. Starting from the work on programmable networks (from active networks to software-defined networking), the chapter introduces emerging concepts and requirements for the computerization of networks. While it breaks some established principles (e.g. end-to-end principle of the Internet or client-server design), this enables to create a seamless core-edge continuum of multiple independent components. Such continuum will require a programmable dataplane, relying on a common data layer. Potential application cases are, for example, the optimization of datacenter design, the next generation of IoT with intelligence everywhere, or computing support for networked AR/VR. We therefore believe 6G should be the opportunity to rethink the network as a programmable platform.

1.3.6 New Perspectives for Network Architectures (Chapters 11 and 12)

Trust has become a major aspect for the adoption of new mobile networking technologies, as they become critical infrastructures. This appears even more important when we consider the increasing interworking of different network domains. Since 2G, interconnection is standardized between mobile network operators for the roaming features. 5G introduces a new step with various private 5G networks that might be partly interconnected with public 5G networks. We believe this will increasingly be the case in future 6G networks, with various players providing connectivity and resources to deliver 6G services. Chapter 11 provides an overview of issues, requirements, and solutions to achieve a trustworthy inter-domain collaboration in software-defined environments. Authors focus on automating the delivery of multi-domain services. They also discuss a set of challenges to address for the networking community to achieve collaborative 6G networks: security of operations, consistency of data and decisions, performance, and scalability.

One of 5G's key achievements is the flexible network architecture defined by disaggregated network functions. This disaggregation has some tradition in the core network but will extend with the adoption of OpenRAN disaggregation principles also within the access network. Chapter 12 introduces more precisely the challenges to build such renewed 6G access networks. Starting from Cloud-RAN and mobile edge computing, authors introduce the rise of intelligence and openness of the RAN components, leading to RAN disaggregation on all its dimensions (radio, compute, management plane, control plane). In line with the OpenRAN architecture, RAN intelligent controllers should be used as the key building block to enable both near real-time and non-real-time 6G services. However, challenges remain to be tackled to shift from traditional tightly coupled RAN to disaggregated RAN, as for example: customized data collection and control, radio resource management, and air interface protocol processing decoupling, but also the need of open API to build an applicative ecosystem on the top of these RAN intelligent controllers.

1.3.7 New Technologies for Network Management and Operation (Chapters 13–15)

As discussed previously, the disaggregation principle with loosely coupled network functions (including in the RAN) will be at the core of the forthcoming 6G networks. This allows the dynamic orchestration of network functions and thus adaptation of the network to specific service requirements and network conditions in an end-to-end manner. In this context, monitoring and management of data will become the new fuel in networking. Chapter 13 addresses major development trends toward the definition of a data-layer-oriented network. It presumes the extraction of a large amount of data, its exchange across different elements, the generation of insight, and its immediate application as customized configurations across the system. A new type of network optimization is obtained based on user behavior instead of the one of the systems complimented by a large level of native automation.

Chapter 14 discusses opportunities given by the adoption of ML at the edge of wireless networks. The authors describe the latest in ML (e.g. Kernel-based learning, deep learning, and reinforcement learning) and suggest application domains within the radio network exploiting the available domain knowledge, such as robust traffic prediction for energy optimization or optimized localization of end systems and beamforming optimization. They proclaim that ML will be the heart of future 6G architectures. Chapter 15 expands on the previous chapter by looking specifically at the adoption and standardization of AI/ML for secure, automated end-to-end slice orchestration and management, both at the edge and at the core of 6G mobile networks. The authors describe the rising use of AI/ML across the control protocol stack, investigating in particular the use of federated learning for optimal communication. They also introduce the challenges for global management of 6G systems that should perform smart resource management, automatic network adjustment, provisioning, and orchestration and rely on real-time data insights to optimize network performance. Standardization should play a key role to achieve this goal, as already started with OpenRAN.

1.3.8 Post-Shannon Perspectives (Chapter 16)

6G network will also be the opportunity to introduce innovative communication mechanisms. Chapter 16 lays out a disruptive view of 6G, addressing major innovations becoming possible by the new post-Shannon theory. To truly achieve the potential gains of post-Shannon communication, we have to break with the concept of a physical layer as a mere transport channel, as the needed algorithms cannot only rely on the softwarized higher layers. A new structure of the physical layer is required to integrate new communication tasks, such as message identification, secure message transmission, and CR (common randomness, i.e. correlated results of a random experiment) generation and extraction with higher layer policies that assign different logical channels for the different services. Chapter 16 provides comprehensive insights on these challenges and evolutions.

We hope you gain interesting insights into the major complementary research aspects of 6G summarized in the Figure 1.1.

We are only at the beginning of a long journey toward 6G, which will hit the market by 2030 and provide the major communications infrastructure until 2040. The fun part is that although it will be a global race to generate new IP, ongoing globalization of our economic markets will necessitate close cooperation among researchers on different continents. We strongly believe that in the end, 6G will be the center of all future communications, connecting all people and things on this planet with infrastructures enabling flexibility and innovation while ensuring trust, reliability, and sustainability.

Have a nice trip, and enjoy reading,
Thomas, Noel, and Emmanuel

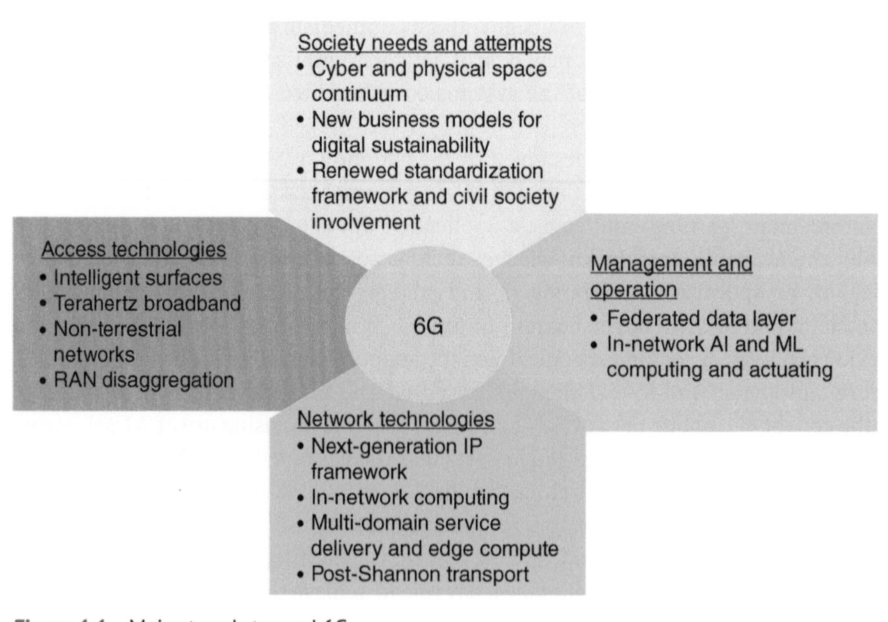

Figure 1.1 Major trends toward 6G.

2

6G Drivers for B2B Market: E2E Services and Use Cases

Marco Giordani[1], *Michele Polese*[2], *Andres Laya*[3], *Emmanuel Bertin*[4], *and Michele Zorzi*[1]

[1] *Department of Information Engineering, University of Padova, Padova, Italy*
[2] *Institute for the Wireless Internet of Things, Northeastern University, Boston, MA, USA*
[3] *Ericsson Research, Stockholm, Sweden*
[4] *Orange Innovation, France*

2.1 Introduction

The 5th generation (5G) of wireless networks was positioned to support, besides the evolution of mobile broadband, new use cases ranging from massive IoT to ultrareliable services. However, in view of future technological innovations, researchers, industrial companies, and standardization bodies have started proposing new use cases and services that, for their generality and complementarity, would not be fully supported by 5G networks and are thus good representatives of future 6G services [1]. While the literature has a larger focus on application domains for the business-to-consumer (B2C) sector, this chapter discusses new beyond-5G drivers for the business-to-business (B2B) markets, such as automotive, manufacturing and logistics, health and government, smart transportation, banking, and financial verticals.

Starting from the 5G-identified verticals, we make the case that, among other services, 6G will (i) fully realize the industrial and manufacturing revolution started with 5G, i.e. the digital transition to a cyber physical facility (Section 2.3.1); (ii) revolutionize business meetings and events by enabling the digitalization and transmission of human sensing and objects through teleportation and digital twins (Sections 2.3.2 and 2.3.3); (iii) support smart (and autonomous) transportation with important implications for the logistics and fleet management sectors (Section 2.3.4); (iv) improve public safety (PS) networking for first responders (Section 2.3.5); (v) transform the healthcare sector through remote interventions and home care to guarantee the most efficient use of healthcare resources and support the reduction of management costs for health facilities (Section 2.3.6); (vi) accelerate the adoption of solutions for digital services in cities, farming, and warehousing, targeting

Shaping Future 6G Networks: Needs, Impacts, and Technologies, First Edition.
Edited by Emmanuel Bertin, Noel Crespi, and Thomas Magedanz.

environmental monitoring, traffic control, and management automation (Section 2.3.7); and (vii) revolutionize the financial sector by supporting novel banking operations and efficient high-frequency trading (HFT) (Section 2.3.8).

This chapter also discusses commonalities and differences among these drivers and outlines the order of magnitude of key performance indicators (KPIs) and requirements to be satisfied. In particular, while 5G-based use cases typically present trade-offs on latency, energy consumption, development and deployment costs, computational complexity, and throughput, 6G will be developed to meet stringent network demands in a holistic fashion, in view of the foreseen economic and business context of the 2030 era. Specifically, 6G paradigms will need to support (i) continuous connectivity, thus enabling coverage expansion compared to 5G in a cost-efficient way to simultaneously reach high capacity, lower latency, and improved reliability; (ii) zero-energy devices, e.g. for Internet of Things (IoT) and sensing applications for devices dispersed in wide areas for which replacing batteries will not be practical; and (iii) network-compute integration to allow better predictions while maximizing performance in terms of latencies under 1 millisecond, low jitter, and high communication resilience.

2.2 Relevance of the B2B market for 6G

First generations of mobile communication networks have mainly targeted the B2C market. Offers targeting the B2B market have been mostly limited to providing connectivity to enterprise employees and providing connectivity to manufactured objects. While the first one has been present since the beginning of mobile communications with resource management offers, the second one has appeared progressively with the rise of Machine-to-Machine (M2M) communications, e.g., for logistics and traceability purposes. However, these use cases used to be quite marginal in the mobile network operator business, which remained focused on delivering connectivity and providing communication means to the mass market.

5G has been a first move to complement the B2C model. The B2B market has been identified from the beginning as an important driver for 5G services, especially through ultra-reliable low latency communications (URLLC) and massive machine type communication (mMTC) features. The deployment models of 5G have also been designed to fit enterprise needs. First, 5G standards enable the provision of end-to-end network slices targeting specific enterprise needs, covering both radio access network and core network. Second, some industry players are going further and have deployed their own 5G private networks, fully dedicated to their specific needs.

It is expected that 6G will amplify this trend, and that B2B market needs will be a strong driver towards 6G[1]. Our societies indeed rely more and more on digital services and on ambient connectivity. Until now, these digital services have been mainly provided by Internet players (i.e., platforms and content providers), while mobile network operators are providing connectivity to this Internet of services. However, when moving forward in the

1 G. Wikström et al, "Ever-present intelligent communication. A research outlook towards 6G," Ericsson White Paper, 2020. [Online] Available: https://www.ericsson.com/en/reports-and-papers/white-papers/a-research-outlook-towards-6g

course of digitalization, social activities and business processes will be progressively transformed. Companies and public organizations will then become key players to implement and benefit from this transformation, and the future 6G requirements should therefore be driven by their pain points and use cases. This also implies a greater involvement of B2B verticals in the 6G standardization process, as outlined in the following chapters of this book.

In the 6G era, innovative mobile network operators will reap benefits from several business models simultaneously: besides selling connectivity services to end-users, telecommunication operators will provide the required connectivity resources to B2B players (including companies, cities, public authorities, etc.) to fulfill their own missions. Connectivity solutions will need to be flexible enough to adapt to more and more heterogeneous Quality of Service (QoS) requirements, along the lines of what 5G is offering with network slicing. We illustrate this point in the following sections by providing relevant KPIs for the described use cases in the B2B market. 6G should therefore be an opportunity for industry players to invent and design new devices and networking strategies to support their own digital transformation needs of processes and social activities.

2.3 Use Cases for the B2B Market

This section reviews the characteristics and requirements of envisioned 6G use cases for the B2B market, as summarized in Table 2.1.

2.3.1 Industry and Manufacturing

Each production facility and industrial area is a different use case with particular requirements. Industrial deployments, particularly in manufacturing, tend to have equipment and machinery with life cycles that extend over a decade. Therefore, the introduction and use of cellular networks for applications with stringent requirements is a long and gradual process that has started in dedicated deployments for a well-defined set of application. Thus, the integration with existing equipment, wired solutions, and radio access technologies (RATs) will still be required for many years into the future to ensure backward compatibility and improve overall performance.

As front-runners expand the applications relying on cellular networks, more complex deployments and applications will emerge. For example: dynamic deployments with URLLC requirements for exploration, mining activities, and wide area monitoring of pipelines, electric grids, and other critical infrastructures and the need for robots to perform cooperative maneuvers that require high precision and coordination. Moreover, for the factory cases with extreme performance requirements, it will be crucial to support URLLC in factory floors with moving objects and reshuffling manufacturing units. Digital twins (discussed in detail in Section 2.3.3) will also play an important role for industrial applications, where a digital twin replicates a real-world object in a digitized environment, which would enable the evaluation of performance and outcomes of industrial applications in safe, digital environments before a solution is deployed in the real world.

Table 2.1 6G use cases for the B2B market.

Use case	Section	Description	Impact for B2B market	Relevant KPIs
Industry and manufacturing	2.3.1	Digital transformation of manufacturing through cyber physical systems and IoT services	• Efficient mining activities • Wide area monitoring of pipelines • Better precision and coordination for cooperative maneuvers	• Reliability • Latency • Spectrum efficiency
Teleportation	2.3.2	Holographic delivery of life-sized three-dimensional stereoscopic experiences in real-time without HMD technologies	• Stimulate telework • Decrease travel time and expenses • Improve labor productivity	• Per-user data rate • Peak data rate • Latency
Digital twin	2.3.3	Digital replica of an object inheriting the same behavior and characteristics of the real object	• Improve the design, engineering, inspection, and maintenance of complex machines and devices • Allow advanced simulations of the product behavior	• Per-user data rate • Peak data rate • Latency
Smart transportation	2.3.4	Evolution of automotive industry to support infotainment, automated driving, and cooperative and intelligent transportation	• Preemptive logistics • Fleet management and telematics • Reduction of goods transportation costs and environmental impact	• Spectrum efficiency • Per-user data rate • Continuous coverage • Vehicle speed • Number of devices • Energy efficiency
Public safety	2.3.5	Collection of services that support fast delivery of information between emergency operators in incident areas and the (remote) command station	• Real-time 3D rendering of the incident area to/from the command station • Remote control operations	• Reliability • Latency • Spectrum efficiency • Vehicle speed • Continuous coverage
Health and well-being	2.3.6	Evolution of healthcare to support telemedicine, healthcare workflow optimizations, and efficient and affordable patient access to health assistance	• Individualized assistance via virtual patient consultation and monitoring involving all senses and health indicators • Efficient use of healthcare resources through preventive care, digitalization, and access to massive data • Reduction of management and administration costs through "care outside hospital" paradigm	• Reliability • Latency • Spectrum efficiency

Table 2.1 (Continued)

Use case	Section	Description	Impact for B2B market	Relevant KPIs
Smart-X IoT	2.3.7	IoT and smart city paradigms targeting life quality improvements, environmental monitoring, traffic control, and city management automation	• Efficient agriculture and farming • Fleet management • Smart warehouse management • Support of zero-energy sensors for home appliances, industrial machines, and robots	• Spectrum efficiency • Per-user data rate • Number of devices • Energy efficiency
Financial world	2.3.8	Evolution of financial sector through high-frequency trading and blockchain technology	• More accessible trading • Elevated security and reduced fees for banking transactions • Robust/secure fraud prevention • Transition toward digital banking	• Latency • Reliability

In 6G we will see improved solutions to support multiple connectivity in the same deployment and application to improve reliability and performance through link diversity and aggregation. While some solutions are already available in 5G such as dual connectivity and carrier aggregation, 6G will cover the need to support and aggregate multiple RATs maintaining the latency and reliability requirements of industrial applications, which will also provide additional access to local spectrum. In addition, the integration of time-sensitive networking (TSN) and deterministic networking (DetNet) standards that started in 5G will be fully utilized to support deterministic data transmission over cellular networks. Finally, network exposure and integration with cloud capabilities are key factors to enable the expansion of 6G for industrial applications, since they will enhance the network capability management and provide support for real-time applications while keeping sensitive operational data on cloud deployments that are kept secured.

2.3.2 Teleportation

Teleportation represents the future of communication, enabling holographic delivery of life-sized three-dimensional (3D) stereoscopic experiences in real-time without head-mounted device (HMD) technologies like augmented reality/virtual reality (AR/VR).

As the business world becomes increasingly automated and ubiquitous, teleportation will stimulate remote telework by allowing flexible virtual interaction during business events and meetings involving geographically distributed colleagues and industries, thus eliminating time and distance barriers. This technology may also decrease carbon footprint (the combustion of fuel for transport accounts for about 30% of global greenhouse gas emissions[2]) and save travel time and expenses, revitalizing small and medium-sized enterprises with limited corporate travel capabilities. Furthermore, replacing in-person business meetings with holographic events may improve labor productivity through more concise interaction and reduced stress for travel planning. Compared to video conferencing technologies, teleportation supports body language and nonverbal communication, thus guaranteeing better audience engagement and reception of information as well as enhanced productivity during professional events. Finally, even though business travel will always represent an essential resource for many company divisions, from sales to marketing, and from production to research and development, it is expected that teleportation will represent a valuable alternative for future corporate business.

Teleportation does not refer only to the digital transmission of physical quantities; it will also enable a clear and reproducible digital representation of all human senses, including smell and taste. This will allow, for example, chemical industries to speed up pharmaceutical product preparation and development via virtualization of drug tests,; healthcare companies to implement noninvasive, real-time diagnostic tools for health monitoring; and agro-food industries to tailor their offerings to consumer preferences while improving quality control of raw materials and increasing the overall efficiency of the food system.

Despite these benefits, remote connections via teleportation will introduce significant demands on the 6G network infrastructure, which are not supported today [2]. Specifically, 5G and previous generations have been typically designed to support audio and 2D-like video communication, where the same data content is broadcast regardless of the viewer's position. In turn, 3D telepresence adds parallax, meaning that the image changes depending on the viewer's position and its interaction with the image itself. This approach will radically change the role of the user (from passive video consumer to interactive consumer of multi-sensory experiences) and lead to a massive increase in the requirements for capturing, transmitting, and interacting with teleportation services. To fully realize an immersive remote experience, all human senses, including touch, smell, and taste, together with video and audio information, will be digitized and transferred across future networks at a data rate up to several terabits per second, which depends on the sensor's resolution and frame rate: for example, a raw uncompressed hologram with colors, full parallax, and 30 fps would require 4.32 Tbps [1]. The latency requirement will also be very challenging to ensure interactive content provisioning and real-time communications. While the 5G paradigm sets the round-trip latency limit in the RAN to 1 ms, 6G technologies will hit the sub-milliseconds to make the holographic experience smoother and more immersive. Finally, hologram-based applications will need to process a massive number of streams originating from sensors at different angles of view, thus involving stringent synchronization requirements.

2 A direct comparison between the transport and information communication technology (ICT) carbon emissions can be found in the report "A quick guide to your digital carbon footprint." [Online] Available: https://www.ericsson.com/en/reports-and-papers/industrylab/reports/a-quick-guide-to-your-digital-carbon-footprint

2.3.3 Digital Twin

A digital twin is a digital replica of an object, generally characterized by a very high level of fidelity that makes it possible to use the digital version as a reliable representation of the behavior and characteristics of the real object [3]. The concept of digital twins has risen to the forefront of the discussion on product life cycle management thanks to improvements in the design and capabilities of sensors (e.g. video cameras, laser scanners, and lidars) and sensor fusion algorithms, which now allow a rich and faithful representation of the real object, as well as thanks to advances in computation capabilities, which enable the real-time manipulation and editing of the digital twin. Moreover, the concept of digital twin is often associated to VR and AR, as the digital representation can be visualized through any immersive visualization technique. Thanks to these properties, digital twins can improve the design, engineering, inspection, and maintenance of complex machines and devices. For example, a machine could be remotely inspected without the need for personnel on the ground, without any loss of realism, and with an improved (digital) access to components that would be hard to reach physically. Similarly, mechanics can monitor the performance and status of different components of a vehicle with a high-fidelity representation without the need for the car to be in the repair center. Additionally, for product development, a digital twin would allow different teams to work on the same product, exploiting a 3D, shared visualization in various remote locations, and can enable advanced simulations of the product behavior.

In these scenarios, the role of the network is to provide a high-throughput, low-latency bit pipe to connect the sensors on the physical product with the computing platforms on which the digital twin is hosted. Several elements contribute to the need for ultrahigh throughput, which would not be supported by 5G technologies, as for the teleportation use case of Section 2.3.2. A digital twin will be generated by a large number of data sources, which need to be distributed around the physical device, and capture different properties, not only the visual aspect. Moreover, the twinning rate, i.e. the rate at which the physical and digital representations are synchronized, could be in the order of hundreds of Hertz, for applications that require a real-time tracking of the evolution of the physical object. Therefore, the data rate required by digital twin use cases can be in the order of tens of gigabits per second, with the need for high-capacity links between the different components of a digital twin system (sensors to database, and database to representation). Similarly, when real-time interaction and control of the physical object through its digital counterpart is required, the latency should be in the sub-millisecond range. However, if real-time control is not of interest, or a lower level of fidelity can be accepted, digital twinning applications can tolerate higher latencies and lower throughput. Therefore, 6G networks should also focus on adaptability and openness to the applications, with open interfaces to enable cooperation between the wireless stack and the higher layers, for example, to optimize the number of sources and the twinning rate according to the capabilities of the network, or, vice versa, to allocate more resources according to the needs of the application.

2.3.4 Smart Transportation

The support of communications in smart transportation is threefold: infotainment, automated driving (AD), and Cooperative Intelligent Transport Systems (C-ITS). Infotainment (sometimes referred to as *navitainment*) comprises information, navigation, and entertainment services for drivers and passengers. These services are expected to evolve into extended reality (XR) experiences for passengers and enhanced high-definition (HD) maps and real-time information services for drivers, and industry players will need to collaborate in order to satisfy the demand for different in-vehicle services. AD will experience a gradual increase in capabilities and market share. It is expected that up to 15% of the new vehicles sold in 2030 will have AD in designated conditions and, while the personal vehicle market is also expected to grow, trucks and delivery vehicles have a stronger business incentive compared to personal vehicles, which will drive faster deployment once technology is available [4]. AD requires high volumes of data to be exchanged between cars and the cloud for HD 3D maps, sensor sharing, and computational offloading. Those are the aspects related to individual vehicles, but the goal of C-ITS is to improve safety and comfort by exchanging information between vehicles and the road infrastructure. Real-time information will include not only measurements and status from sensors but also path planning and cooperative maneuvers, that are particularly relevant for unmanned aerial vehicles. An important consideration is vulnerable road users (VRUs) such as pedestrians, cyclists, and road workers that can be increasingly protected with solutions based on positions and path crossing alerts enabled by the communication between smartphones (or other personal devices) and vehicles. The former aspects are mostly related to the mobility safety and experience but, in the future, other use cases such as preemptive logistics, fleet management, and telematics will expand and have a key role in society. These services are expected to be implemented by global players in the coming years and, even if they have less stringent requirements on data rate and latency, network coverage and secure private cloud platforms that leverage on network capabilities will be essential for fleet operators and vehicle manufacturers.

While some of these functionalities can be supported in 5G networks, 6G will play a key role in increasing the flexibility to expand coverage and enable services in all locations and conditions. Continuous coverage will be key if AD should be able to rely on connectivity. Moreover, even lower latencies can enable the use of services at higher traveling speeds. Also, the expected timelines for many of these services in the mass market match the 6G expected release plans. With respect to C-ITS requirement, data can be exchanged as collective perception messages (CPM) [5] where an average payload of 900 bytes generated at 1–10 Hz can be assumed depending on sensors, speeds, and traffic density. The download requirements will depend on the number of vehicles and other relevant user equipments (UEs) in proximity. A very important requirement will be the possibility to enable accurate positioning for moving objects, where 1 m–10 cm is the commonly referred range depending on the use case (which corresponds to 30–3 ms latencies at 120 km/h).

In this perspective, even a significant increase in the channel capacity may not be enough to satisfy the boldest service requirements of future automotive applications. One possible solution is to realize a fully distributed user-centric architecture in which end terminals make autonomous decisions, "disaggregated" from the network. This approach removes the burden of communication overhead to and from centralized network entities, thus achieving quasi-real-time latency, e.g. yielding more responsive driving decisions.

2.3.5 Public Safety

Communications are a primary enabler of critical PS operations. First responders need to be aware of their surroundings and of the activities of the other personnel in the field. Moreover, communications are essential to deliver information and orders throughout the chain of command, i.e. between emergency operators in the incident area and the command station that is often remote. While traditionally the technologies for PS communications have focused on voice, data services can significantly improve the experience and safety of first responders. Notably, enhanced monitoring capabilities could allow a real-time 3D rendering of the incident scenario at the command station or in head-mounted headsets for the first responders. This can be done through video, from body cameras, or from flying platforms and with additional sensors such as lidars, 3D cameras, and thermal cameras, among others. Moreover, health and position sensors on PS operators could continuously stream telemetry data to other first responders and to the command station. Finally, the communications will not only be human-to-human but also extend to machine-type traffic, to networking among vehicles (e.g. ambulances, fire trucks), and to remotely controlled devices. Remote control operations are indeed fundamental in several PS scenarios, where robots (e.g. wheelbarrow robots) are used to remotely defuse bombs, inspect incident locations, and perform operations in conditions that would be dangerous for first responders (e.g. during chemical leaks).

Given the importance of the related scenarios and use cases, PS networking has thus been at the forefront of standardization and research efforts throughout different generations of cellular networks, with notable examples in the device-to-device communications and proximity services introduced in long term evolution (LTE) Release 12 [6] and the development of FirstNet using LTE technologies. Following this trend, 5G research has focused on how to improve the throughput of data services in emergency scenarios, relying on the new spectrum bands (i.e. mmWaves) and mobile communication platforms (i.e. vehicular communications and drones). As discussed in [1], however, it is not clear whether 5G technologies will be capable of delivering the improved quality of service (QoS) (e.g. the ultrahigh throughput) with the high reliability level and the ubiquitous coverage required to support PS services.

Therefore, there is a case for further developing promising 5G innovations and bringing them to full fruition in 6G networks, focusing on reliability and coverage, with possible improvements in throughput and latency. Notably, the integration of non-terrestrial (e.g. with satellites, balloons, and unmanned aerial vehicles (UAVs) and terrestrial networks in 6G will increase the coverage of the network, allowing connectivity of a staggering 10^7 devices per square kilometer. PS communications will also benefit from the increased throughput, to provide teleportation-like experience between the command station and the incident site. Moreover, orchestration and remote control of robots requires *end-to-end* ultralow latency, thus pushing the over-the-air latency requirement into the sub-milliseconds region and placing tight constraints on the latency budget of the rest of the network. An important requirement of PS networking is related to the sustainability and autonomy of the infrastructure, which should strive to consume as little power as possible to improve battery life in off-grid infrastructures and mobile devices. To this end, 6G is expected to increase the energy efficiency by a factor of 10 with respect to 5G, with improvements in both the device battery lifetime and the overall network consumption.

2.3.6 Health and Well-being

The global increase in the cost of providing healthcare services to a continuously ageing and growing population is rapidly becoming unsustainable. In this context, 6G is positioned to foster the healthcare revolution by eliminating time and space barriers through telemedicine, achieving healthcare workflow optimizations, and guaranteeing patient access to increasingly more efficient and affordable health assistance.

On one side, 6G connectivity solutions should enable the transition from a traditional provider–patient relationship toward a "care outside hospital" paradigm, where primary care services will be delivered by health professionals directly to the patients at home. Moving care outside clinics and health facilities will not only promote more individualized and personalized assistance but also empower preventive care while avoiding that fragile patients with limited mobility capabilities need to travel. From a business-oriented perspective, home care guarantees the most efficient use of healthcare resources (e.g. preventive care can drastically reduce the need for expensive treatments for patients with chronic conditions) and a significant reduction of management and administration costs for institutional care centers [7].

Cost savings in the healthcare industry will also increase the reach and accessibility to healthcare assistance to the most unprivileged and least developed countries in the world, thus making it possible for an estimated billion extra patients globally to receive quality treatment [8]. The goal is to achieve healthier life years and more efficient health and social care for a larger population [8, pp. 5–6].

Moreover, the development of 6G technologies, together with the digitalization of healthcare services, will allow more granular and higher quality data to be collected on patients, thereby improving clinical analyses and reducing health costs associated with treatments of diseases.

Furthermore, 6G innovations should drive the design and adoption of new use cases in the healthcare sector. VR- and AR-based technologies will facilitate remote patient monitoring, while artificial intelligence and tactile sensing will enable even more invasive healthcare assistance through robotic telesurgery, i.e. remote surgery where surgeon and patient are geographically separated. Robotics and automation advancements will empower connected ambulances, while holographic solutions combined with the transmission of important health indicators (collected through wireless body sensors) will make it possible to improve healthcare assistance via virtual patient consultation and monitoring.

Due to technology limitations in today's wireless networks, future healthcare applications are calling for the design of new wireless communication systems that support continuous interaction with mobile end users. Besides the high cost and the lack of medical professionals and infrastructures in today's healthcare industry, the current major limitation is the lack of real-time tactile feedback. Moreover, the explosion of advanced eHealth and mobile Health (mHealth) services challenges the ability to meet their stringent QoS requirements. For reliable remote surgery, for example, the latency demands will be in the order of sub-milliseconds, which are not yet achievable with upcoming 5G innovations. Even for less latency-critical use cases, e.g., digital healthcare assistance enabled through VR/AR technologies, combined with holographic communication, will pose very strict requirements in terms of end-to-end throughput will need to be satisfied (for 3D MediVision products, a resolution of 1920 × 1080 pixel and a frame rate of 120 fps for 3D displays will require multi-Gbps data rates to be supported [9]). Extremely high reliability (>99.99999%) will also be needed due to the potentially catastrophic consequences of a communication failure. It is estimated that the increased spectrum availability, combined with the refined intelligence of 6G networks, will guarantee these KPIs, together with 5–10× gains in spectral efficiency [1].

In this context, integrating networks and applications emerges as a viable approach to support resource-demanding services by exchanging information in such a way that specific requirements are satisfied. For example, different network configurations and related system parameters (including – but not limited to – the choice about the optimal deployment, power allocation, interface selection) should be adopted based on network's available resources and supported capabilities and dynamically (and iteratively) updated until network requirements are satisfied.

2.3.7 Smart-X IoT

The IoT and smart city paradigms have emerged as typical use cases for 4G and 5G networks, as well as for a wide set of noncellular communication technologies (e.g. LoRaWAN, LP-WAN, among others). 6G will further expand these platforms, to ease the transition toward digital services in a wide range of business areas. Notably, Smart-X (where X refers to everything) use cases will require ubiquitous coverage, with a support for more than 10^7 connected devices per square kilometer, and high energy efficiency, so that zero-energy devices can be globally deployed to gather environmental awareness, track complex processes (e.g. shipments, production lines), and then actuate data-driven policies. Future Smart-X scenarios will heavily rely on artificial intelligence and machine learning techniques to infer trends and behaviors, which will need to be driven by a constant stream of data from the network.

An area in which the importance of connectivity-based digital services is central is that of agriculture and farming. A sensor-enabled data pipeline is key to monitor the efficiency and results obtained when growing crop in very large areas. Moreover, the availability of data in this area makes it possible to track the evolution of terrain needs and characteristics and infer future trends in crop availability. Similarly, for animal breeding, connectivity-based approaches allow farmers to monitor the position and health status of each single animal in the herd.

Data-driven predictive analytics and monitoring are useful also for fleet management, as discussed in Section 2.3.4, and for smart warehouse management. In particular, with Smart-X approaches, the status, position, and value of goods can be tracked with cheap sensors from the source (e.g. the factory in which the good is manufactured) to the warehouse in which it is stocked and, eventually, to the final destination. Moreover, for products such as home appliances, industrial machines, and robots, among others, embedded zero-energy sensors can measure and report failures or anomalous behaviors toward an integrated factory-to-customer assistance and maintenance pipeline.

To this end, as previously discussed, it is important to provide low-cost, low-energy and highly available connectivity in 6G networks. Notably, Smart-X applications need global coverage and the possibility of transmitting and receiving data with the same connection throughout the whole world. Consider, for example, the current production paradigm where a product may be manufactured in China, shipped to Europe (through either a ship or an airplane), further processed, and then delivered to a customer in the United States. To enable a seamless, global tracking and monitoring, the sensors on the device should be able to authenticate and securely transmit data, irrespective of the area in which the device is located. Moreover, given the scale of the deployment of Smart-X sensors, it will be important to design 6G networks so that such sensors can be cheap, consume very little energy, and be easily disposable and/or reusable for different applications.

2.3.8 Financial World

6G is positioned to revolutionize the financial sector by allowing companies to launch new products and services not previously possible, move into new markets, and increase productivity.

Along these lines, 6G can enable efficient HFT, a new method of trading where powerful computer programs can transact millions of financial decisions in fractions of a seconds. For these applications, receiving data even a millisecond sooner can represent a clear advantage over the competitors and generate profits. To this aim, financial institutions typically tend to buy computing facilities as close as possible to the trading and exchange offices to get the trading transactions close to the speed of light, which is, however, a very expensive and often impractical approach. In this context, 6G innovations in the wireless architecture design could offer a better (and more accessible) solution for achieving ultralow-latency communication in comparison to fiber optic equipment and on-site deployments, especially in those (rural and/or remote) areas where wired connectivity cannot be easily provided.

The blockchain technology has also gained momentum in the financial industry as a solution to decentralize and eliminate intermediaries in financial feeds while guaranteeing transparency and anonymity. In this perspective, blockchain can be considered as an additional component technology to support financial instruments in 6G networks [10]. 6G innovations, in fact, can provide elevated security features, e.g. through quantum computation, as well as reduce processing fees and improve resilience and resistance to external attacks.

Moreover, in the financial world, 6G can facilitate merchants and banks to deploy transformative and highly personalized customer service experiences, including virtual tellers in banking, high-resolution digital signage with facial recognition in retail, and AR/VR-enabled e-commerce (which would allow potential customers to explore virtual showrooms, improve their shopping experience and, in the end, encourage them to complete a purchase). By supporting high-capacity data communications in real-time, 6G can also support time-sensitive banking operations for both ordinary customers and banks, and accelerate inclusion of small financial institutions, e.g. from emerging markets, by transitioning from (expensive) banking facilities to more accessible, ready-to-use digital banking experiences. Furthermore, 6G can help implement more robust and secure fraud prevention without consumer intervention or other expensive direct activities.

Large-scale financial operations could further stress already congested communication networks, which will struggle to guarantee low-latency connectivity with very high degrees of reliability. For example, for HFT, latency requirements should be in the order of sub-millisecond (a 1-millisecond advantage in trading applications may be worth $100 million a year to a major brokerage firm, according to some estimates [11]), in contrast to the more relaxed (though already challenging) 1 ms 5G target. For blockchain operations, in turn, security, as well as energy consumption, will be key concerns. Finally, digitalization in the financial industry and the introduction of artificial-intelligence-based procedures will result in very large amounts of data to be exchanged and processed, which will likely saturate 5G capacity: the per-user data rate may need to reach the Gbps range, at least one order of magnitude up from the 100 Mbps of 5G.

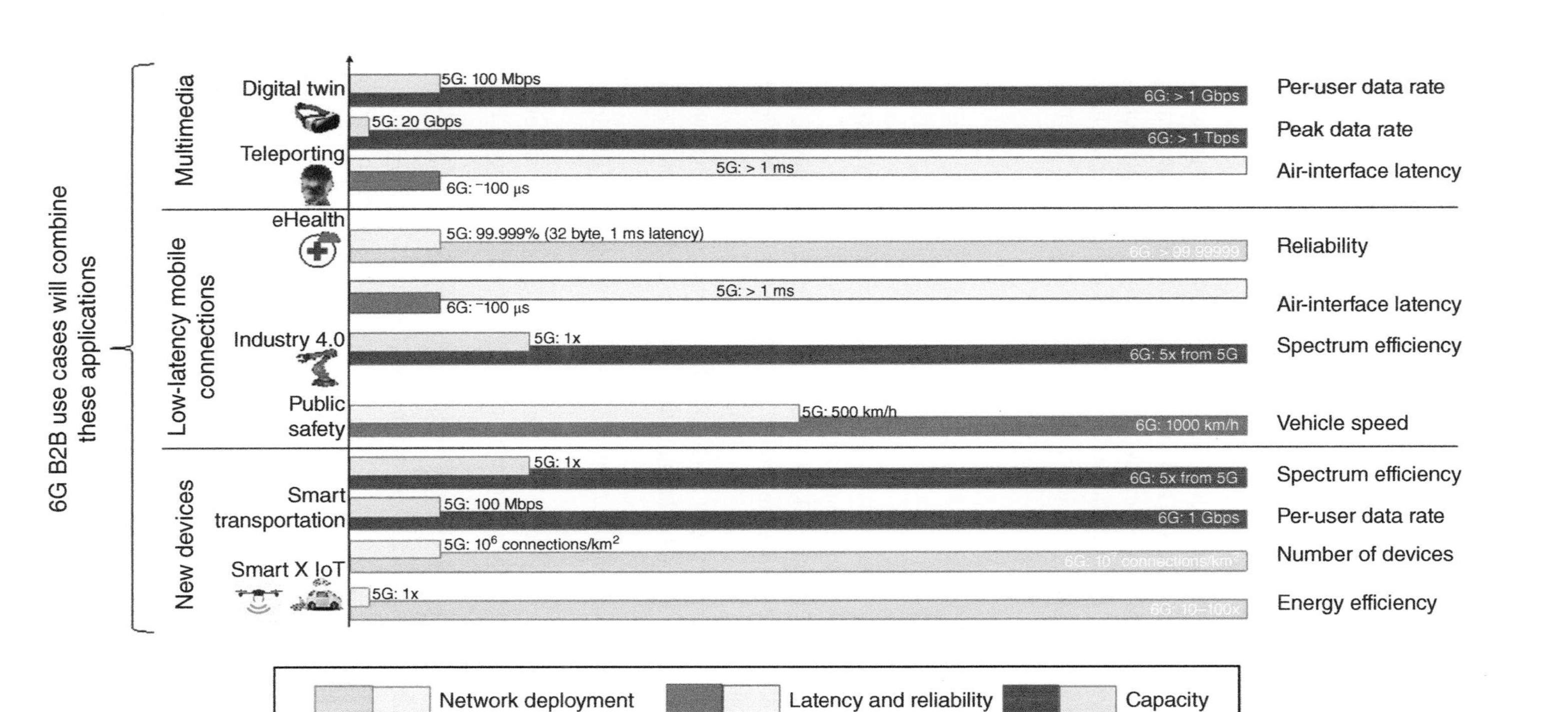

Figure 2.1 Representation of multiple KPIs of 6G use cases and improvements with respect to 5G.

2.4 Conclusions

In this chapter, we analyzed possible business-oriented use cases, scenarios, and relative KPIs for future 6G networks, as summarized in Figure 2.1. Notably, even though the market verticals we presented have been already considered in 4G and 5G networks, we focused on applications of these verticals that would not be supported by 5G networks in terms of (often combined) throughput, latency, coverage, and reliability requirements. After a brief introduction, we discussed the importance of services that (i) enable remote presence or inspection with high fidelity, including sensory information (teleportation, digital twin); (ii) globally connect goods, vehicles, and machines (Smart-X IoT, smart transportation and industry); and (iii) provide secure continuous connectivity for critical information (remote healthcare interventions, PS, financial world). We believe that this chapter can provide a basis for a requirement-driven development of future 6G networks, to enable key digital services for the connected society of 2030 and beyond.

References

1 Giordani, M., Polese, M., Mezzavilla, M. et al. (2020). Toward 6G networks: use cases and technologies. *IEEE Communications Magazine* 58 (3): 55–61.

2 Clemm, A., Vega, M.T., Ravuri, H.K. et al. (2020). Toward truly immersive holographic-type communication: challenges and solutions. *IEEE Communications Magazine* 58 (1): 93–99.

3 Jones, D., Snider, C., Nassehi, A. et al. (2020). Characterising the digital twin: a systematic literature review. *CIRP Journal of Manufacturing Science and Technology* 29 (Part A): 36–52.

4 ITF (2019). *ITF Transport Outlook 2019*. Paris: OECD Publishing https://doi.org/10.1787/transp_outlook-en-2019-en.

5 ETSI TR 103 562 V2.1.1, Intelligent Transport Systems (ITS); Vehicular Communications; Basic Set of Applications; Analysis of the Collective Perception Service (CPS); Release 2, 2019.

6 Lin, X., Andrews, J.G., Ghosh, A., and Ratasuk, R. (2014). An overview of 3GPP device-to-device proximity services. *IEEE Communications Magazine* 52 (4): 40–48.

7 Pouttu, A., Burkhardt, F., Patachia, C. et al. (2020). 6G white paper on validation and trials for verticals towards 2030. White Paper.

8 Laya, A. (2017). The internet of things in health, social care, and wellbeing. Ph.D. dissertation. KTH Royal Institute of Technology.

9 Zhang, Q., Liu, J., and Zhao, G. (2018). Towards 5G enabled tactile robotic telesurgery. arXiv preprint arXiv:1803.03586.

10 Hewa, T., Gür, G., Kalla, A. et al. (2020). The role of blockchain in 6G: challenges, opportunities and research directions. 2nd 6G Wireless Summit.

11 Martin, R. (2007). Wall Street's quest to process data at the speed of light. *InformationWeek*.

3

6G: The Path Toward Standardization

Guy Redmill[1] *and Emmanuel Bertin*[2]

[1] *Redmill Communications Ltd, London, UK*
[2] *Orange Innovation, France*

3.1 Introduction

Historically, standardization has been fundamental to the success of each new generation – or "G" – of mobile network technology. It is the process through which we arrive at a set – or sets – of replicable guidelines governing technology, interoperability, and performance for a specific technology that realizes specific technical goals and that enables commercial delivery and operation, from a diverse community of providers.

This success is clear. There are now more than 5 billion mobile people using different generations of network technology, while the industry contributed more than $4 trillion to the global economy in 2019 [1]. In addition, millions more devices are connected via these networks, serving a wide range of Internet of Things (IoT) and machine-to-machine (M2M) applications.

For recent "Gs," a broadly similar process has been followed, which has resulted in the release of standards, which can be adopted and followed by industry stakeholders. These standards provide templates that enable participants to contribute and to develop the solutions required to build a new network.

However, there are a number of factors that suggest the path to 6G may diverge from the most recent path taken to the delivery of a global standard. First, there are historic precedents that highlight alternative paths to realizing the common goal of a particular generation of mobile technology.

Second, regardless of the ultimate realization of 6G, the stakeholder ecosystem has changed dramatically as the mobile landscape has evolved. The traditional mobile value chain has expanded to include new classes of operators, new vendor consortia, as well as new spectrum holders (with license conditions that may differ from the previous model of country-wide coverage obligations), and has attracted a wide range of new vertical actors, largely drawn by the new capabilities 5G can unlock. 6G will likely lead to further expansion of this ecosystem.

Shaping Future 6G Networks: Needs, Impacts, and Technologies, First Edition.
Edited by Emmanuel Bertin, Noel Crespi, and Thomas Magedanz.

Third, new political pressures and fault lines have emerged, which have already affected the standardization work of the 3rd Generation Partnership Project, commonly known as 3GPP. Fourth, while a broad community of solution providers has been envisaged, in practice, this has narrowed dramatically, leaving many operators dependent on just a handful of suppliers. Efforts are underway to change this, which could lead to deviation from the current standards path – accelerating or impeding progress toward 6G. Finally, 5G is enabled by an entirely new operating model that, in turn, could change the way in which stakeholders interact to define new standards in the future.

This chapter explores the evolving landscape and considers possible future standardization models for 6G – and beyond – based on interviews with selected stakeholders and lessons drawn from the evolution of mobile network technology to date.

3.2 Standardization: A Long-Term View

> How did we reach the point at which 5G could be realized, via what is effectively a single body that represents global mobile standards, and which has become the preeminent voice in the industry? Historically, this was not the case for each previous generation of mobile technology.

There have been a number of successful standardization initiatives that have resulted in the release of each new generation of mobile technology – 2G, 3G, 4G, and, more recently, 5G. However, there has often been debate about how each such generation should be realized, which has led to variation in the implementation of some previous "Gs" and the emergence of parallel standards that have aimed to deliver the same outcome. In this section, we will briefly review current activities and explore previous approaches.

For a number of years, each new "G" has begun with the release of what is known as an "International Mobile Telecommunications" (IMT) recommendation. IMTs are defined by the International Telecommunications Union (ITU), a specialist agency of the United Nations. It is important to note that an IMT is simply a recommendation. It does not define how it should be realized, but it does define expected performance levels:

> The term International Mobile Telecommunications (IMT) is the generic term used by the ITU community to designate broadband mobile systems. It encompasses IMT-2000, IMT-Advanced and IMT-2020 collectively, [2].

At the ITU, work is already underway toward 6G under the overall banner of IMT-2030. Stakeholders such as solution vendors, operators, research institutions, and other agencies are collaborating to define 6G and what it may mean in practice. As part of these activities, FG-30 (Focus Group on Technologies for Network 2030) is creating a definition of what a 6G network should deliver:

> The Focus Group intends to study the capabilities of networks for the year 2030 and beyond, when it is expected to support novel forward-looking scenarios. . .The study aims to answer specific questions on what kinds of network architecture and the enabling mechanisms suitable for such novel scenarios [3].

This will eventually result in a new IMT requirements definition, which will be available for the industry.

3.3 IMTs Have Driven Multiple Approaches to Previous Mobile Generations

Previous generations of mobile technology have also been kick-started by the agreement of a relevant IMT. While IMT-2020 led to 5G and an agreed, unified approach spearheaded by the 3GPP, earlier IMTs have given rise to different approaches to achieve the same desired outcomes.

For example, IMT-Advanced provided requirements for the 4th generation of mobile technology (4G), but there were several approaches that could have met these. Long-term evolution – otherwise known as "LTE" – was just one example; Mobile WiMAX and Ultra Mobile Broadband, among others, were, for several years, considered as viable alternatives, although only two remained as candidate systems – LTE and Mobile WiMAX [4]. Similarly, the earlier IMT-2000 recommendations, which gave rise to 3G, also generated several different candidate solutions, some of which saw commercial operation.

3GPP was originally founded in 1998, with the aim of bringing together a number of foundational partners to collaborate on common standards to meet the requirements of IMT-2000 and the realization of what became known as 3G [5]. 3GPP based its efforts on the evolution of the existing GSM (the Global System for Mobile Communications) standards for 2G, originally driven through the European Telecommunications Standards Institute, or ETSI. The partnership agreement brought together organizations that had previously, either individually or via different collaboration groups, contributed to earlier generations of mobile technology.

3GPP's efforts resulted in a set of standards for 3G – the Universal Mobile Telecommunications System (UMTS) – whose work has ultimately led to today's 5G. However, a parallel organization (3GPP2) shadowed the work of the original 3GPP during this period, considering the perspectives of a different group of stakeholders, some of which proposed alternative candidates for 3G, such as CDMA2000, to meet the same IMT-2000 requirements.

CDMA2000 was adopted in the United States, South Korea, Japan, China, and Canada, for example, but saw little deployment outside these countries, as most nations favored the UMTS standards defined by 3GPP. 3GPP2 subsequently pursued initiatives to create standards for the 4th generation of mobile; however, these met with little success, because, as far as IMT-Advanced is concerned, LTE won the race and became the default commercial vehicle for 4G.

The alternatives also failed to gather significant commercial momentum (although some such as Mobile WiMAX did reach deployment in several countries) and, eventually, fell by the wayside. Different approaches to IMTs are illustrated in Figure 3.1.

As a result, 3GPP2 effectively ceased operations in around 2013 and is now a dormant organization. This latter point may be highly significant for the future, as we shall see.

Even before this, while GSM had been promoted for 2G by ETSI, other approaches had been developed by alternative standards organizations, resulting in multiple 2G standards that enjoyed widespread adoption for a long period of time.

5G is thus something of an anomaly in the history of the mobile industry, as it represented a coordinated, single global approach to the realization of the IMT-2020 demands.

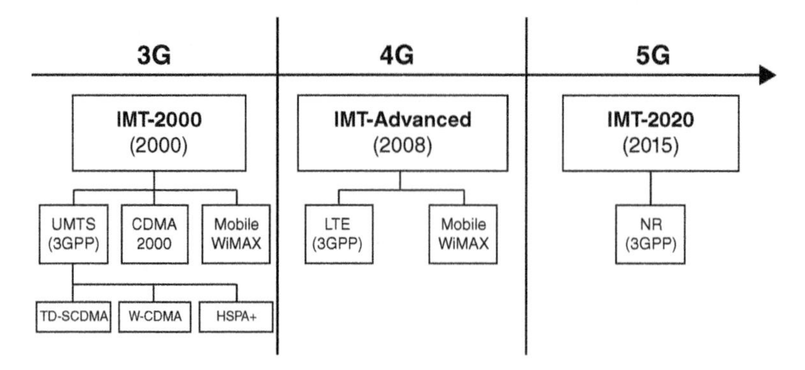

Figure 3.1 IMTs and different generations of mobile network technology.

For each previous generation, there was not, at the outset, necessarily an obvious winner or a single unified approach. Will 6G see a return to such a situation, or will a global effort prevail? There are many reasons to think that 6G could well lead to fragmentation in mobile standardization. Why?

3.4 Stakeholder Ecosystem Fragmentation and Explosion

> 3GPP is currently the main body pushing forward with the evolution of 5G standards, but it has expanded considerably in recent years. What has driven that expansion and what might this mean for the long path towards 6G, currently being driven by early-stage discussions in working groups of the ITU?

Since 1998, 3GPP has expanded considerably, growing from an initial five "organizational partners" and six "market representation partners." Organizational partners are different telecommunications standard development organizations (SDOs). Currently, there are seven SDOs: Association of Radio Industries and Businesses (ARIB) and Telecommunication Technology Committee (TTC) (Japan), Telecommunications Standards Development Society India (TSDSI) (India), European Telecommunications Standards Institute (ETSI) (Europe), China Communications Standards Association (CCSA) (China), Alliance for Telecommunications Industry Solutions (ATIS) (USA), and Telecommunications Technology Association (TTA) (Korea) [6]. Market representation partners are groups representing industry and sector interests. There are 21 market representation partners, including the GSMA. Each of these has, in turn, many of its own members.

Significantly, the number of market representatives has continued to grow in recent years, with new organizations that represent the specific interests of vertical industries entering the fray. Examples include 5GACIA (the 5G Alliance for Connected Industries and Automation), 5GAA (5G Automotive Association), 5G Americas, PSCE (Public Safety Communication Europe Forum), CSAE (China Society of Automotive Engineers), and others. Each of these is thus now equivalent, in status at least, with the GSMA, which has been seen as the most significant industry group for many years. No longer – which could mean new pressures emerge that disrupt the unified approach to realizing IMTs.

Finally, there are also "observer" members, which may be new national or international SDOs that are candidates for future elevation to organizational partner status. While 3GPP has accomplished much in becoming the primary organization responsible for leading mobile standardization initiatives, it is an increasingly complicated and rapidly growing body.

Of course, on the one hand this growth speaks for the strength of the industry and the alignment that has been achieved, particularly since IMT-Advanced was brought to fruition. On the other, it points also to an increasingly diverse spectrum of views. 3G and 4G developments were largely driven by a combination of vendor, operator, and regulator interests, with additional contributions from integrators and academia, and with the goal of realizing mobile telecommunications standards primarily for people and devices. It was a relatively simple and well-defined community. 5G has now brought a dramatic change to this ecosystem, expanding the focus to new sectors that can benefit from wireless communications. Why, and what might it mean for 6G?

5G is the first generation of mobile technology to have been designed, from the outset, to support multiple services. This was supported by a phased implementation. First, enhanced mobile broadband (eMBB) services were launched by many operators, using 5G spectrum but also leveraging existing core assets. eMBB services are largely dedicated to personal use, for consumer and business customers, and support the standard range of mobile capabilities (data, voice, and messaging services) but with enhanced performance.

However, the second phase brought new services that can be run on 5G networks in parallel with eMBB – massive IoT (MIoT) and Ultra-Reliable Low-Latency Communications (URLLC). It is these new services that have driven the most profound changes to the mobile ecosystem. That is because, while eMBB could be said to be "people-driven," MIoT and URLLC are largely "device-driven" [7].

The new capabilities that 5G (Phase 2) brings are designed to service new levels of performance for connected things and processes. So, while there are undoubtedly many exciting and novel applications and services that can be enabled for people, there are expected to be many, many more for devices and applications, across multiple industries and sectors.

Indeed, it is the new use cases enabled by MIoT and URLLC that are driving the expansion of the mobile ecosystem: new stakeholders have been attracted by the potential to use wireless connectivity to service a growing range of applications.

Broadly speaking, these can be divided into two categories: public and private. According to consulting firm, Arthur D. Little, use cases extend to areas such as "mobility and public transport; public safety; healthcare; energy and utilities; education; education and retail; media and entertainment; and, industry and agriculture" [8].

Within each category, an unprecedented number of use cases (when compared with previous generations of mobile technology) can be found – and will continue to be discovered, as the market matures. As a result, a growing number of new stakeholder groups have emerged to manage contributions to mobile standard evolution from interested parties.

These have aligned over both public and private sector interests. A UK-based software industry trade organization with more than 850 members, techUK, has noted that "the ecosystem is likely to expand rapidly" [9]. Similarly, diverse groups exist in numerous other countries. There is also a growing number of industry-specific groups that seek to represent the interests of members from particular sectors.

As such, it seems equally likely that competing interests will emerge. Indeed, 6G is already targeted at an even broader set of use cases, based on the introduction of new features and performance capabilities. As noted, the ITU is working on a definition of these new requirements, in the shape of IMT-2030. While this does not yet amount to 6G – and, indeed, may not [10] – this will shape new enhancements to networks and should be key considerations toward the definition of 6G. Specifically:

> to support various applications driven by devices. . .and devices to be developed in the future, the Network2030 infrastructure is expected to include fixed and wireless networks, cloud and space communications infrastructures. . . We expect virtualisation, memory and computing technologies in addition to Artificial Intelligence (AI)/ Machine Learning (ML) [to] continue to impact Network2030 [11].

As such, the emerging requirements for IMT-2030 – and, in all likelihood a new series of documents that will support evolution toward 6G – already embrace yet more stakeholders and technologies. A key conclusion from this is that standardization for 6G will, by necessity, involve a wider group of stakeholders participating and contributing their unique perspectives than has been seen before. For the time being, 3GPP is the first to align with IMT-2030, as can be seen in Figure 3.2.

But, it is clearly reasonable to ask if the current standardization model, based on an increasingly diverse and expanded 3GPP, can sustain and absorb contributions from such a broad ecosystem to arrive at a single set of standards that realizes the requirements demanded in 6G – and whether it can remain the sole voice of the industry. To complicate matters further, new political pressures have emerged that may further influence this ecosystem.

3.5 Shifting Sands: Will Politics Influence Future Standardization Activities?

> The issue of dependence on high-risk vendors has been thrown to the fore under the Trump presidency. Other countries are also beginning programs to de-risk their networks and limit the involvement of specific vendors.

Broadly speaking, the 3G world could be defined in two camps: US-led initiatives contributed to the evolution of CDMA2000, while 3GPP-led initiatives saw UMTS brought to

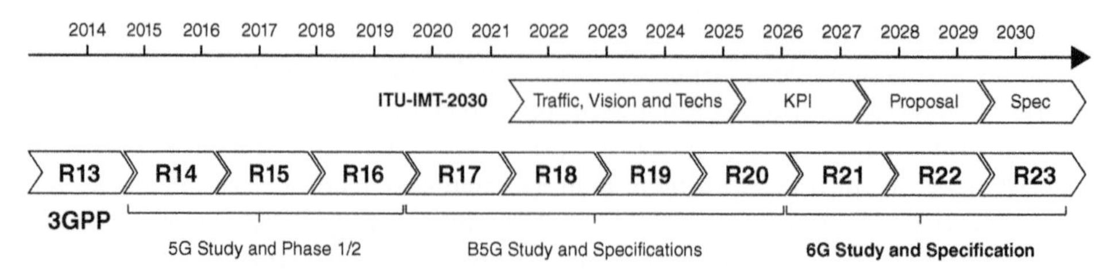

Figure 3.2 3GPP alignment with IMT-2030.

market. Since then, the evolution of mobile standards based on successive IMTs has been dominated by 3GPP, with 3GPP2 entering into hibernation. For 5G, 3GPP has been the sole standards body responsible for converting the target IMT performance goals into practical standards that can be implemented by stakeholders.

However, there has been significant, sustained debate on the current situation, mostly driven by the post-2016 White House [12]. Former President Trump had called for US leadership in the mobile industry to be reestablished while also demanding progress toward 6G [13]. Partly this has served to highlight US successes in 5G deployments, but it has also been part of a long-lasting series of attacks against what the administration sees as the commercial and political rivals of the United States, most notably China [14].

As a result, unprecedented tensions have emerged that have begun to have an impact on market players, with other governments echoing some of the former president's remarks [15]. In the United Kingdom, this has led to the identification of what has been defined as "high-risk vendors" (HRVs) and a clear policy to ensure that operators choose alternatives to such actors [16].

The advice goes so far as to define functional elements that are deemed to be "high-risk," thereby providing a template for operators to follow. Specifically, the advice notes that "the cyber security risk of using HRVs in the network functions set out below cannot be managed. . .if effective risk management of HRVs is to be undertaken, their products and services should not be used in the following network functions" [16].

A detailed list of such elements is provided. For 5G networks, this includes – but is not limited to – "5G core database functions, 5G core-related services," such as "Authentication Server Function" and "Network Slice Function," to name but two in a lengthy, comprehensive, and diverse set of instructions [16]. In other words, operators in the United Kingdom are now obliged to diversify their supply chains, an issue to which we shall return.

Similarly, the European Union has also started to move in a similar direction, with the European Commission reported as expressing the view that "progress is urgently needed to mitigate the risk of dependency on high risk suppliers" [17]. While this climate could settle through the creation of new trade agreements or political changes, it has also created considerable uncertainty, not to say unease, in the market and what seemed firm, long-standing global commercial relationships have been severely disrupted. These shifting sands will likely have an impact on the future standardization of 6G.

3GPP has become a global success story, drawing contributions from stakeholders from around the world. Chinese vendors, in particular, have made important and valuable contributions, alongside counterparts from elsewhere. While this has undoubtedly boosted the industry and accelerated 5G realization, it did not please the then current US leadership – so much so, that attention has now turned to the international standards that support the development of the mobile industry. According to *The Economist*, companies with manufacturing operations in America have "been frozen out of some standard-setting as an accidental consequence of the American government's attack" [18]. This report adds:

> The effect has been particularly acute at standards bodies that convene outside America, where the organisers are less inclined to make arrangements to accommodate firms that are subject to export-control rules. At those meetings, in some instances, Huawei and other Chinese companies have had a voice where American companies have not.

> Some, such as 3GPP, a body that deals with 5G, and IEEE, an engineering body, have declared themselves to be "open" meetings, in an attempt to remove liability from firms with American operations. But uncertainty persists [18].

Moreover, this "uncertainty" has led to "talk of competing bodies being set up outside America, to make truly global discussion possible" [18]. Whatever the outcomes of the political noise – bluster, negotiation, a return to the status quo – the ground is already set for a possible return to the situation pre-2013, in which 3GPP2 acted as a counterpart to 3GPP.

Plus, the very ITU operational model of setting requirements, not *standards*, for mobile communications, and which allocates spectrum, but which does not license it, could easily enable multiple bodies to set independent standards paths to realize the same goals.

ITU recommendations explicitly allow for this possibility, which is not without precedent, as we have seen. They may be said to encourage diversity. Indeed, it has been argued in this chapter that recent harmonization of global standardization activities for mobile broadband are a deviation from the past – and cannot be said to represent an enduring norm.

Allowing for such uncertainty from political and governmental stakeholders is one thing, but disruptive factors from within the industry, both in commercial approaches and as a consequence of the current standards, are quite another. There are disruptive influences that must be considered as we look toward 6G standardization. To put these in context, however, we must first look at the benefits of standardization towards a single template.

3.6 Standards, the Supply Chain, and the Emergence of Open Models

> The global mobile supply chain has become increasingly limited, with high entry costs leaving operators dependent on a handful of vendors. As we have seen, there are political moves to change supplier models, but there are other industry initiatives that could also have a lasting impact on the ecosystem.

Standards are designed to be replicable; that's to say, any vendor should be able to follow a template and develop a solution that can be deployed in the appropriate network domain or as part of infrastructure.

They have delivered universal mobile access across multiple generations of technology, which have gradually converged on a single template for the realization of 5G. Again, this is very positive for the market, as it allows manufacturers and suppliers to build solutions, while enabling operators to follow a clear path toward service delivery.

However, there is a complication: standards can also limit choice and have led to unintended consequences. There are several reasons for this. First, standards can create challenges for market entry for new actors; it is not easy to implement, test, and validate established standards.

Second, it is expensive to ensure continuing compliance with what are often moving targets, subject to regular evolution and iteration. Third, they can increase dependency on a small number of suppliers. It can be argued that standards which, on the one hand enable interoperability, have actually become a barrier to an open and competitive market.

The barriers to entry emerge because it is extremely costly to develop appropriate solutions that implement the desired functionality. Moreover, operators insist on strict performance requirements, partly to meet customer expectations and partly to meet the needs of regulators and law enforcement agencies [19]. While fulfilling the requirements of a standard is necessary to enter the market, it is far from sufficient. A new entrant must ensure that they can meet the criteria defined by target operator customers, which proves a challenge for many.

As a result, the number of vendors has declined, through consolidation and acquisition – or even withdrawal from sectors of the market – leaving operators to:

> choose from a relatively limited menu of products from a small number of suppliers [19].

Numerous analysts and vendors have reached similar conclusions.

Thus, while interoperability and multi-vendor strategies have been goals for many operators, in practice, few have achieved this, relying on a small handful of suppliers. Dependence on a single supplier increases risk and exposure to the shifting sands of political change.

This has not gone unrecognized. The flaws in the current model have led to new initiatives that seek to change the situation, a process that began long before political tensions were wound up in 2020. For example, the open radio access network Alliance, or "O-RAN Alliance," is an effort to:

> clearly define requirements and help build a supply chain ecosystem to realise its objectives [20].

The Alliance, which was launched in 2018, seeks to:

> drive new levels of openness in the radio access network of next generation wireless systems [20].

This is significant because the RAN accounts for up to 70% of a mobile operator's total infrastructure [21].

Others, such as the Open RAN Policy Coalition, pursue similar goals, aiming to:

> promote policies that will advance the adoption of open and interoperable solutions in the Radio Access Network (RAN) as a means to create innovation, spur competition and expand the supply chain for advanced wireless technologies including 5G [22].

In this context, the term "open" is used to denote the opening of interfaces:

> While 3GPP defines the new flexible standards. . . the O-RAN Alliance specifies reference designs consisting of virtualised network elements using open and standardised interfaces [23].

As such, current "open" initiatives are tightly bound to 3GPP standards [24]. They do not seek to replace 3GPP standards, in other words, but rather to augment them. So, while there may, in time, be divergence from 3GPP efforts, such divergence is not the *raison d'être* for such organizations. They are focused on diversity and innovation, *not* a radical new approach to standards.

That's because 3GPP defines functionality, while implementation has been the preserve of the vendor. By enabling further decomposition and promoting off-the-shelf hardware and processing solutions, it is hoped that more stakeholders can deliver discrete components, ultimately resulting in a more diverse supply chain – each element could be said to represent a lower barrier of entry.

The recent policy decisions by national governments that we have already discussed highlight the importance of promoting such efforts, and there is growing eagerness among the operator community to adopt products that result and which reduce dependency on a small number of powerful vendors.

In summary, 3GPP standards have had the perverse effect of limiting supply to the operator community by creating difficult barriers for new entrants to surmount, but initiatives that seek to break this model are gaining ground and are likely to lead to a more diverse vendor community – which is increasingly politically desirable to de-risk supply chains and to avoid dependence on HRVs.

Moves are afoot to find alternative actors and even new national champions to deliver critical infrastructure. Some also seek early deployment of 6G, which may end reliance on 3GPP, as we have seen. And, other bodies could also create new open standards that can be adopted by the industry at large or by discrete sectors.

From where else may such pressures come? Another avenue to explore is the nature of 5G itself. There are aspects of 5G that move beyond traditional models and that suggest a future non-3GPP-based approach. We shall explore this briefly.

3.7 New Operating Models

> 5G brings a shift from tightly specified protocol interaction between different systems and entities within the mobile network, to interaction based on APIs. While these may be specified, they may also evolve rapidly. The API-based network may also lead to fragmentation, as different approaches to APIs for specific integration requirements emerge.

Until 5G, mobile network architecture had followed a broadly similar template. Functional entities are defined, with interfaces specified that allow communication and information transfer between them, according to rules and criteria. The interfaces have been implemented using standardized protocols. As such, a vendor can develop a solution to meet functional criteria and then use the required interfaces to connect to the relevant adjacent nodes. Protocols used have included the Signaling System No. 7 (SS7) family of interfaces, session initiation protocol (SIP) and Diameter, as well as others such as H.248, and so on.

Access to mobile entities has also been possible from external systems. While there have been different approaches to this, the general principle is that application programming interfaces (APIs) can be exposed, enabling third-party systems and processes to control, at some level, services and procedures within the mobile core. Examples of such initiatives include Parlay, among others. Although there is a long history of such initiatives, adoption and uptake were relatively slow and constrained, but, today, RESTful API exposure is relatively commonplace.

5G is different, because API exposure has become fundamental to both internal and external systems. Third-party access will be enabled by RESTful APIs, but this now also extends to internal communications. This is due to the adoption of the new service-based architecture (SBA). Functions that are internal to the 5G core – network functions (NFs) – will connect to others via service-based interfaces (SBIs), which will also present RESTful APIs [25]. "It is this change that enables the network programmability, thereby opening up new opportunities for growth and innovation beyond simply accelerating connectivity" [26].

This is an important shift – not only for how 5G networks will be built and for how innovation from operators and third parties will be enabled but also because it points to the likely growth in openness of all subsequent generations of mobile technology. With APIs available internally and externally, there may be less need for the standardization of such interfaces and, instead, more emphasis on a functional definition, not a specified definition.

As such, it can be seen that 5G networks are inherently more open than any previous generation and that 6G is likely to adapt and build on the same principles. Vendors and solution providers will be able to create their own APIs – for internal and external consumption. As such, other interest groups will, in all likelihood, align around the needs of vertical industries (as has already happened) and hence drive APIs and SBIs defined according to their needs, in addition to any defined by a globally focused standards organization – or any others that emerge, for that matter. They may derive their own interpretations of IMT-2030 in order to accelerate time to market and alongside efforts from 3GPP.

3.8 Research – What Is the Industry Saying?

Stakeholders believe that 6G can be standardized by 2030 and that 3GPP will play a leading role. However, it is increasingly difficult to be heard, with the result that independent voices could lead to new approaches, while growing political pressure is also recognized – and could be disruptive.

Thus far, we have considered factors that are shaping 5G and are likely to influence the ways in which 6G is likely to take shape. We have explored how standardization happens, based on high-level definition of performance goals in the shape of ITU-T IMTs; the increasingly diverse stakeholder ecosystem; political influences; emerging models for open solutions; and changes inherent to 5G architecture that are likely to have significant and lasting impact on all subsequent generations of mobile technology. We also conducted interviews with a number of industry stakeholders to solicit external opinions.

The interviews were conducted in July 2020. Representatives from academia, standards bodies, and the wider industry were presented with the same set of questions. Key findings are discussed next.

3.9 Can We Define and Deliver a New Generation of Standards by 2030?

Respondents believe that we can, presumably based on the success of moves toward 5G. However, when asked if the definition of 6G should be led by traditional stakeholders – in other words, operators and vendors – the answer is also clear: users should be driving requirements.

Given that the ecosystem has grown so dramatically and given that it can be expected to diversify still further, this raises questions. Can such a user-driven definition proceed at the appropriate pace? Vertical sectors are clearly understood as likely to provide the most compelling requirements for 6G – but one issue that is singled out is the fact that, while actors in an industrial sector may know what they need, they may not understand how to convert that into a wireless technology. In other words, some form of mediation between aspirations and physical reality will be required.

Moreover, is there room in the standardization process for all such voices? As one respondent noted, 3GPP "is too complex and [it is] difficult to get engaged as an enterprise." Others disagree, expressing full confidence in the ability of 3GPP to deliver. And yet, the political storm clouds have been noted – external political pressure is now recognized as a significant new influence that may impact standardization, positively or negatively.

Curiously, for all the support shown for 3GPP (and, indeed, its efforts must be applauded) and for all that respondents recognize the need for new stakeholders to join the process, not all are willing to do so. Only 75% of those surveyed declared an interest in participation. Of course, participation in the activities of an SDO is time consuming and an investment – for many representatives, it seems to be almost a full-time job. Some stakeholders may simply ignore the process and accept the outcomes that result. Others may find their own path.

Finally, it should be noted that 3GPP, the ITU, and specific industry associations, spanning both those for mobile operators, such as the GSMA, and those for verticals, such as 5GACIA, are seen as the preferred forums for the development of 6G standards by our respondents. However, this does not preclude the emergence of new forums in the future.

3.10 Conclusion

Alignment around standards has been key to the success of previous generations of mobile technology. This has delivered definitions of what is required for the delivery of the capabilities demanded by different stakeholders, based on the origination of IMTs by the ITU. While multiple approaches have been followed, convergence toward a single SDO took place for the realization of 5G.

But 5G itself changes this dynamic, as it is explicitly designed to fulfil multiple services. Consequently, it has given rise to an expanded ecosystem. 6G will accelerate this process, as more stakeholders are drawn to the discussions. 5G created this possibility because it is multiservice and driven by API interaction – which both widens the ecosystem further and opens the gate to further innovation for specific needs that deviate from agreed specifications.

In addition, regional and political pressures have been exposed. These have already proven to be disruptive and are unlikely to ease in the foreseeable future. Because of the lengthy time that it takes to create standards, efforts must start soon after IMT-2030 has been approved – which means that alternative paths, if they are to emerge, must do so soon.

As a result, it is by no means certain that there will be a single 6G framework. Indeed, it seems much more likely that, while there will be a uniform set of requirements, there may also be different approaches and initiatives that enable these to be met. That means that as things stand, while there will be a clear, 3GPP-driven path, there may also be others to meet the same goals.

To summarize, we have reviewed key factors that will influence standardization of 6G:

1) The IMT process – there is now no reason why 3GPP should be sole arbiter.
2) A new, highly fragmented ecosystem – which means there may simply be too many stakeholders for 3GPP to deliver a single set of new standards.
3) Mounting political pressure to find alternatives to HRVs – which suggests geographic realignment and – just possibly, the revival of 3GPP2 (or something very much like it) as a counterweight to 3GPP, which is seen as being too open to satisfy some governments.
4) Economic pressure for some economies to move early toward 6G performance, once IMT-2030 is available, with the result that national standardization efforts could be pursued – faster than the planned 3GPP timeline (see Figure 3.2).
5) New industry groups promoting open products that may support specific national initiatives and national champions and which could deviate from the current standards path, in order to accelerate 6G deployment.
6) New open APIs rather than protocols – there isn't a standard, which suggests future innovation and increasing diversity.
7) Recognition from within the industry that political pressure is now a major factor and that some industry voices may struggle to be heard, perhaps leading to yet more bodies to help ensure their needs are met, as previously noted.

There is one additional factor that we have not discussed here: spectrum. The spectrum for 6G will be identified by the ITU. This may be from existing frequencies and may be licensed or unlicensed. Different countries will then allocate spectrum for national and private operators.

However, it is worth noting that spectrum allocation for 5G has already proven to be disruptive, with new entrants in some countries (e.g. United Kingdom, United States, and Germany) being able to access spectrum for highly localized deployments. It is possible that further disruption to classic spectrum allocation models will also take place, which may change the radio access requirements that are at the heart of any mobile system, and thus lead to further divergence from 3GPP activities.

In conclusion, by asking questions about future standardization, we are really asking whether we need a single set of standards – as we have today – to fulfil the requirements of the next IMT. The fact is that we do not need such a single standard; we happened to have one for 5G, but the answer must be "not necessarily" for 6G – provided that any standard meets the requirements of IMT-2030 and provided there is sufficient political and industry support to ensure that alternative approaches gain commercial momentum.

Whether we will see true fragmentation remains moot. That depends on willingness to assume the burden of creating a solution to the challenge of realizing IMT-2030, but the effort required to do so will be assisted by the new openness of the network and the sheer power of both governments and new industrial players, which have not previously been considered as factors. They are now, which the industry, analysts, and media clearly recognize. 3GPP will provide a solution for 6G, but it is increasingly likely that there will be multiple efforts to achieve the same goals. As a result, all stakeholders must consider this deviation – and whether they want to support existing bodies or to form new coalitions for the realization of 6G.

References

1 GSMA (2019). Annual Review 2019. GSMA.

2 ITU-R (2020). FAQ on IMT. April ed.

3 ITU (2018). https://www.itu.int/en/ITU-T/focusgroups/net2030/Pages/default.aspx

4 Wikipedia (2020). *IMT Advanced*. Wikipedia.

5 3GPP (1998). *Third Generation Partnership Project Agreement*. Sophia Antipolis: 3GPP.

6 3GPP (2020). Sophia Antipolis: 3GPP. https://www.3gpp.org/about-3gpp/membership.

7 Kim, C. (2019). *5G and Massive IoT: Legacy Technologies will Bridge the Gap for Now*. London: Informa Omdia.

8 Melanie, N., Maximilian, S., and Karim, T. (2020). *Is Your City Ready to Go Digital?* Luxembourg SA: Arthur D Little https://www.adlittle.com/en/insights/viewpoints/your-city-ready-go-digital.

9 TechUK (2020). London. https://www.techuk.org/resource/atomico-s-state-of-european-tech-report-lifts-the-lid-on-a-turbulent-year-for-the-sector.html.

10 Chris Adams (2019). IMT-2030 starts today. . . (and don't call it 6G). AccessPartnership. https://www.accesspartnership.com/imt-2030-starts-today-and-dont-call-it-6g/.

11 ITU Focus Group on Technologies for Network 2030 (2020). FG NET-2030 Technical Specification on Network2030 Architectural Framework. NET2030-I-132, ITU Sub-G3.

12 Whitehouse Briefings Statements (2019). https://trumpwhitehouse.archives.gov/briefings-statements/remarks-president-trump-united-states-5g-deployment/

13 South China Morning Post (2019). https://www.scmp.com/tech/big-tech/article/2187190/donald-trump-says-he-wants-us-lead-5g-and-even-6g-wireless-technology.

14 The Atlantic Council (2020). https://www.atlanticcouncil.org/blogs/new-atlanticist/the-battle-for-5g-leadership-is-global-and-the-us-is-behind-the-white-houses-new-strategy-aims-to-correct-that/.

15 The Financial Times (2020). UK review of Huawei eyes impact of US sanctions. https://www.ft.com/content/9e581ace-69ec-4a42-81c3-c28d2bb40aa1.

16 NCSC (2020). *Advice on the Use of Equipment from High Risk Vendors in UK Telecoms Networks*, 1.0e. London: NCSC https://www.ncsc.gov.uk/guidance/ncsc-advice-on-the-use-of-equipment-from-high-risk-vendors-in-uk-telecoms-networks.

17 Donkin, C. (2020). EC demands urgent action on 5G supply chains. Mobile World Live.

18 The Economist (2020). The fight with Huawei means America can't shape tech rules. https://www.economist.com/united-states/2020/04/23/the-fight-with-huawei-means-america-cant-shape-tech-rules

19 Chappell, E.A. (2020). *Accelerating Innovation in the Telecommunications Arena*, 1e. London: Telecom TV.

20 O-RAN Alliance (2018), cited in "https://www.gruppotim.it/tit/it/technology/standards-focus/dicembre-2018/looking-around.html"

21 Mobile Europe and European Communications (2020). *Keys to the Kingdon: OpenRAN Seeks to Break Open Vendor Lock-In*, Q2e. SJP Business Media.

22 Open RAN (2020). Policy Coalition. https://www.openranpolicy.org/open-ran-a-year-in-review/.

23 Jordan, E. (2020). *Open RAN 101–A Timeline of Open RAN Journey in the Industry: Why, What, When, How? (Reader Forum)*. Austin: RCR Wireless https://www.rcrwireless.com/20200715/opinion/readerforum/open-ran-101-a-timeline-of-open-ran-journey-in-the-industry-reader-forum.

24 ABI Research (2020). Open RAN: market reality and misconceptions. White paper at https://go.abiresearch.com/lp-open-ran-market-reality-and-misconceptions.

25 Mayer, G. (2018). *Restful APIs for the 5G Service Based Architecture. Journal of ICT Standardization* 6 Combined Special Issue 1 & 2 ed.

26 Jan Friman, M.E.P.C.J.M.J.S. (2019). *Service Exposure: A Critical Capability in a 5G World*, 4e. Ericsson Technology Review.

4

Greening 6G: New Horizons

Zhisheng Niu[1], Sheng Zhou[1], and Noel Crespi[2]

[1] *Tsinghua University, Beijing, China*
[2] *IMT, Telecom SudParis, Institut Polytechnique de Paris, Paris, France*

4.1 Introduction

Until today, the evolution of mobile communications from 1G to 5G has been driven mainly by spectrum efficiency, which unfortunately has led to a very high level of energy consumption. This is, on one hand, caused by the ever increasing complexity of signal processing for the new multiplexing schemes from TDMA (time-division multiple access) to CDMA (code-division multiple access) and then to OFDMA (orthogonal frequency division multiple access) and MIMO (multi-input multi-output) technologies. But, on the other hand, it is more caused by the cell densification, through which the precious spectrum can be reused many times, and therefore a vast number of base stations (BSs) need to be deployed. Even worse, it is foreseen that 6G needs to further deploy a large number of edge computing facilities to support task offloading and machine learning (ML) algorithms for smarter, ultrareliable low-latency communications (URLLC). To this end, not only the radio access networks (RANs) face with high energy cost but the computing on edge servers will also have huge energy barriers. As a result, a new horizon and a holistic approach need to be taken in order to make future 6G networks greener and sustainable.

The chapter first identifies the dominating factors of the energy consumption in 6G networks in Section 4.2 from a system point of view. Then a holistic approach for greening 6G RAN will be introduced in Section 4.3 from the viewpoints of network planning to radio resource management and further to service provisioning. In particular, we will highlight the role of artificial intelligence (AI) in the radio resource management, which has been considered as one of the key approaches for making 6G RAN smarter and greener. In Section 4.4, a new horizon of greening AI in 6G networks will be discussed by additionally considering the computing energy for AI training and inference. Conclusions together with some discussions will be summarized in Section 4.5.

Shaping Future 6G Networks: Needs, Impacts, and Technologies, First Edition.
Edited by Emmanuel Bertin, Noel Crespi, and Thomas Magedanz.

4.2 Energy Spreadsheet of 6G Network and Its Energy Model

4.2.1 Radio Access Network Energy Consumption Model

A precise energy consumption model of radio access technologies is important to assess the network energy efficiency (EE). For a typical BS in 4G or 5G mobile networks, the energy consumption mainly comes from the power amplifier and the circuits for signal processing. In [1–6], the energy consumption (P_A) of one or more power amplifiers on a BS is modeled as a function of bandwidth B, power amplifier efficiency ξ, the total transmit power P_t, and the static energy consumption P_B, i.e. $P_A = \frac{BP_t}{\xi} + P_B$. Here, the static energy consumption includes the power supply and may also include air conditioning, which heavily depends on the type and the scale of the BSs.

Moreover, the calculation of the circuit energy consumption, mainly the energy consumption of signal processing, is more complex. In particular, the number of antennas plays a vital role, as massive multi-antenna transmission at high frequency band is expected to address the need for very high spectral efficiency in wireless access networks of 5G and very likely in 6G too. In [4], the circuit energy consumption P_C is modeled as $P_C = \sum_{n=1}^{N} P_{RF}^{(n)}$, where $P_{RF}^{(n)}$ denotes the circuit energy consumption on the radio frequency (RF) chain attached to the nth antenna and N represents the number of antennas. Assuming that the energy consumption of each RF chain is identical [5], the circuit energy consumption PC is proportional to the number of RF chains, i.e. $P_C = NP_{RF}$. A more detailed model of circuit energy consumption can be found in [3], in which the energy consumption for channel estimation and zero-forcing multiuser detection is quantified with the number of complex operations, and in particular:

$$P_C = KP_{dec} + NP_r + 2NKC_0B + 4NK^2\frac{C_0}{T_c} + 8K^3\frac{C_0}{3T_c},$$

where K, N, C_0, B, Tc, P_{dec}, and P_r denote number of single antenna users, number of antennas on the BS, energy required to compute a single complex operation, bandwidth, channel coherence time, average power consumed at the BS for decoding each user's coded information stream, and average power consumed in each RF chain.

Overall, the energy consumption of a single BS can be expressed as

$$P_{BS} = \begin{cases} \frac{BP_t}{\xi} + P_B + P_C, & \text{if } P_t > 0 \\ P_{sleep}, & \text{if } P_t = 0 \end{cases},$$

where P_{sleep} is the energy consumption when the BS is in sleep mode, which can be modeled by $P_{sleep} = \delta P_B$, where a typical value of $\delta = 0.29$ is reported in [7]. To gain some insights of the total energy consumption of a typical BS, Table 4.1 shows the power consumption of typical 4G and 5G macro BSs working in 2.5 GHz band, as measured by China Mobile [8]. As seen from the table, a 5G macro BS has almost fourfold power consumption of a 4G BS, mainly due to the substantially increased number of antennas and the transmit power, and also the signal processing power due to the large bandwidth. Although the throughput is

Table 4.1 Energy consumption of a macro 4G and a macro 5G BS (the 4G LTE BS has 8 antennas while the 5G NR BS has 64 antennas).

	Total Tx power	Bandwidth	Baseband	Radio unit	Total power	Throughput
4G LTE	40 W	20 MHz	150 W	950 W	1100 W	120 Mbps
5G NR	240 W	100 MHz	220 W	4077 W	4297 W	1963 Mbps

enhanced by almost 16 times, leading to higher EE per BS, the network-wise EE can still be worse in 5G since the deployment of 5G BSs will be much denser.

TeraHz-band communication is also foreseen as one of the key radio technologies of 6G. In [9], estimations reveal that the energy consumption of TeraHz receiver can be prohibitive for receivers, especially the handsets, meaning that energy can be a major bottleneck of adopting TeraHz communications in 6G.

4.2.2 Edge Computing and Learning: Energy Consumption Models and Their Impacts

With the paradigm shift of the mobile networks from "connected things" to "connected intelligence," 6G requires vigorous support from edge computing and edge learning, and they can be more energy consuming than BSs [10, 11]. Therefore, it is important to study the corresponding energy consumption models, in order to understand the bottlenecks and possible solutions in reducing the energy consumption of edge computing and learning systems. The energy consumption can be modeled from different levels of perspective, e.g. from the digital circuit level, the processor level (central processing unit [CPU], graphics processing unit [GPU], field-programmable gate array [FPGA], etc.), and the network infrastructure level (BS or edge server). Several energy computation models used in edge computing and edge learning systems are summarized as follows.

4.2.2.1 Energy Consumption Models in Edge Computing

In edge computing, two processor-level energy consumption models are widely used. The first model gives the energy consumption of processing a task, which requires w CPU cycles. According to [12] and references therein, the energy consumption per CPU cycle is κf^2, where κ is a coefficient related to the hardware architecture and f is the CPU clock frequency. Therefore, the total CPU energy consumption to complete the task is $\kappa w f^2$. The second model is inspired by the observation that the CPU energy consumption is linear with the utilization ratio [13]. Therefore, given the maximum energy consumption of a fully utilized CPU E_{max} and the minimum energy consumption of an idle CPU E_{min}, the energy consumption of the CPU with utilization ratio u is $E_{min}+u(E_{max}-E_{min})$. Although these two models are not very accurate, they enjoy low complexity and good generality and thus are widely used [12].

4.2.2.2 Energy Consumption Models in Edge Learning

Since the research on edge learning is still in its infancy and the computation of AI tasks varies, we provide potential energy consumption models for edge learning. From the digital circuit-level perspective, the fundamental computation unit of DNNs is the multiply-and-accumulate operation, which first accesses the data in the memory and then computes the result [14]. The majority of the energy consumption of multiply-and-accumulate operations is caused by two actions: data access and computation, and thus summing up the number of each action times the energy consumption per action gives the energy consumption [15]. Although digital circuit-level modeling can be accurate, it is complex due to many hardware-specific parameters.

Analogous to edge computing, the processor-level model is given by $\frac{W}{E_f}$, where W is the total number of float-point-operations (FLOPs) of the task, and E_f is the EE of the processor, which can be measured by the number of FLOPs that are processed with unit energy, and the typical values of E_f for different processors are reported in [16]. The computing capabilities of processors are usually measured by the number of FLOPs that can be processed within unit time, and one state-of-the-art GPU reaches 100 TFLOPs per second [17]. FPGA-based DNN accelerators have lower computing capability compared to the high-end GPUs or data center chips but having relatively better EE [16]. Furthermore, from the perspective of network infrastructure level, cooling consumes a major portion of the energy, and an edge server still consumes substantial amount of energy even if it is idle [18]. Therefore, queueing theory can be used to model the task processing process of a BS equipped with edge servers, and the power consumption is given by $\rho P_{on} + kP_{idle}$, where ρ is the average number of ON servers, P_{on} is the power consumption of an ON server, k is the average number of idle servers, and P_{idle} is the power consumption of an idle server.

4.3 Greening 6G Radio Access Networks

4.3.1 Energy-Efficient Network Planning

The network-wise energy consumption tightly relates to the number of BSs or equivalently the BS density when the BSs are fully loaded, which is a major concern during the planning phase, and so does the network capacity. There have been studies on the optimal BS deployment to maximize the EE of the network in two-tier heterogeneous networks [19] and ultradense networks [20, 21]. However, in the operation phase, many BSs are underutilized, and thus dynamic BS sleeping has been shown effective to reduce the network energy consumption, by putting part of the BSs into sleep mode and offloading the users to the active BSs [22–24]. In 5G and 6G networks, facing with more directional transmissions led by large antenna arrays and higher frequency bands, and in particular with the introduction of intelligent metasurfaces, new potentials and issues in energy-efficient network planning arise.

4.3.1.1 BS Deployment Densification with Directional Transmissions

When beamforming is used to perform directional transmissions, the coverage area of a BS can be time varying and situation dependent, especially under BS densification. It is thus challenging to optimize the BS deployment to maximize the network EE as well as guaranteeing the network coverage. The EE of different deployment strategies including macro BS densification, extremely dense femtocells, and dynamic distributed antenna systems is studied in [25]. The results indicate that densely deployed femtocells are much more spectrum efficient and energy efficient than densifying the outdoor macro BSs when indoor

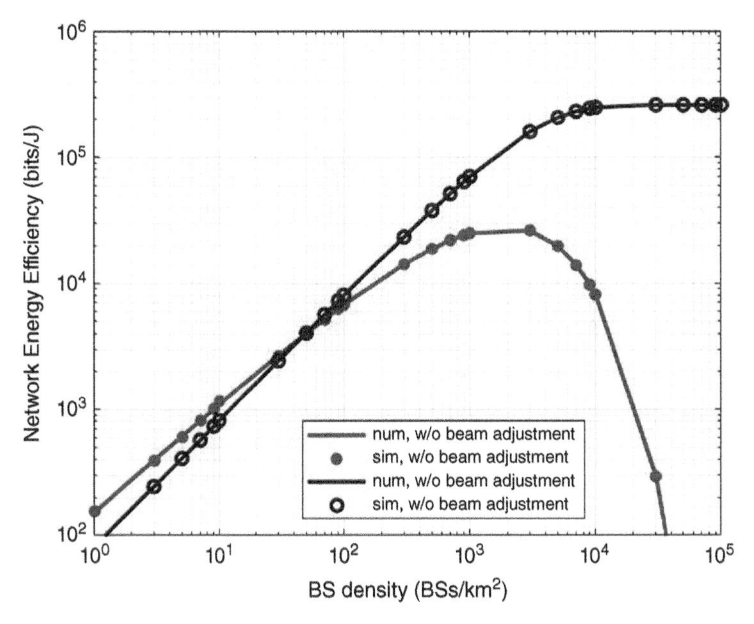

Figure 4.1 The EE of ultradense networks with directional transmissions. The system bandwidth is set to 20 MHz, and the per BS power consumption is 1000 W.

capacity demand is vital. In terms of directional transmissions, its coverage performance under BS densification is analyzed and the result reveals a beamwidth dependent optimal BS density [26] and shows the potential gain from beamwidth adaptation for the spectrum efficiency as well as the EE. As shown in Figure 4.1, even with directional transmissions, the network EE will eventually decrease dramatically as the network density keeps increasing. But, if the beamwidth is adjusted with respect to the network density, the network EE does not drop but saturates with large BS density.

4.3.1.2 Network with Reconfigurable Intelligent Surfaces (RISs)

The RIS, a synthetic surface equipped with massive number of passive reflecting elements, is considered promising for better coverage and EE in 6G systems [27], and detailed introductions and discussions can be found in Chapter 7. When the line-of-sight path between a BS and a user is blocked, the RIS can provide an extra path by reconfiguring the propagation characteristics of the radio wave. It requires much less energy compared with relays due to its passive reflecting nature. Deploying RIS certainly decreases the network outage probability and enhances the capacity of the network [28]. In other words, the network performance can be maintained on a certain level while the energy consumption is saved by replacing part of the BSs with the RISs. To address this, the energy consumption model of the RISs is required to evaluate the network EE when BSs and RISs coexist. Even though RISs are supposed to be passive devices, its control and interaction with BSs or controller still consume some energy, which should be carefully judged. In addition, the equivalence between the BS density and RIS density, i.e. how many RISs can replace one BS, needs further verification, and the energy-efficient operation strategy of RIS is yet to be discovered.

4.3.2 Energy-Efficient Radio Resource Management

The remarkable growth of BS density and the massive number of devices and antennas in 6G networks demand for scalable and low-complexity methods for optimizing the network EE. Conventional optimization methods face the curse-of-dimensionality problem. Luckily, the recent development in the area of deep learning provides a new approach. Compared with conventional optimization methods, AI-aided methods have the following advantages:

4.3.2.1 Model-Free

AI-aided methods can directly deal with the massive data generated in 6G networks. This shift from model-oriented to data-oriented avoids the introduction of explicit model as well as the corresponding deviations. For instance, the network can have a better knowledge of the traffic distribution over time and space with the help of big data and deep learning. This knowledge can be utilized for power allocation, BS sleep control [29], and so forth to improve the EE. Moreover, deep reinforcement learning (DRL) can achieve model-free control by interacting with the environments. In Figure 4.2, taking BS sleeping as an example, the learning agents deployed in the RANs can observe the traffic states and make BS sleeping decisions according to a learning model. After the decisions are conducted, the rewards related to the energy consumption and quality of service (QoS) are fed back to the agents and utilized for updating the learning model. Repeating this cycle, the agents can gradually learn better actions and ultimately improve the EE in an online manner.

4.3.2.2 Less Computation Complexity

AI-aided methods can reduce the computation complexity by replacing the optimization algorithms with an online inference of a learning model, which is trained offline. Specifically, a supervised learning framework maps the observed system states to the optimal solutions. This is achievable since neural networks are universal approximators on any measurable functions providing adequate and appropriate training data [30]. Here the training data can be obtained by solving the optimization problems with traditional methods. Notice that the training process is conducted offline, and the online inference process is usually computational friendly.

With the aforementioned advantages, deep learning has been widely used in many applications to optimize the network EE. From the prospective of network planning and deployment, the work in [31] proposes a reinforcement learning-based node clustering method to improve the EE in wireless sensor networks. Authors in [29] exploit DRL to schedule BS sleeping. A bunch of works consider AI-aided resource allocation to optimize the EE of the networks. The work in [32] adopts deep neural network to achieve online energy-efficient power control. Reinforcement learning is applied in [33] for power control in favor of energy-efficient wireless communication. Apart from them, deploying AI in physical layers can also potentially improve the EE. For example, authors in [34] propose an energy-efficient hybrid precoding method for massive MIMO inspired by ML.

Nevertheless, the drawbacks of deep learning cannot be neglected, bringing about many promising research directions. First, deep learning introduces extra computation cost, and it is doubtful that the overall energy cost is actually reduced when deploying AI algorithms. Therefore, a comprehensive study on the energy cost for AI-aided energy-efficient

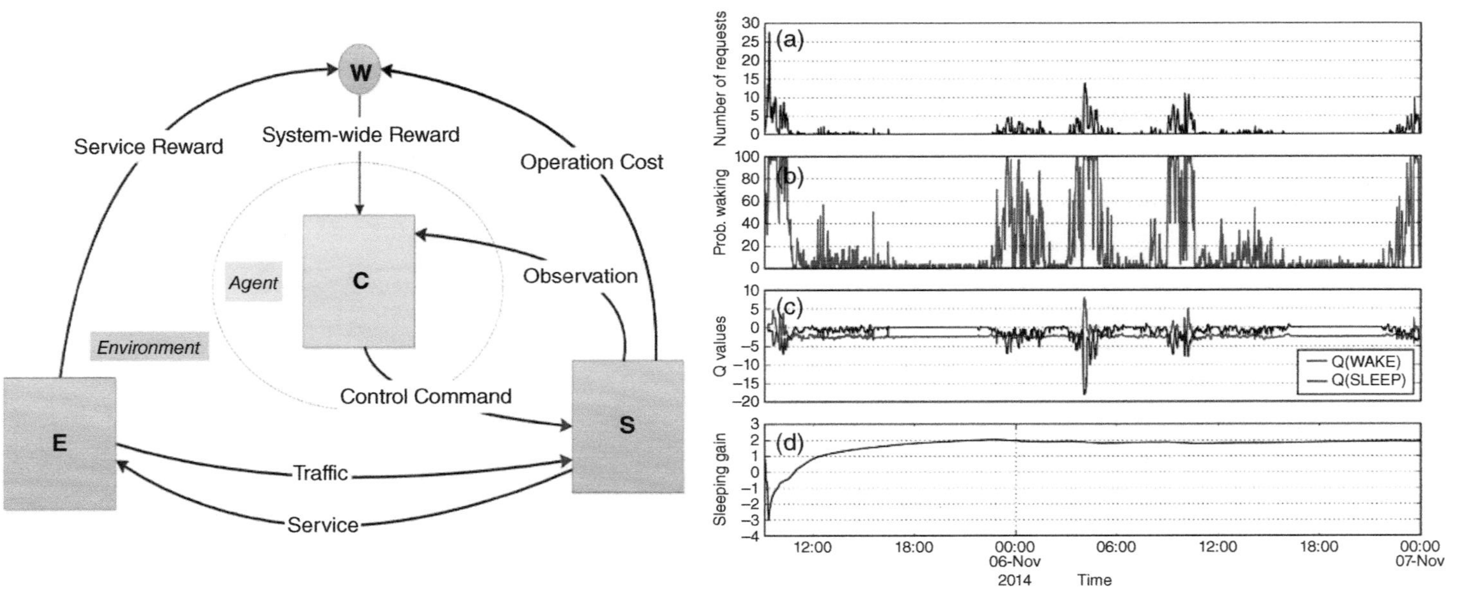

Figure 4.2 (*left*) Reinforcement learning formulation of BS sleeping control in [29]. The controller (C) serves as the agent, while the traffic emulator (E), traffic server (S), and reward combiner (W) together serve as the environment. (*right*) Simulated sleeping patterns of the proposed framework. (a) Number of requests generated in each time step, (b) percentage of waking time steps, (c) Q values for waking and sleeping actions, and (d) sleeping gain above the baseline always-on agent. *Source:* Liu et al. [29].

communication is needed. Second, the physical environment and network topology can be dynamic, meaning that the pretrained learning model can be outdated. Further studies on learning scalability and knowledge transfer are important for field deployments. Last but not least, deep learning is a black box and the performance may not be guaranteed. Explanations on how these AI algorithms make decisions on improving the EE are desirable.

4.3.3 Energy-Efficient Service Provisioning with NFV and SFC

The rapidly increasing demands for delay-sensitive and computation-intensive tasks prompt the continuous expansion of mobile communication networks. However, expanding network services is energy consuming under the traditional network paradigm, since the service providers need to deploy more dedicated hardware or infrastructures for new services. Besides, once the service requests change, the service providers may have to switch corresponding physical devices. Fortunately, network function virtualization (NFV) has been proposed to improve the network flexibility and scalability. Exploiting NFV, the dedicated hardware can be replaced by programmable software modules running on general-purpose servers, i.e. virtual network functions (VNFs). To this end, the traffic flows need to pass through different VNFs in a specific sequence to complete the network services, i.e. service function chaining (SFC). With NFV and SFC, the operators can decouple the service tasks from underlying physical entities, which can improve the EE via VNF placing and traffic routing [35, 36]. Figure 4.3 shows two potential approaches to enhance EE with NFV and SFC: VNF consolidation and exploiting the renewable energy. Related research issues are discussed, respectively.

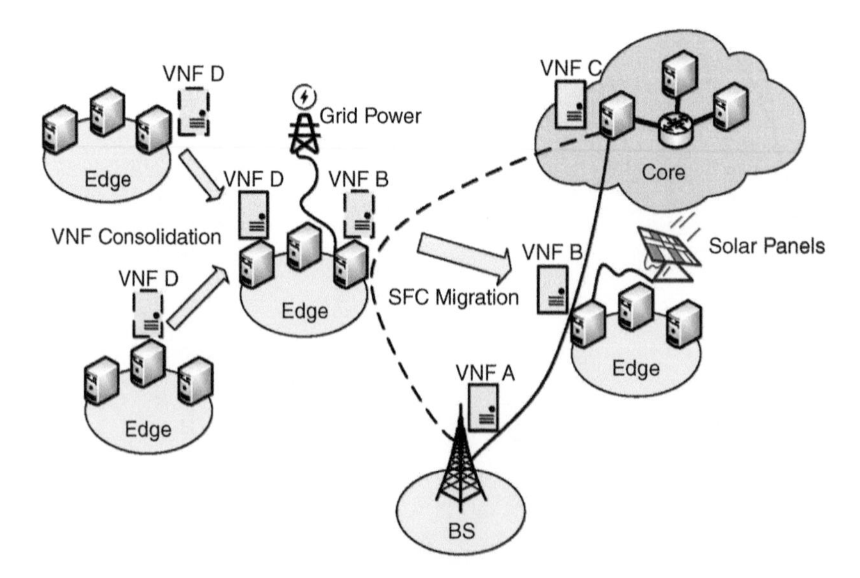

Figure 4.3 Improve the EE via VNF consolidation and SFC migration.

4.3.3.1 VNF Consolidation

The VNFs are software applications running on universal hardware and can be shared by multiple tasks. In the edge cloud, the energy consumption mainly comes from maintaining these universal servers and processing the tasks. Since the traffic load has uneven distribution over time and geographic domains, the network operators can consolidate the VNFs in fewer BS or servers to reduce the number of physical machines so that the energy to maintain the servers can be saved. In [37], the authors propose a solution for minimizing the energy consumption of SFC migration. However, they assumed that the traffic load is predictable and service demands are known in advance. Planning the SFC with respect to task arrivals with high randomness and burstiness brings challenges for service migration strategy designs, yet on the other hand, provides opportunities for energy savings from the fluctuations of the traffic. Notice that the migration of VNFs can also result in QoS degradation (e.g. packet loss, extra delay). Therefore, the SFC planning and function migration should achieve proper trade-off between energy saving and QoS degradation.

4.3.3.2 Exploiting Renewable Energy

The renewable energy (e.g. solar or wind energy) is environmentally friendly and can be considered almost with no cost. However, the availability of renewable energy is highly unstable and its utilization efficiency is often low. By leveraging SFC, the network operators can flexibly adjust the locations of deployed VNFs without changing the deployment of physical network entities. Therefore, there are opportunities to maximize the utilization of renewable energy by migrating the network services among different locations, according to the characteristics of renewable energy arrivals.

4.4 Greening Artificial Intelligence (AI) in 6G Network

As shown in Section 4.3, ML and AI algorithms will be widely used in future 6G networks, in particular for emerging applications such as autonomous driving, augmented and virtual reality, Internet of things (IoT), etc., which are typically communication- and computation-intensive as well as delay- and reliability-sensitive. With the development of mobile edge computing (MEC), the ML-based training and inference tasks from these applications can be executed at the edge of the wireless network in real-time, namely, *edge training* and *edge inference*, respectively. However, the benefit comes at costs. While meeting the timeliness requirements, i.e. to train ML models and infer tasks accurately under stringent delay constraints, a great deal of energy is consumed by edge devices and servers for computation and communication, as described in Section 4.2.2. To prolong the lifetime of devices and reduce the operation costs at the network side, energy-efficient training and inference becomes a key research issue.

In traditional MEC systems, the total workloads of a task, including the input and output data sizes and the required CPU cycles, are typically given in advance, and the energy consumption of edge devices is minimized by jointly optimizing the offloading decision, transmit power allocation, and CPU frequency adjustment [12]. However, the accuracy of an ML model and its communication and computation workloads are in fact tightly correlated, indicating that the workloads of a training or inference task can also be optimized based on the timeliness requirements and energy budget. We can exploit the dependency between accuracy, delay, and energy consumption to achieve green edge training and inference.

4.4.1 Energy-Efficient Edge Training

There are two major edge training architectures: centralized edge training and federated learning (FL). In centralized training, edge devices such as IoT sensors collect and upload real-time data to the edge server for timely training. From the perspective of devices, uploading high volume of fresh data is both bandwidth hungry and energy thirsty. The key idea is to reduce the total number of bits for transmission. Notice that different data samples contribute unequally for training, and using important data samples helps the ML model converge faster. Therefore, data filtering and compression techniques can be adopted to improve the EE [11, 38]. Specifically, the importance of each data sample can be evaluated by each device before uploading, according to a metric of uncertainty [11], which indicates the confidence in correct prediction using the current ML model, or training loss [38]. Then, given the timeliness requirements and bandwidth constraints, edge devices filter the important data samples to upload. Data samples can be further compressed with different ratios, i.e. allocating more bits to important data [38]. However, the theoretical relation between data importance, model accuracy, and energy consumption still remain open, yet it is important for optimizing EE.

From the perspective of the network side, the centralized training at the edge server involves high computation workloads to carry out training algorithms such as stochastic gradient descent (SGD), and it takes many training rounds before the model convergence. To reduce the total computation workloads so as to save energy, simplifications can be jointly implemented from data, model, and training algorithms, that is, to use less data samples to train ML models with low-complexity gradient estimation algorithms. Meanwhile, it is also important to achieve a certain accuracy threshold with minimum number of rounds, which lies in the area of training acceleration. In [39], by using stochastic mini-batch dropping, selective layer update, and predictive sign gradient descent methods, 80% energy saving can be realized in training a convolutional neural network for image classification without apparent accuracy degradation.

On the contrary, in FL, a number of edge devices jointly train a shared ML model based on their local data, while an edge server or a BS coordinates the training process and schedule devices for global model aggregation. The edge server or BS only needs to periodically broadcast the global model, while each edge device consumes energy for both local model training and model uploading. Similar to centralized training, the main idea to reduce the energy consumption for local training is to simplify the training process from data, model, and training algorithm aspects. Besides, the edge server is responsible for efficient device scheduling. Particularly, when the local data is non-independent and identically distributed (non-*i.i.d.*), some devices with more important data samples may help the global model converge faster, and thus the BS should schedule these devices more frequently to reduce the total number of rounds. However, uploading the complex ML model periodically also consumes bandwidth and energy for devices. Gradient compression techniques such as quantization and sparsification are beneficial [40, 41]. Here the key is to design

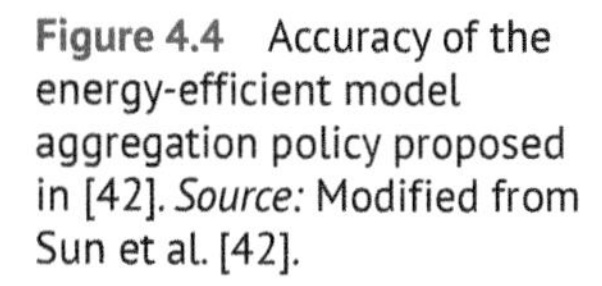

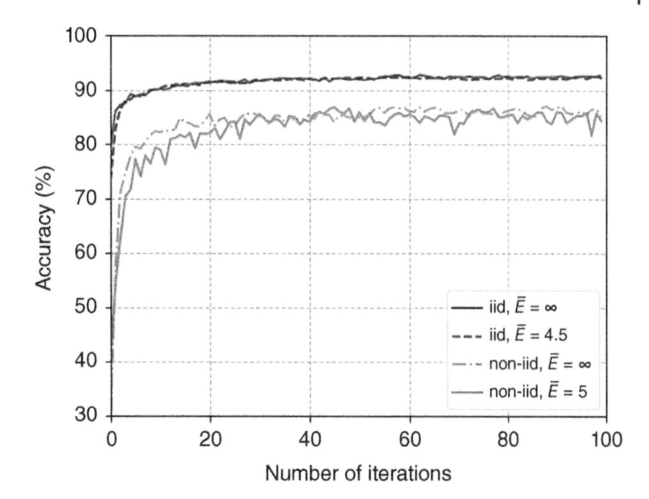

Figure 4.4 Accuracy of the energy-efficient model aggregation policy proposed in [42]. *Source:* Modified from Sun et al. [42].

effective compression schemes so as to keep the important information of the gradients while removing the redundant information. Intuitively, a high compression ratio saves the delay and energy for model uploading in each round but degrades the convergence speed and increase the total number of training rounds, which may in turn lead to high energy consumption.

Last but not least, joint computation and communication resource management is also a key research direction for energy-efficient edge training. Existing works mainly consider the energy minimization for model aggregation [42, 43], the total training delay minimization, [44] and the delay-energy trade-off [45]. Figure 4.4 shows the training accuracy of the energy-efficient model aggregation policy proposed in [42] using MNIST dataset. Both *i.i.d.* and non-*i.i.d.* data distributions are considered. The term $\bar{E}$ represents the average energy constraint in Joules, and $\bar{E} = \infty$ refers to the optimal policy where devices have infinite energy. By using the Lyapunov-based device scheduling policy for wireless model aggregation, the training accuracy is close to optimal. Resource management can help to reduce the energy consumption in each training round and should be combined with the aforementioned techniques to make the whole training process greener without the compromise of accuracy.

4.4.2 Distributed Edge Co-inference and the Energy Trade-off

In edge inference, a well-trained ML model is deployed at edge servers and devices, where each inference task can be offloaded entirely or inferred by server and device cooperatively, namely, co-inference. When an inference task is entirely offloaded, edge devices use energy for data uploading, while edge servers use energy mainly for inference computation. When co-inference is enabled, part of the inference computation is carried out at edge devices.

To reduce the computation energy, the key is to reduce the complexity of ML models. Structure pruning is considered as a promising solution [46, 47], where some less important parts of a neural network, such as neurons or convolution channels, are pruned. The trade-off between the inference accuracy and energy consumption should be balanced to yield energy-efficient edge inference.

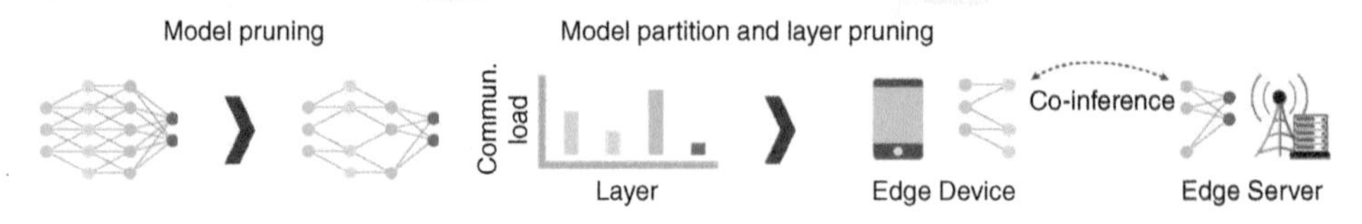

Figure 4.5 An illustration of joint pruning and model partition for co-inference.

To reduce the communication energy, the key idea is to transmit fewer bits of each inference task via data compression, based on the fact that the input data of inference tasks have information redundancy [48]. Using a high compression ratio helps reduce transmission delay and energy consumption but may cause inference failures due to too much information loss. Thus, the compression ratio is the main optimization variable here. Considering a multi-device inference system, bandwidth and power allocation should be jointly optimized to further improve the EE while meeting the timeliness requirements of tasks.

In the co-inference scenario, the computation and communication energy are correlated. As shown in Figure 4.5, given the energy budgets and the processing speeds of edge devices and edge servers, as well as the available bandwidth, we can decide how much neurons should be pruned and where to partition the ML models. The last layer implemented on edge devices can be further pruned to reduce the communication costs [47].

4.5 Conclusions

In summary, future 6G networks will get much denser, deploy much more antennas both on BSs and UEs, and more seriously deploy extra edge servers for AI-aided applications. As a result, it will lead to much higher total energy consumption than 5G networks, which might become the main burden of developing 6G networks. To solve this problem, a holistic approach of applying AI to every corner of 6G networks so that the total energy consumption can be reduced and new horizons such as distributed MLs so that ML algorithms can be widely applied within the network need to be taken. In other words, 6G and AI can help each other and therefore should work collaboratively, i.e. greening 6G via AI and greening AI via 6G.

References

1 Cui, S., Goldsmith, A.J., and Bahai, A. (2004). Energy-efficiency of MIMO and cooperative MIMO techniques in sensor networks. *IEEE J. Sel. Areas Commun.* 22 (6): 1089–1098.

2 Li, H., Song, L., and Debbah, M. (2014). EE of large-scale multiple antenna systems with transmit antenna selection. *IEEE Trans. Commun.* 62 (2): 638–647.

3 Mohammed, S.K. (2014). Impact of transceiver energy consumption on the EE of zero-forcing detector in massive MIMO systems. *IEEE Trans. Commun.* 62 (11): 3874–3890.

4 Liu, Z., Li, J., and Sun, D. (2017). Circuit energy consumption-unaware EE optimization for massive MIMO systems. *IEEE Wireless Commun. Lett.* 6 (3): 370–373.

5 Fiorani, M., Tombaz, S., Martensson, J. et al. (2016). Modeling energy performance of C-RAN with optical transport in 5G network scenarios. *IEEE J. Opt. Commun. Netw.* 8 (11): B21–B34.

6 Arnold, O., Richter, F., Fettweis, G., and Blume, O. (2010). Energy consumption modeling of different BS types in heterogeneous cellular networks. *Proceedings of Future Network and Mobile Summit.*

7 Athley, F., Tombaz, S., Semaan, E. et al. (2015). Providing extreme mobile broadband using higher frequency bands, beamforming, and carrier aggregation. *Proceedings of IEEE Annual International Symposium on Personal, Indoor, and Mobile Radio Communications (PIMRC)*, Hong Kong, China.

8 Chih-Lin, I., Han, S., and Bian, S. (2020). Energy-efficient 5G for a greener future. *Nat. Electron.* 3 (4): 182–184.

9 Skrimponis, P., Dutta, S., Mezzavilla, M. et al. (2020). Power consumption analysis for mobile mmWave and Sub-THz Receivers. *Proceedings of 2020 2nd 6G Wireless Summit (6G SUMMIT)*, Levi, Finland.

10 Letaief, K.B., Chen, W., Shi, Y. et al. (2019). The roadmap to 6G: AI empowered wireless networks. *IEEE Commun. Mag.* 57 (8): 84–90.

11 Zhu, G., Liu, D., Du, Y. et al. (2020). Toward an intelligent edge: wireless communication meets machine learning. *IEEE Commun. Mag.* 58 (1): 19–25.

12 Mao, Y., You, C., Zhang, J. et al. (2017). A survey on mobile edge computing: the communication perspective. *IEEE Commun. Surv. Tutor.* 19 (4): 2322–2358, Fourthquarter.

13 Rivoire, S., Ranganathan, P., and Kozyrakis, C. (2008). A comparison of high-level full-system power models. *Proceedings of USENIX Workshop on Power Aware Computing and Systems (HotPower), held at the Symposium on Operating Systems Design and Implementation (OSDI 2008)*, San Diego, California, USA.

14 Sze, V., Chen, Y., Yang, T., and Emer, J.S. (2017). Efficient processing of deep neural networks: a tutorial and survey. *Proc. IEEE* 105 (12): 2295–2329.

15 Yang, H., Zhu, Y., and Liu, J. (2019). Energy-constrained compression for deep neural networks via weighted sparse projection and layer input masking. *Proceedings of International Conference on Learning Representations (ICLR).*

16 Guo, K., Zeng, S., Yu, J. et al. (2019). A survey of FPGA-based neural network inference accelerators. *ACM Trans. Reconfigurable Technol. Syst.* 12 (1): 1–16.

17 Reuther, A., Michaleas, P., Jones, M. et al. (2019). Survey and benchmarking of machine learning accelerators. *2019 IEEE High Performance Extreme Computing Conference (HPEC).*

18 Dayarathna, M., Wen, Y., and Fan, R. (2016). Data center energy consumption modeling: a survey. *IEEE Commun. Surv. Tutor.* 18 (1): 732–794, Firstquarter.

19 Cao, D., Zhou, S., and Niu, Z. (2013). Improving the EE of two-tier heterogeneous cellular networks through partial spectrum reuse. *IEEE Trans. Wireless Commun.* 12 (8): 4129–4141.

20 Zhang, T., Zhao, J., An, L., and Liu, D. (2016). EE of BS deployment in ultra dense HetNets: a stochastic geometry analysis. *IEEE Wireless Commun. Lett.* 5 (2): 184–187.

21 Kamel, M., Hamouda, W., and Youssef, A. (2016). Ultra-dense networks: a survey. *IEEE Commun. Surv. Tutor.* 18 (4): 2522–2545, Fourthquarter.

22 Liu, C., Natarajan, B., and Xia, H. (2016). Small cell BS sleep strategies for EE. *IEEE Trans. Veh. Technol.* 65 (3): 1652–1661.

23 Zheng, J., Cai, Y., Chen, X. et al. (2015). Optimal BS sleeping in green cellular networks: a distributed cooperative framework based on game theory. *IEEE Trans. Wirel. Commun.* 14 (8): 4391–4406.

24 Zhang, S., Gong, J., Zhou, S., and Niu, Z. (2015). How many small cells can be turned off via vertical offloading under a separation architecture? *IEEE Trans. Wirel. Commun.* 14 (10): 5440–5453.

25 Yunas, S.F., Valkama, M., and Niemelä, J. (2015). Spectral and EE of ultra-dense networks under different deployment strategies. *IEEE Commun. Mag.* 53 (1): 90–100.

26 Xu, Y. and Zhou, S. (2020). On the coverage and capacity of ultra-dense networks with directional transmissions. *IEEE Wireless Commun. Lett.* 9 (3): 271–275.

27 Basar, E., Di Renzo, M., De Rosny, J. et al. (2019). Wireless communications through RISs. *IEEE Access* 7: 116753–116773.

28 Renzo, M. and Song, J. (2019). Reflection probability in wireless networks with metasurface-coated environmental objects: an approach based on random spatial processes. *J. Wireless Com. Network 99.*

29 Liu, J., Krishnamachari, B., Zhou, S., and Niu, Z. (2018). DeepNap: data-driven base station sleeping operations through deep reinforcement learning. *IEEE Internet Things J.* 5 (6): 4273–4282.

30 Hornik, K., Stinchcombe, M., and White, H. (1989). Multilayer feedforward networks are universal approximators. *Neural Netw.* 2 (5): 359–366.

31 Mustapha, I., Ali, B.M., Rasid, M.F.A. et al. (2015). An energy-efficient spectrum-aware reinforcement learning-based clustering algorithm for cognitive radio sensor networks. *Sensors* 15: 19783–19818.

32 Mastronarde, N. and van der Schaar, M. (2011). Fast reinforcement learning for energy-efficient wireless communication. *IEEE Trans. Signal Process.* 59 (12): 6262–6266.

33 Zappone, A., Debbah, M., and Altman, Z. (2018). Online energy-efficient power control in wireless networks by deep neural networks. *Proceedings of 2018 IEEE 19th International Workshop on Signal Processing Advances in Wireless Communications (SPAWC)*, Kalamata, 1–5.

34 Gao, X., Dai, L., Sun, Y. et al. (2017). Machine learning inspired energy-efficient hybrid precoding for mmWave massive MIMO systems. *Proceedings of 2017 IEEE International Conference on Communications (ICC)*, Paris, 1–6.

35 Tajiki, M.M., Salsano, S., Chiaraviglio, L. et al. (2019). Joint energy efficient and QoS-aware path allocation and VNF placement for service function chaining. *IEEE Trans. Netw. Service Manag.* 16 (1): 374–388.

36 Farkiani, B., Bakhshi, B., and MirHassani, S.A. (2019). A fast near-optimal approach for energy-aware SFC deployment. *IEEE Trans. Netw. Service Manag.* 16 (4): 1360–1373.

37 Eramo, V., Miucci, E., Ammar, M., and Lavacca, F.G. (2017). An approach for service function chain routing and virtual function network instance migration in network function virtualization architectures. *IEEE/ACM Trans. Netw.* 25 (4): 2008–2025.

38 Huang, X. and Zhou, S. (2020). Adaptive transmission for edge learning via training loss estimation. *Proceedings of IEEE International Conference on Communications (ICC)*, Dublin, Ireland.

39 Wang, Y, Jiang, Z., Chen, X. et al. (2019). E2-Train: training state-of-the-art CNNs with over 80% energy savings. *Proceedings of Advances in Neural Information Processing Systems (NIPS).*

40 Du, Y., Yang, S., and Huang, K. (2020). High-dimensional stochastic gradient quantization for communication-efficient edge learning. *IEEE Trans. Signal Process.* 68: 2128–2142.

41 Mohammadi Amiri, M. and Gunduz, D. (2020). Machine learning at the wireless edge: distributed stochastic gradient descent over-the-air. *IEEE Trans. Signal Process.* 68: 2155–2169.

42 Sun, Y., Zhou, S., and Gunduz, D. (2020). Energy-aware analog aggregation for federated learning with redundant data. *Proceedings of IEEE International Conference on Communications (ICC)*, Dublin, Ireland.

43 Zeng, Q., Du, Y., Huang, K., and Leung, K.K. (2020). Energy-efficient radio resource allocation for federated edge learning. *2020 IEEE International Conference on Communications Workshops (ICC Workshops).*

44 Shi, W., Zhou, S., Niu, Z. et al. (2021). Joint device scheduling and resource allocation for latency constrained wireless federated learning. *IEEE Trans. Wireless Commun.* 20 (1): 453–467.

45 Yang, Z., Chen, M., Saad, W. et al. (2021). Energy efficient federated learning over wireless communication networks. *IEEE Trans. Wireless Commun.* 20 (3): 1935–1949.

46 Liu, Z., Sun, M., Zhou, T. et al. (2019). Rethinking the value of network pruning. *Proceedigs of The International Conference on Learning Representations*, New Orleans, USA.

47 Shi, W., Hou, Y., Zhou, S. et al. (2019). Improving device-edge cooperative inference of deep learning via 2-step pruning. *Proceedings of IEEE INFOCOM Workshops*, Paris, France.

48 Huang, X. and Zhou, S. (2020). Dynamic compression ratio selection for edge inference systems with hard deadlines. *IEEE Internet Things J.* 7 (9): 8800–8810.

5

"Your 6G or Your Life": How Can Another G Be Sustainable?

Isabelle Dabadie[1]*, Marc Vautier*[2]*, and Emmanuel Bertin*[2]

[1] *Laboratoire de recherche en sciences de gestion Panthéon-Assas (LARGEPA), Université Paris 2 Panthéon-Assas, Paris, France*
[2] *Orange Innovation, France*

5.1 Introduction

The accumulation of environmental crises (climate change, destruction of biodiversity, pollution, etc.) and likely future energy crises will all have very great and painful economic and social consequences in coming years. They will therefore force us to accelerate the changes initiated in recent years in all areas of industrial activity in order to minimize environmental impacts and contribute positively to the ecological transition.

The environmental impact of the Information and Communication Technology (ICT) sector, which includes telecommunications activities, is well documented in the literature [1]. According to an assessment by the European Commission [2], its energy and environmental footprint corresponds to "a range of 5–9% of the world's electricity consumption and accounts for more than 2% of all emissions." These 2% are similar to the global CO_2 emissions of the civil aviation sector [3]. Of course, ICT does not operate in isolation and also has an impact on all other sectors by its cross-cutting nature. ICT is present in all daily activities in the professional and personal spheres. Taking a medium-term view (to about 2030) of electricity consumption in particular, ICT represents a high-growth industry [4] that will have huge environmental impacts (CO_2 emissions, raw materials consumption, pollution, etc.), while also being one of the solutions to help other industries reduce their own impacts. The forthcoming 5G technological solutions will therefore have to provide answers to these challenges. This is also true for 6G, the future generation of cellular networks, a technology that will be rolled out in 2030–2035, in a world that will probably be very different from the one we live in today.

The telecoms industry in general, and designers of new generations of cellular networks in particular, are strongly committed to helping reach the UN IPCC (United Nations Intergovernmental Panel on Climate Change)[1] objectives to limit climate change

1 https://www.ipcc.ch/

Shaping Future 6G Networks: Needs, Impacts, and Technologies, First Edition.
Edited by Emmanuel Bertin, Noel Crespi, and Thomas Magedanz.

and contribute to the well-being of future generations. They must be careful, however, not to offset these efforts by generating higher impacts through mobile network upgrades and growth and should always seek to minimize the overall impact of their activity. In order to achieve this, innovation for sustainability in the telecoms industry has so far been oriented mainly toward resource efficiency. Unfortunately, this has not proven to be sufficient, in part due to the so-called rebound effect (when increasing energy efficiency ends up driving up energy consumption). While there are also other factors than this rebound effect that account for the rise in cellular communications consumption (new technology allowing development of new services, social changes, demography, etc.), the point here is to draw the attention of 6G actors to a central question: besides energy efficiency, how can new-generation cellular networks contribute to achieving a long-term global reduction in the resources used (energy and raw materials) and CO_2 emissions related to their own activity? The risk is that by focusing on energy efficiency, we take our eye off the big picture, which has been one of a steady increase in environmental impacts over the years.

In order to address this question, we will focus on the case of business-to-consumer (B2C) markets and the particular position of service providers in the game, although our approach could also apply to business-to-business (B2B) markets. If energy efficiency is not sufficient to curb CO_2 emissions, the solution may lie in reducing traffic. Another thorny issue then needs to be tackled: how can a network operator go down this road without the risk of disrupting its business? The case of service providers is interesting as it allows the complexity of the task to be shown, while raising broader questions of relevance to all industry players and particularly equipment manufacturers.

This chapter will begin by describing the environmental crises and challenges that 6G players will have to contribute to solving. Solutions will then be described, while highlighting the obstacles to be overcome by network operators. The necessity of a paradigm shift will then be raised, along with the opportunities that it can bring. Finally, the subject will be broadened to consider other levels of action and question the role of policy makers. While we acknowledge that this chapter will raise questions to which there are no simple responses, we hope that by addressing practical problems, it can stimulate innovation and help to identify new courses of action for a sustainable 6G.

5.2 A World in Crisis

5.2.1 Ecological Crisis

Within the framework of the UN IPCC objectives to limit the increase in global surface temperature to 1.5–2 °C by 2050 (with reference to the preindustrial period), entropic CO_2 emissions must be neutral in 2050, i.e. all entropic emissions must be fully compensated by natural absorptions, or stored. For a country like France, these objectives correspond to each man, woman, and child reducing their CO_2 emissions from 11 to 2 tons per year per person [5] over this period, i.e. dividing emissions by 6 within 30 years. Of these 11 tons, 6–7 are emissions on French territory [6], with the rest being related to imports of manufactured products. One-quarter of these emissions is the direct responsibility of the population, while the remaining three-quarters are the responsibility of companies, local authorities, and states.

In parallel with these ambitious CO_2 emission reduction targets, the world population will grow by 30% by 2050, according to UN forecasts, from the current 8 billion to 10 or 11 billion people. Over the period during which CO_2 emissions must decrease to become neutral, planet Earth will therefore have to accommodate some three billion extra people who will need to be fed, clothed, and equipped, and who will want to consume and be mobile locally but also internationally, which is to say, in the end, consume energy and raw materials and thus emit greenhouse gases. This growth will take place mainly on the African continent, which does not emit much CO_2 at present but will make an increasingly important contribution, without counting the political consequences of such growth.

5.2.2 Energy Crises

Another challenge concerns the long-term availability of "fuel" in our consumer societies, meaning cheap and easily accessible fossil fuels. The major economic crises in developed countries in recent decades have often been preceded by an energy crisis that served as a trigger. The crisis of the 1980s was preceded by the oil shocks of the 1970s. In 2007, the price of a barrel of oil rose to over $150, raising the cost of living for Americans, impacting their ability to repay loans, and triggering the subprime crisis with the consequences we all know. In 2018, the International Energy Agency in its World Energy Outlook announced that the peak of conventional oil production had been reached in or around 2010, and since then it has been the production of shale oil that has been offsetting this decrease. Even without the COVID-related economic crisis of 2020, an overall decline in oil production would have been expected in the coming years (by 2030). In addition, we do not have the renewable energy sources to offset the decline in fossil fuel output. In 2018, for example, only 50% of the growth in energy demand was covered by new renewable energy sources, while the other half came from coal. And even if we assume that the development of renewable energy does make it possible to meet all demands in the medium term, the deployment costs will still be very high. The energy return on energy injected (EROI) key performance indicator (KPI), which reflects the ratio between the amount of energy needed to extract a certain amount of energy, gave higher values of between 100 and 30 for conventional oil extraction a few decades ago. This value is currently below 30 and is much lower for renewable energy (~10) [7]. This means that it will always be more expensive to deploy renewable energy than other forms of energy. Finally, history shows that when new energy sources are discovered, they do not replace the existing ones but are added to them. To sum up, the era of the economic development model based on virtually free and abundant energy is coming to an end.

5.2.3 Technological Innovation and Rebound Effect: A Dead End?

To limit the consequences of these crises, the obvious solution is to reduce energy consumption and CO_2 emissions rapidly and drastically. Innovating to enhance energy efficiency would therefore appear to be a necessity. Unfortunately, this solution suffers from a significant shortcoming known as the "rebound effect."

In standard economic conditions, companies have always sought to be more efficient. They try to reduce the amount of energy and materials they need to manufacture products and deliver services. They also strive to optimize their various industrial processes. The energy savings resulting from these efforts have potential positive environmental impacts. However, the efficiency gains also generate cost reductions and in turn have potential adverse consequences from an environmental point of view. Because they spur demand for the products and services the company provides – which have become more attractive because they are cheaper or offer new benefits – they contribute to global growth in consumption of energy and raw materials. This is called the "rebound effect," a concept developed by economist William Stanley Jevons in 1865 in reference to the energy performance of steam engines. Economic players at the time believed that an improvement in energy efficiency would lead to a decrease in coal consumption. Jevons, however, showed that these gains led to an overall increase in coal consumption, as technological progress led to improved performance and lower production costs of steam engines and thus to their multiplication. This led to an increase in uses and thus ultimately to an increase in energy consumption. A similar example can be found in the aviation industry: between 1968 and 2005, fuel consumption per passenger kilometer was reduced by approximately 35%, while overall greenhouse gas emissions more than doubled (multiplied by 2.3) [8].

This observation calls for some nuance, however. The extent of the impact of this rebound effect is subject to academic debate. According to Gillingham et al. [9], "rebound effects are small and are therefore no excuse for inaction. People may drive fuel-efficient cars more and they may buy other goods, but on balance more-efficient cars will save energy" [9]. The rebound effect is also a complex phenomenon, including direct, indirect, and macroeconomic effects. Our purpose is not to investigate it further here but instead to draw the attention of telecom industry players working on 6G or other ICT solutions to an important issue. Although it is difficult to define the extent of the rebound effect, it is necessary to acknowledge the fact that while working on energy efficiency is necessary to achieve a global reduction in energy consumption and CO_2 emissions, it is not sufficient.

In ICT, each new generation has brought efficiency gains. Between 4G and 5G, for example, energy efficiency will be improved by a factor of between 10 in the midterm and 20 in the longer term. This energy efficiency improvement will also contribute to traffic growth, and the expected energy consumption gain will probably not be as large as expected, due to the rebound effect. As a consequence, one obvious solution to achieve the energy consumption and CO_2 emission reduction objectives would be to stabilize or reduce data traffic on these future 6G networks but also the number of connected devices. While this raises a number of questions that we will address later (i.e. how can industry players wish for degrowth in their own industry?), we also have to take account of the fact that ICT can contribute to reducing global environmental impacts in different sectors (e.g. transport, housing). This topic is also subject to debate and should always be addressed with precaution, as we know that technology can prove to be for better or for worse in this domain. In order to contribute to the ecological transition, ICT should be developed with human well-being and environmental protection in mind (including electric and magnetic field impacts) and strive to achieve carbon neutrality and sparing use of all resources. To sum up, for business models based on growth in consumption, achieving an absolute reduction in energy consumption and associated CO_2 emissions requires a dual approach: improving the overall efficiency of the products and services that are developed, while encouraging customers to use them sparingly. The question to be addressed is therefore: how can service providers pull off this tricky task and position themselves in the game?

In order to address this issue, the approach we would like to focus on here is related to consumption behaviors in B2C markets. In cellular communications, the challenge is to limit the environmental impact without being a barrier to all the benefits they can bring for a socio-ecological transition.

5.3 A Dilemma for Service Operators

5.3.1 Incentives to Reduce Consumption: Shooting Ourselves in the Foot?

For a mobile service operator, one straightforward path toward contributing to reducing CO_2 emissions in the mobile industry could consist in simultaneously increasing the energy efficiency of its infrastructure and reducing the amount of traffic going through it. Simple mathematics can demonstrate this. Improving energy efficiency would reduce the amount of energy necessary to deliver a particular service (e.g. video streaming), while stabilizing or reducing the use of this particular service would avoid canceling out that effort. As a matter of fact, with a 30% increase in energy efficiency in video streaming delivery and a 50% increase in video streaming consumption, the balance would be negative.

The first part of the equation seems relatively easy to achieve (cf. Chapter 4). Thanks to the efforts made in terms of energy efficiency, some telecom operators have managed to keep the energy consumption of their network at the same levels for several years despite traffic doubling every two years.[2] This result has also been obtained because no new generation or network extension occurred during the same period of time. While going down this path requires significant research and development investments, such efforts have resulted in lower operating costs (reduced energy bills) and increased profitability, and this win-win situation is very encouraging. Every operator that wishes to reduce the absolute energy consumption of its infrastructures should continue to work on overall efficiency (energy/material/process). This is an ongoing movement, and every generation of technology brings significant efficiency gains. In the context of the current ecological crisis, this objective has become a major priority for all industry players.

Unfortunately, the second part of the plan, which consists in stabilizing or reducing traffic over the networks, raises more complex issues. In recent years, large telecom operators have seen the global traffic on their networks double every two years – with a growth mainly due to video consumption from providers like Netflix or YouTube – and there is no reason to expect the development in traffic to be different on other networks. As a result, we can anticipate difficulties on two levels. First, reducing the consumption of services provided by the operators would most likely result in a decrease in average revenue per user (ARPU) and a subsequent decrease in company profitability. Second, encouraging customers to use the services offered to them more responsibly, while, as indicated above, other industry players (e.g. content providers) are encouraging unbridled consumption,

2 https://www.orange.com/en/Group/Non-financial-reporting

would be no small feat and we may wonder which company would dare envisioning such a strategic move. Although it may seem risky, this option, which could be very efficient from an environmental point of view and may prove to be a driver of positive innovation, should be investigated further.

5.3.2 Incentives to Reduce Overconsumption: Practical Solutions

While optimizing network infrastructure in order to minimize the environmental impact of service operations is a strategic move that a number of operators have already made, as mentioned above, involving consumers and influencing their behavior in favor of reasonable consumption of products and services is more complex. A number of solutions do exist, however.

One of the very first steps consists in informing customers of the impacts of their uses (energy consumption and CO_2 emissions). As a matter of fact, not only it is necessary to raise awareness, but it has also been proven that the mere fact of providing feedback on energy savings automatically drives consumption down in significant proportions [10]. A second step consists in providing consumers with offers that simply meet their needs and are designed to keep pace with changes in these needs, without being oversized, as is often the case.

The design of the products and services supporting these offers must be based on a number of principles of frugality or minimalism:

- Minimalism in features: limit the number of features that are offered or allow customers to choose just the one(s) they need. Do we really need, for example, up to four Ethernet and several USB interfaces on an Internet gateway? One or two Ethernet interfaces would probably be enough.
- Minimalism in scale: adjust the offers as precisely as possible to the exact customer needs. If a customer wants to subscribe only to a volume of data of 20 Go/month, do not sell them 50 Go/month, even for the same price.
- Minimalism in use: do not encourage overconsumption, just keep up with changes in the needs of the customer. If the customer needs more data a short time after subscription, then the switch should be simple and easy for them to make.

The implementation of such offers could be based on modular design and scalable service plans. These kinds of offers are still scarce on a market in which customers are usually offered all-inclusive plans or equipment with promises of "colossal power" and endless possibilities. However, we have observed some recent changes in the buzzwords used to promote products in ICT – we are hearing new claims of products that are designed to bring customers "just what they need" – and have also noticed the emergence of disruptive minimalist devices, "designed to be used as little as possible,"[3] which come with an optional minimalist service plan tailored to fit the phone.[4] In other sectors, some industry players are also taking a stand against overconsumption and encouraging their own customers to reduce the use of the very services they deliver. This is the case in the energy sector in

3 https://www.thelightphone.com/ (accessed 1 September 2020)
4 https://www.thelightphone.com/plans (accessed 1 September 2020)

which some providers accompany their customers not only toward energy efficiency but also toward the adoption of more responsible behaviors. A more striking example recently came from the aviation sector, in which, on top of its initiatives to make flying more sustainable, an airline company encouraged its customers to "fly responsibly" by preferring trains to planes for short trips or packing a lighter suitcase to reduce fuel consumption when they choose to fly.[5] Time will tell whether – or when – there is a market for such offers and whether people respond positively to such cues to limit their consumption to what they really need.

5.3.3 Opportunities... and Risks

Going down this path is a great opportunity to innovate. The cellular communication industry has always been largely focused on performance and cost. Service providers keep on fighting to provide their customers with better coverage, higher speed, and more minutes, messages, data, and features for the best price. Cell phone plan comparison websites are a good illustration of this situation. The names of the offers marketed by service providers also reflect this tendency to focus on price and performance, with plan names such as "5Go 4G+" and "120Go 4G+." Other operators focus on the endless possibilities offered to their customers, with plans named "Start unlimited," "Play more unlimited," "Unlimited plus," etc., including music or video content and additional features such as cloud storage. But what if we took a different approach, considering the ecological crisis, changing lifestyles and perceptions (e.g. some people believe that less can sometimes be more), by developing new offers that simply fit the real needs of end-users, rather than trying to add to already-unlimited possibilities? Instead of coming up with technological innovations that are supposed to reduce environmental impacts but end up increasing them by spurring the development of new uses, how about thinking about inventing more frugal lifestyles and developing the technological solutions that facilitate them.

Service providers could also derive potential economic benefits from an approach based on matching customer needs precisely, rather than exceeding them. If the network infrastructure and end user devices, marketing channels, and whole supply chain are designed to deliver modular solutions in which customers pick just what they need, this could allow the oversizing of offers to be limited in a cost-effective manner and constitute a source of cost reductions for the operator.

Encouraging customers to moderate their digital consumption is not without risks for mobile operators, however, when their business model is based on selling products and services designed to allow people to communicate ever more quickly, anytime and anywhere, with access to ever-richer content. One option consists in relying on regulators to address this issue and change the rules for providers and/or for consumers. The role of regulation, which might prove necessary, will be addressed later in this chapter. However, considering that we do not know how or when this may happen, we consider an alternative path here, in which some operators position themselves as game changers. We can fear three types of negative consequences of such strategies. First, if we assume that, as suggested earlier, mobile operators do choose to replace their traditional "all you can use" flat rate offers with

5 https://flyresponsibly.klm.com (accessed 1 September 2020)

pay-per-use contracts, subscribers will most likely reduce their consumption and expect a proportional decrease in their monthly bill, and this will have a direct impact on the operators' revenue streams and profitability. The same will probably be true if they design "minimalist" packages or tailor-made offers meant not to exceed customers' needs. A second kind of negative impact may be foreseen in terms of brand image. In a domain in which the performance of cell phone plans – measured in download speeds and amounts of minutes, messages, and data available – is a key attribute of products and services, coming up with offers aimed at reducing the amount of data exchanged over the network could be interpreted as a decrease in quality. Resorting to deliberate exaggeration, we might imagine the reaction of customers who are offered a "slow" connection offer. The cognitive dissonance could be such that it would tarnish the image of the service provider. It would also probably – and this is the third risk we anticipate – result in pure and simple failure of such offers, even if less radical and only including incentives to use less data or renew communication devices less frequently.

5.4 A Necessary Paradigm Shift

5.4.1 The Status Quo Is Risky, Too

While going down this path does definitely seem to have its risks, there is also risk in the status quo. At macro-level, continuing to encourage the "unlimited use" mindset increases the risk of worsened environmental impacts – water, air, and soil pollution; increasing CO_2 emissions; natural resource depletion; and biodiversity loss – despite the efforts made to improve the eco-efficiency of products and services. At meso-level, if organizations allow the ecosystem to which they belong to be altered, their performance will be impacted sooner or later. At micro-level, which is the level we will focus on here, continuing down the path of unbridled consumption may come short of understanding consumers' concerns in a changing world. While telecom services are more necessary than ever in a digital, hyper-connected society – a necessity that the recent COVID-19 crisis has dramatically highlighted – people are also becoming aware of the impact of their digital consumption on the environment on a wider scale, as well as on their individual well-being. These changes are reflected in a recent trend toward "digital minimalism" [11], based on the promise of regaining control over our digital lives by "knowing how much is just enough." While still marginal, this inclination may be related to long-standing trends, such as the "simple living" movement [12]. It can also be seen as a desire for deceleration of consumption and "escape from a sped-up pace of life" [13]. This trend can also be related to a growing concern about the influence of Internet platforms that are driving users toward always more engagement, as highlighted by Tristan Harris at his hearing at the US Senate in June 2019.[6]

Consumers are increasingly aware of the need to reduce the environmental impact of their consumption. Yet because a large part of what makes mobile communications possible is invisible to them, most end users consider it immaterial and are unaware of the

6 https://www.commerce.senate.gov/2019/6/optimizing-for-engagement-understanding-the-use-of-persuasive-technology-on-internet-platforms (accessed 1 September 2020)

environmental impact. They ignore the significant share that digital communications represent in global CO_2 emissions. While they may be aware of the environmental impacts related to the manufacturing, use, and end-of-life of the devices they have in their hands (e.g. their smartphone or tablet), they often ignore the impacts of the infrastructure needed to support the services available on these devices. Awareness is growing, however, and data centers, which can be considered as the tip of the iceberg, have been pinpointed as a major source of CO_2 emissions. People are also slowly coming to realize that sending messages, photos, downloading music, and video streams have a significant impact on the environment. A recent campaign in France, led by a think tank named The Shift Project, has drawn attention to "the unsustainable use of online video" [14], but at the same time, consumers are being urged by most industry players – network operators, device manufacturers, and content providers – to indulge in unrestrained use of telecom services to enjoy unlimited video streaming offers. In such a situation, not only do consumers who want to opt for a more reasonable behavior feel trapped between paradoxical demands, but they also suffer from a lack of offers on the market allowing them to do so. As an example, while it is possible to buy basic "voice-only" mobile phones, there is not a single service plan including only voice and simple text messages available on the French market, as we write this book.[7] In this context, acknowledging consumer aspirations for a streamlined digital life and accompanying them toward more frugal digital consumption appear a necessity.

5.4.2 Creating Value with 6G in the New Paradigm

Encouraged by the ecological necessity and signs of change in consumer lifestyles and expectations, there is an opportunity for the telecoms industry to shift to a new paradigm centered on frugality, a paradigm in which operators would not be focused solely on always offering more, but instead on answering the precise needs of their customers, while always keeping up with any evolution of their needs. Considering that 6G will be delivered in 2030–2035, this new generation of technology must focus on creating value for a world in which the scarcity of resources will most likely be a more acute reality and in which consumers will be seeking solutions that are both sustainable and designed for their well-being.

A number of industry players have clearly identified a need among some people to decelerate or take control of a digital life that can sometimes become overwhelming. Most smartphone manufacturers provide applications that allow consumers to control or self-limit their use of their devices. Some manufacturers go a step further and a few offers for "minimalist" phones have been launched on the market in the past few years.[8] Some of them are brand new designs, while others are renewed versions of old generations of cell phones, and some operators are providing plans tailored to these phones. However, these remain niche offers and there is still a lot that needs to be invented.

A key challenge for cellular operators therefore lies in understanding complex consumer behaviors and expectations, their needs in terms of connectivity, their concerns

7 https://www.echosdunet.net/dossiers/forfait-mobile-sans-internet (accessed 1 September 2020)
8 https://www.telegraph.co.uk/recommended/tech/best-basic-phones-dumbphones/ (accessed 1 September 2020)

for the impact of their consumption on the environment, and their (positive or negative) representations of frugality. Such an understanding is a prerequisite to build new value propositions based on something other than "unlimited" consumption. Such value propositions, which should create value for consumers as well as service providers, are yet to be invented. This is another major challenge. One possible route could consist in adopting a different marketing approach to consumer value, leaving out the traditional performance/price ratio and adopting a holistic, multidimensional approach to consumption value [15]. Instead of focusing on utilitarian value – unlimited data, maximum speed, etc. – it would consist of considering other sources of value. Frugal offers would provide ethical value to consumers who are eager to stay connected without creating negative environmental or social impacts. Hedonic value could be derived from the satisfaction of simple, unsophisticated, long-lasting devices with which consumers could build relationships over the long term [16]. The possibility of displaying this form of responsible consumption would also create social value. Finally, the possibility of regaining control over our digital life would alleviate the burden that hyper-connectivity or information overload creates on many of us, thus offering a form of spiritual value.

5.4.3 Empowering Consumers to Achieve the "2T CO_2/Year/Person" Objective

The "more and more" approach, which has reached the limits imposed by our planet, may also be "too much" from a consumer perspective. The time may have come to consider that, even for ICT, less can be more and to question how much is enough. Consumers should not be left alone in this process, and the whole industry should play its role in accompanying them on this path toward sustainable digital consumption. This implies gaining a deep understanding of consumer behaviors and, on a larger scale, of cultural changes in a world in transition, faced with ecological deadlines. This also requires new value propositions that take account of these complex expectations and find the words to communicate them, while raising awareness of the necessity of adopting sustainable consumption behaviors.

We have mentioned the fact that reaching the UN IPCC objectives to limit the increase in global surface temperature by 1.5–2 °C by 2050 would mean that each French citizen should bring their CO_2 emissions down under the threshold of 2 T/year/person by then. At the same time, due to technological progress – including the advent of 5G and 6G – as well as sociological changes, we expect an increased need for wireless connectivity. As a result, consumers will see greater value in products and services designed to allow them to stay connected while maintaining their CO_2 emissions under the limit of 2 T/year/person. This is the opportunity that all 6G players should seize: design a new generation of cellular technology that will empower consumers to reach the UN IPCC objectives and contribute to their well-being by easing the mental burden vis-à-vis the digital world.

5.5 Summary and Prospects

5.5.1 Two Drivers, Three Levels of Action

To sum up, we identify two main drivers for achieving the objective of an overall reduction in energy consumption and associated CO_2 emissions linked to the use of ICT services:

efficiency and control of consumption and uses. For each of these drivers, action can be taken at three levels: the regulator, companies, and consumers. We have discussed changing consumer expectations and behaviors and seen all the challenges and opportunities that this represents for companies, but the regulatory level could play a complementary role.

5.5.2 Which Regulation for Future Use of Technologies?

National and international regulations have had a positive impact on reducing environmental impacts in various industrial areas, such as chemicals (Registration, Evaluation, Authorisation and Restriction of Chemicals [REACH][9]), the use of hazardous substances in electronics (Restriction of Hazardous Substances in Electrical and Electronic Equipment [RoHS][10]), and the management of waste from electrical and electronic equipment (WEEE[11]). In the field of ICT, the energy consumption of home network products, such as TV decoders or Internet access gateways, are subject to regulations or voluntary agreements in Europe, such as the Complex Set Top Boxes Voluntary Agreement[12] for TV decoders and the Broadband Code of Conduct[13] for gateways, either to implement low power modes or to limit energy consumption of functional blocks. These regulations have a significant impact on the energy consumption of products, for example, through the implementation of standby mode (0.5 W) on TV set-top boxes when they are not in use.

If regulation has shown itself to be relevant in improving the energy efficiency of end user devices, it could be just as relevant for the consumption and use part, if policy makers should decide to regulate the use of mobile services in order to limit bandwidth consumption.

Just as we were writing this book and raising the possibility of a ban on unlimited cell phone plans, a debate opposed the French regulator for electronic communications Autorité de Régulation des Communications Electroniques et des Postes (ARCEP) and French governmental organizations over the possibility of encouraging operators to offer limited plans for both fixed and mobile Internet access.[14] The evolution of policies in this domain is hard to predict, and the issue is a worldwide one and quite complex. As a result, both paths – limitation of bandwidth consumption imposed by regulators on one hand and self-limitation driven by operators and end users on the other hand – should be considered and worked on in parallel.

5.5.3 Hopes and Prospects for a Sustainable 6G

Your 6G or your life? We should not have to choose! 6G must be sustainable. A key challenge in the development of the next generation of cellular networks is therefore to allow all their expected environmental benefits – within the telecoms industry, with increased energy efficiency, as well as in other sectors, for example, with the implementation of

9 https://echa.europa.eu/regulations/reach/legislation
10 https://ec.europa.eu/environment/waste/rohs_eee/index_en.htm
11 https://ec.europa.eu/environment/waste/weee/index_en.htm
12 http://cstb.eu/
13 https://ec.europa.eu/jrc/en/energy-efficiency/code-conduct/broadband
14 https://www.capital.fr/economie-politique/larcep-dit-non-a-toute-limitation-des-forfaits-internet-illimites-1375392

smarter transportation systems – without getting caught out by a rebound effect that could undermine all the efforts made to contribute to the ecological transition. While there is no easy solution to implement it, the formula to deliver a sustainable 6G is quite simple: increase energy efficiency while keeping consumption and uses under control. While the first part comes from technological innovation, the second one can only be achieved with change in consumption behaviors. 6G players could play a significant role in accompanying consumers down this road, as we expect them to promote the products and services that will meet their wireless communications needs while helping them stay within the 2T CO_2/year/person threshold. This is an opportunity and a challenge, which we hope to see stimulate innovation and give rise to solutions with positive impacts on sustainability.

References

1. Malmodin, J. and Lundén, D. (2018). The energy and carbon footprint of the global ICT and E&M sectors 2010–2015. *Sustainability* 10 (3027) https://www.researchgate.net/publication/327248403_The_Energy_and_Carbon_Footprint_of_the_Global_ICT_and_EM_Sectors_2010-2015 (
2. European Commission (2020). Communication from the commission to the European parliament, the council, the European economic and social committee and the committee of the regions. Shaping Europe's digital future, Brussels. https://ec.europa.eu/info/sites/info/files/communication-shaping-europes-digital-future-feb2020_en.pdf.
3. Wikipedia (2020). Environmental impact of aviation. https://en.wikipedia.org/w/index.php?title=Environmental_impact_of_aviation&oldid=974811845 ().
4. Andrae, A.S.G. (2019). Predictions on the way to 2030 of internet's electricity use. In: *The Lost Decade? Planning the Future*. Ålborg University Copenhapen.
5. Carbone 4 (2020). 2-infra challenge methodological guide. http://www.carbone4.com/wp-content/uploads/2020/07/Carbone4_2-infra_challenge_methodological_guide_july2020.pdf.
6. Commissariat Général au Développement Durable and I4CE – Institute for Climate Economics (2019). Chiffres clés du climat. https://www.i4ce.org/download/chiffres-cles-du-climat-2019-france-europe-et-monde/.
7. Weißbach, D., Ruprecht, G., Huke, A. et al. (2013). Energy intensities, EROIs (energy returned on invested), and energy payback times of electricity generating power plants. *Energy* 52: 210–221.
8. Kharina, A. and Rutherford, D. (2015). *Fuel Efficiency Trends for New Commercial Jet Aircraft: 1960 to 2014*. Washington: ICCT – The International Council on Clean Transportation. https://trid.trb.org/view/1372314 (accessed 1 September 2020).
9. Gillingham, K., Kotchen, M.J., Rapson, D.S. et al. (2013). The rebound effect is overplayed. *Nature* 493 (7433): 475–476.
10. Faruqui, A., Sergici, S., and Sharif, A. (2010). The impact of informational feedback on energy consumption – a survey of the experimental evidence. *Energy* 35 (4): 1598–1608.
11. Newport, C. (2019). *Digital Minimalism: On Living Better with Less Technology*. New York: Portfolio/Penguin.

12 Elgin, D. (1981). *Voluntary Simplicity: Toward a Way of Life that is Outwardly Simple, Inwardly Rich.* New York: Quill.

13 Husemann, K.C. and Eckhardt, G.M. (2019). Consumer deceleration. *Journal of Consumer Research* 45 (6): 1142–1163.

14 Efoui-Hess, M. (2019). Climate crisis: the unsustainable use of online video. The Shift Project. https://theshiftproject.org/wp-content/uploads/2019/07/2019-02.pdf.

15 Holbrook, M.B. (1999). Introduction to consumer value. In: *Consumer Value: A Framework for Analysis and Research*, 1–28. London: Routledge.

16 Dabadie, I. and Robert-Demontrond, P. (2021). What being an owner can also mean: A socio-anthropological study on the development of a forgotten relationship to objects. *Recherche et Applications en Marketing (English Edition).* DOI: 10.1177/20515707211014436.

6

Catching the 6G Wave by Using Metamaterials: A Reconfigurable Intelligent Surface Paradigm*

Marco Di Renzo and Alexis I. Aravanis

CNRS, CentraleSupelec, L2S, University of Paris-Saclay, Gif-sur-Yvette, France

6.1 Smart Radio Environments Empowered by Reconfigurable Intelligent Surfaces

The ever increasing demand for broadband access has pushed wireless networks away from the archetypal wireless network paradigm, where (mainly) outdoor users were served by centralized network entities. As opposed to this approach, emerging B5G network architectures are expected to employ atomized and softwarized network infrastructures in a dispersed, device-centric manner. Moreover, the unabated increase of indoor traffic [1] is pushing for even more disruptive network designs able to provide high-quality service even to indoor users, surrounded by urban blockages. Furthermore, the emphasis of the envisaged 6G networks on high frequency bands (e.g. D-band and THz) makes the adverse effect of blockages more acute [2]. In this setup, a fundamentally distinct wireless ecosystem needs to arise, not only to be able overcome the challenges posed by wireless blockages but more importantly to satisfy the need for increased spatial bandwidth, i.e. the need of B5G networks to deliver bits per second per cubic meter rather than simply bits per second [2], by exploiting the spatial domain of the wireless environment.

In this course, future networks must challenge the entrenched status where the wireless environment is perceived as an invariable "unintentional adversary," to which the system needs to adapt, and where only the endpoints of the communication network can be optimized. This is an extremely inefficient paradigm, where base stations (BSs) transmit radio waves of the order of magnitude of Watts while user equipment detects signals of the order of magnitude of μWatts or even nWatts, with the rest of the energy being dissipated over the wireless environment (i.e. the channel), while creating interference to other network elements. In order to challenge this status, an antipodal approach is required, where the

* The work of M. Di Renzo and A. I. Aravanis was supported in part by the European Commission through the H2020 ARIADNE project under grant agreement 675806 and through the H2020 MSCA IF Pathfinder project under grant agreement 891030.

Shaping Future 6G Networks: Needs, Impacts, and Technologies, First Edition.
Edited by Emmanuel Bertin, Noel Crespi, and Thomas Magedanz.

wireless environment (i.e. the channel) needs to dynamically adapt to the wireless network operation, giving rise to what is referred to as a *smart radio environment*.

A smart radio environment is a controllable wireless environment, where software-defined materials or software-controlled metasurfaces can be overlaid on top of environmental objects or blockages to transform them into controllable network entities, conducive to wireless propagation [3]. This transformation of the wireless networks into a smart radio environment is made possible by the advancement in the area of electromagnetic (EM) metamaterials [4], which has given rise to a new technology allowing the control and manipulation of the radio waves traversing through the wireless channels [3]. This technology has allowed for the design and fabrication of large software-controlled metasurfaces [5–7], known as *Reconfigurable Intelligent Surfaces (RISs)* [3], which are able to recycle radio waves without the need for generating new signals but rather controlling the radio waves from within the wireless channel.

6.1.1 Reconfigurable Intelligent Surfaces

An RIS is an intelligent surface able to manipulate the impinging radio waves at will. As opposed to a regular surface that reflects impinging waves by adhering to Snell's law of specular reflections, (where the angle of the incident wave is equal to the angle of the reflected wave) an RIS is an "intelligent" surface in the sense that it is able to reflect and transmit EM waves in a desired direction. This anomalous reflection, also referred to as *geometric scattering*, is not bound by the rules of reflection or refraction but guides reflected waves toward directions that can be considered anomalous with respect to Snell's law, and this reflection is governed by the so-called generalized Snell's law [8]. The capability of RISs to shape and direct the EM waves in an intelligent way gives rise to a number of interesting applications that are expected to play a crucial role in 6G [9]. These applications, depicted in Figure 6.1 [10], include but are not limited to (a) anomalous reflection, (b) beamforming to focal points [11, 12], and (c) joint encoding on the RIS reflection pattern [13].

RISs are capable of applying the aforementioned customized transformations to the reflected radio waves, by changing the phases of the scattering particles that comprise the RIS [4], in a way that allows for the constructive interference of all the particle-engendered

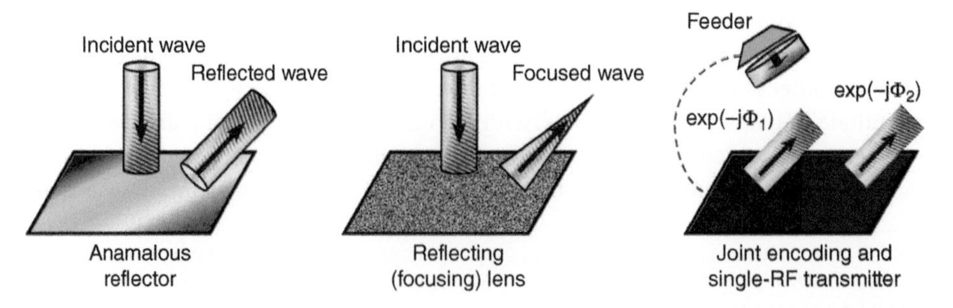

Figure 6.1 Potential applications of "intelligent" reflectors. *Source:* Di Renzo et al. [10]. Licensed under CC BY 4.0.

multipath components at the desired location. These constituent subwavelength scattering particles (aka meta-atoms) are dielectric or metallic [14], forming a subwavelength array of planar or curved conformation. This malleable conformation of RISs allows for the coating of environmental objects with software-controlled, artificial, thin films of RISs, facilitating the dynamic manipulation of the incident radio waves in a fully reconfigurable manner [15]. This allows the efficient management of the transmitted radio waves toward (a) improving network coverage, by circumventing physical obstacles [3]; (b) enhancing the signal power at the receiver, exploiting the scaling law of the power reflected from the RIS scattering particles [11, 16]; (c) reducing BS power consumption by substituting power-hungry MIMO (multiple-input and multiple-output) RF (radio frequency) chains with RIS-empowered multi-stream transmitters on a single-RF chain (media-based modulation) [16]; or (d) cloaking antennas from one another in particular frequency bands allowing for their ultra-tight packing [17].

Numerous companies are already productizing RIS-based solutions, engendering the next generation of RIS-empowered wireless networks. In the present chapter, we will elaborate on many of those industrial activities, while outlining the state of the art. In this course and in order to visualize the RIS-empowered network operation, Figure 6.2 [18] demonstrates a Metawave-implemented solution (Metawave is one of the companies actively involved in the engineering of RIS-based solutions [18]) toward improving network coverage by coating building facades and street furniture with RISs, steering radio waves toward locations of poor coverage. The subsequent interconnection and control of all network RISs (that are used for coating the network environment) through radio access intelligent controllers can indeed transform wireless networks into customizable smart radio environments. In this setup, the spatial domain arises as a fully malleable pillar of flexibility playing an active role in transferring and processing information [3].

Figure 6.2 RIS-enhanced network coverage. *Source:* Metawave Corporation.

6.2 Types of RISs, Advantages, and Limitations

RISs can be manufactured in practice either as large arrays of inexpensive antennas, usually spaced half of wavelength apart (e.g. Figures 6.6 and 6.7), or as large metamaterial-based surfaces (i.e. metasurface-based RISs), whose scattering elements have sizes and inter-distances much smaller than the wavelength (e.g. Figures 6.8 and 6.9). These two approaches can be effectively viewed as "discrete" and "continuous" implementations of RISs, respectively. In the discrete approach, the phases of the antenna elements are optimized individually, whereas in the continuous approach, the phases of all scattering particles are optimized collectively. This collective optimization of all scattering particles, in the latter case, is imposed by the very nature of metasurface-based RISs, as will be detailed in the following paragraphs.

To elaborate on metasurface-based RISs, metasurfaces are electrically thin and electrically large structures, which means that their thickness is considerably smaller and their transverse size is considerably larger than the wavelength. The mode of operation and the reflection properties of these metasurfaces are defined by the inter-distance and the size of their constituent scattering particles (aka unit cells). Metasurface-based RISs constitute the third generation of metasurfaces, coming as the evolution of first generation of uniform metasurfaces (i.e. surfaces whose unit cells are uniformly distributed), and of the second generation of nonuniform metasurfaces. This second generation of nonuniform metasurfaces extended the range of applicability of the first generation, by introducing a variability in the spatial domain (over the uniform spatial pattern) to support ultrawideband, beamforming, multibeam, and multifrequency operation. The current third generation of metasurface-based RISs introduced an additional variability in the time domain, through the employment of switches, varactors, and diodes, to allow for the dynamic reconfiguration of the reflected or refracted waves.

Having outlined the concept of metasurface-based RISs and since the fundamental of antenna arrays is already known to the community, we can proceed with the axiomatic formulation of the aforementioned classification of RISs into "discrete" and "continuous," based on the concept of homogeneity. In particular, RISs can be axiomatically categorized based on the size of their unit cells to homogenizable and inhomogenizable. Homogenizable RISs are surfaces that can be globally or locally described by effective material parameters, like surface impedance, susceptibility, and polarizability [19]. This means that RISs characterized by homogeneity can be practically considered continuous and so can be the ensuing analysis on the reflected waves, as opposed to the discrete RISs of inexpensive antennas mentioned earlier. Figure 6.3 [19] demonstrates the threshold between homogenizable and inhomogenizable surfaces, with surfaces of unit size (denoted by α) smaller than half a wavelength, like uniform metasurfaces, high impedance surfaces (HDS), or some frequency selective surfaces (FSS) being homogenizable, whereas surfaces of unit size greater than half a wavelength being inhomogenizable. This homogeneity-based taxonomy of RISs is further refined in Figure 6.3, by taking into account not only the unit size (characterizing the homogeneity of the structure) but also the period of the unit cells, i.e. their inter-distances. In particular, surfaces of unit size smaller than half a wavelength, and of inter-distance (denoted by D) also smaller than half a wavelength, are globally homogenizable, whereas surfaces of unit size smaller than half a wavelength but of inter-distance higher

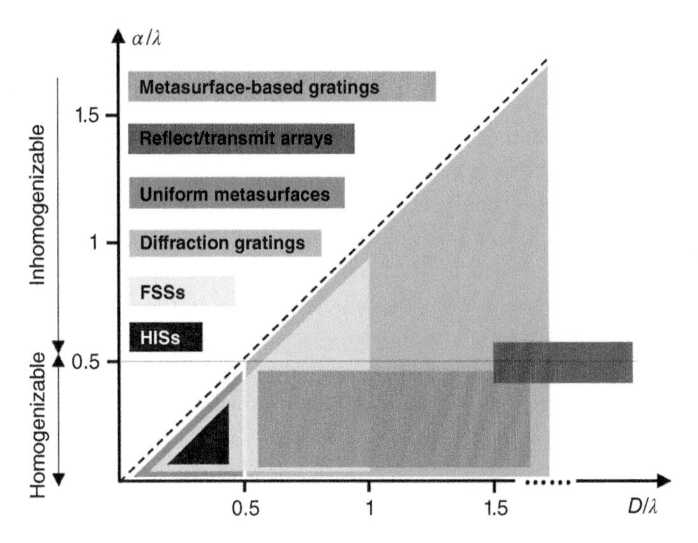

Figure 6.3 Planar structures classified by homogenization property. Here, a is the unit size and D is the structure period (inter-distance). *Source:* Wang [19].

than half a wavelength maintain the homogeneity, but only at a local level. This means that structures such as metasurface-based gratings, which are clustering unit cells in super-cells of varying inter-distances, maintain the homogeneity, not at a global but at a local super-cell level [19].

The aforementioned homogeneity of RISs is extremely important to provide properties such as larger angles of reflection, and perfect anomalous reflection or refraction [20–22], as opposed to the imperfect anomalous reflection and refraction of discrete, inhomogenizable structures. In particular, as already mentioned, the unit cells of discrete RISs (that are inhomogenizable due to the half wavelength spacing of the reflect/transmit arrays as shown in Figure 6.3) are optimized independently of each other, whereas the unit cells of continuous RISs follow a coupled optimization. These two optimization approaches, described in the literature as local and nonlocal design, respectively, are imposed by the very nature of the two implementations and endow different properties to discrete and continuous RISs.

In the case of the nonlocal design of continuous RISs, the coupled optimization of unit cells allows the transfer of energy from one unit cell to another. In particular, energy from an incident wave can be transferred by exciting evanescent superficial (not propagating) waves, traveling on the surface of the RIS [23]. Hence, the power received from one unit cell can be transferred to a different unit cell, giving rise to a more flexible power budget. To elaborate, the local design of the discrete RISs imposes a standalone power constraint for each unit cell, where the impinging power on each unit cell must be less or equal to the power reflected by the same unit cell, since RISs are in general passive structures. However, the nonlocal design allows for the transfer of energy between unit cells, giving rise to a globally passive structure that can, however, be active at a local level. That is, the reflected power by a unit cell can be higher than the power that impinged on the unit cell, due to the amplification achieved by the aggregation of power from the neighboring unit cells. This gives rise to a flexible power budget of higher power efficiency, that allows continuous RISs to provide larger angles of reflection and perfect anomalous reflection, exploiting the fact that the structures are passive on a global but not on a local level.

At this point it should be noted that even though second-generation metasurfaces are completely passive structures, the aforementioned characterization of metasurface-based RISs as globally passive structures is a relatively abusive term, since in reality an RIS is only a nearly passive structure. That is, since minimal power and digital signal processing capabilities are needed in order for the RIS to interact with the environment and in order for the surface to be reconfigured dynamically over time. In this course, RISs are equipped with processing units, microcontrollers, and radio frequency chains that allow them to either report changes in their environment if they are equipped with appropriate sensing elements or receive commands during the control and configuration phase in response to the changing wireless channel. These commands are subsequently implemented by the RIS tunable elements (i.e. switches, varactors, and diodes), resulting in the consumption of power. However, after the control and configuration phase of the RISs is over, no power amplification is used after the RISs enter the normal operation phase. Due to the passive behavior at the normal operation phase and the only minimal power requirements at the control phase, RISs can more accurately be characterized as nearly passive structures.

6.2.1 Advantages and Limitations

Having outlined the types of RISs and their modes of operation, we can proceed with the comparison of RISs with other existing technologies to understand the advantages, limitations, and trade-offs of this emerging technology. In this direction, an experimental case study is examined as performed by the startup company Pivotal [24, 25] that develops metasurface-based RIS solutions. In this case study, an RIS architecture (developed by the company under the framework of their "holographic beamforming" technology and described in detail in a number of white papers [24, 25]) is compared with the MIMO and phased array architectures in terms of cost, size, and implementation complexity and is presented in Table 6.1 [24].

The block diagrams of Table 6.1 demonstrate that the complexity of the MIMO array is very high, since a high number of antennas is required along with an equal number of RF chains and a complex signal processing unit, while the complexity increases significantly with the number of antenna elements. In the case of the phased array, the complexity decreases since instead of an RF chain per antenna element, a phase shifter is employed and the signal processing unit is much less complex. As opposed to these two systems, Table 6.1 demonstrates that the RIS system requires a much higher number of unit cell elements; however, the circuits supporting each element are much less complex and much less expensive. Hence, this gives rise to an implementation-related trade-off where the successful application of RISs requires a large number of RIS elements supported, however, by circuits characterized by extremely low complexity compared to legacy technologies. This conclusion becomes also evident in Table 6.2 [25], where an RIS system and a phased array achieving the same quality of service (QoS) are compared. Table 6.2 demonstrates that the number of RIS elements is much higher than the corresponding elements of the phased array (640 *vs.* 256); however, their power consumption is much lower than that of the phased array elements (12.9W *vs.* 39.6W).

Table 6.1 Summary of key differences among RIS, phased array, and MIMO beamformers.

Architecture	Block diagram	Cost	Size	Challenges
RIS	Simple BBU, Radio & Control	Super-sampled, commercial off the shelf (COTS) design, low price	Thin, comfortable	Single beam per polarization per sub-aperture
Phased array	Simple BBU, Radio & Control; ϕ ϕ ϕ ϕ	Distributed phase shifters and amplifiers, moderate price	Trades cost for thickness, thin are too expensive	Thermal challenges, difficulty due to distributed amplification, multibeam at increased cost
MIMO	Complex BBU, Timing & Control System; Radio Radio Radio Radio	Radios behind every element, complex baseband unit (BBU) entails high price and power consumption	Thick, thickness can be reduced by hiding BBU in baseband cabinet	No frequency-division duplex (FDD), unworkable at mmW, spectral efficiency scales poorly with cost

Source: Eric J. Black, PhD, CTO, Pivotal Commware, Holographic Beam Forming and MIMO, 2021.

Table 6.2 Difference between RIS and active phased array power consumption.

	Phased array	RIS	Unit
Number of unit cells	256	640	#
Antenna gain	28	26	dB
Number of RF chains	256	1	#
Transmit power per chain	6.2	2512	mW
Total RF transmit power	1.58	2.51	W
Power-added efficiency	4.0%	25.0%	%
Direct current (DC) draw for RF	39.6	10.0	W
RIS controller	0	2.9	W
Total DC power	39.6	12.9	W

Source: Pivotal Staff, Pivotal Commware, Holographic Beam Forming and Phased Arrays, 2021.

Evidently, compared with other transmission technologies, e.g. phased arrays, multi-antenna transmitters, and relays, RISs require the highest number of scattering elements, but each of them needs to be backed by the fewest and least costly components, while no dedicated power amplifier per element is needed. This trade-off between the number of elements and the power, cost, and complexity of implementation per element needs to be investigated in detail in future works. However, the aforementioned case study has demonstrated that RISs arise as prime candidates for realizing what is called a reduced cost, size, weight and power consumption (C-SWaP) design, i.e. a transmitter design at reduced cost, size, weight, and power [24], through a software-defined RIS-enabled architecture. Moreover, such architectures can achieve a sustainable wireless design of low EM field exposure, by recycling transmitted waves while the transmitter can be manufactured by recyclable materials, that (as will be demonstrated in the following section) can be seamlessly integrated in urban network environments by being physically and aesthetically unobtrusive.

The previous case study has already demonstrated that in order to achieve performance comparable or even superior to that of other technologies a high number of RIS elements is required. Two more key factors determining the performance of RISs that need to be taken into account prior to their productization and their incorporation in wireless networks are the size and transmit distance of the RISs. In particular, it has been demonstrated that for an indicative transmission frequency of f = 28 GHz and an RIS length of 1.5 m = 140 λ (i.e. an electrically large RIS), an RIS can provide a similar performance or even outperform a full duplex (FD) relay without even employing a power amplifier. The data rates achieved by the RIS, the FD relay, and a half duplex (HD) relay for different transmission distances are demonstrated in Figure 6.4 [10]. The results obtained in

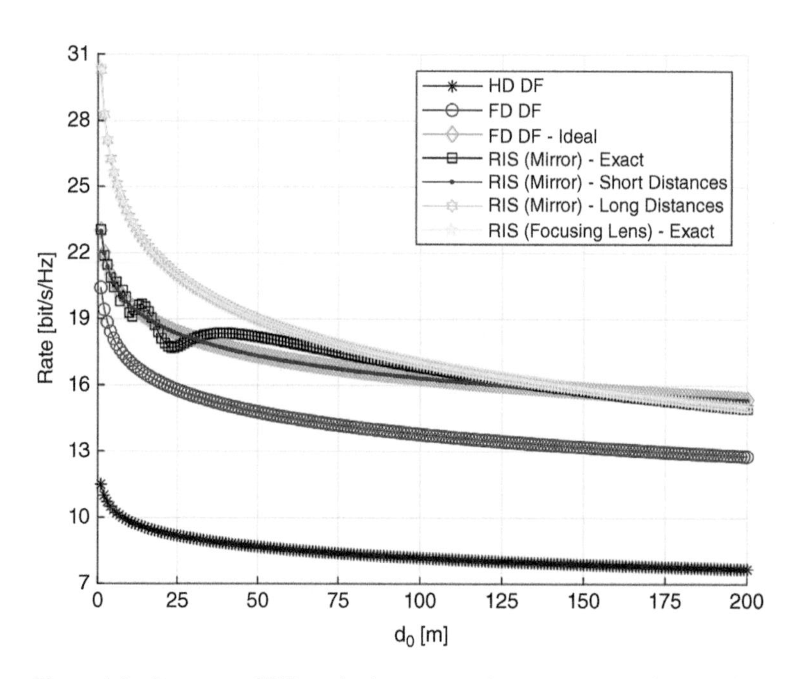

Figure 6.4 Data rate of RISs and relays versus the transmission distance. *Source:* Di Renzo et al. [10]. Licensed under CC BY 4.0.

Figure 6.4 are obtained for inter-distances in the range of $\lambda/5$ and a number of unit cells equal to 700 (or $\lambda/2$ and 280, respectively). An interesting finding is that for distances up to 25–50 m the RIS behaves as an anomalous mirror (i.e. the transmitter is close enough to the RIS to perceive it as electrically large), whereas for distances greater than 75–100 m the RIS behaves as a local diffuse scatterer (i.e. the transmitter resides that far from the RIS; that in practice the transmitter sees the RIS as electrically small), while for distances greater than 150 m the RIS exhibits a performance that is inferior to that of the FD relay.

This finding demonstrates the importance of the transmit distances and of the size of the RIS, since at higher distances RISs are perceived as electrically small and therefore larger RISs may be required in order to outperform legacy technologies. However, the fact that as already mentioned and as will be demonstrated in the following section modern RISs can be aesthetically unobtrusive and seamlessly integrable on glass building facades, billboards, etc. allows for the employment of electrically large structures (depending of course on the operating frequency) giving rise to a performance comparable to this of Figure 6.4. Moreover, due to the aforementioned C-SWaP design, the use of RISs is expected to be pervasive with ultradense deployments of RISs that are expected to minimize transmit distances of RIS reflected waves, allowing network operators to effectively manage this trade-off between RIS sizes and transmit distances.

This trade-off is also visualized in Figure 6.5 where the bitrate achieved by the employment of an RIS is compared with the bitrate achieved by an FD and an HD relay for the

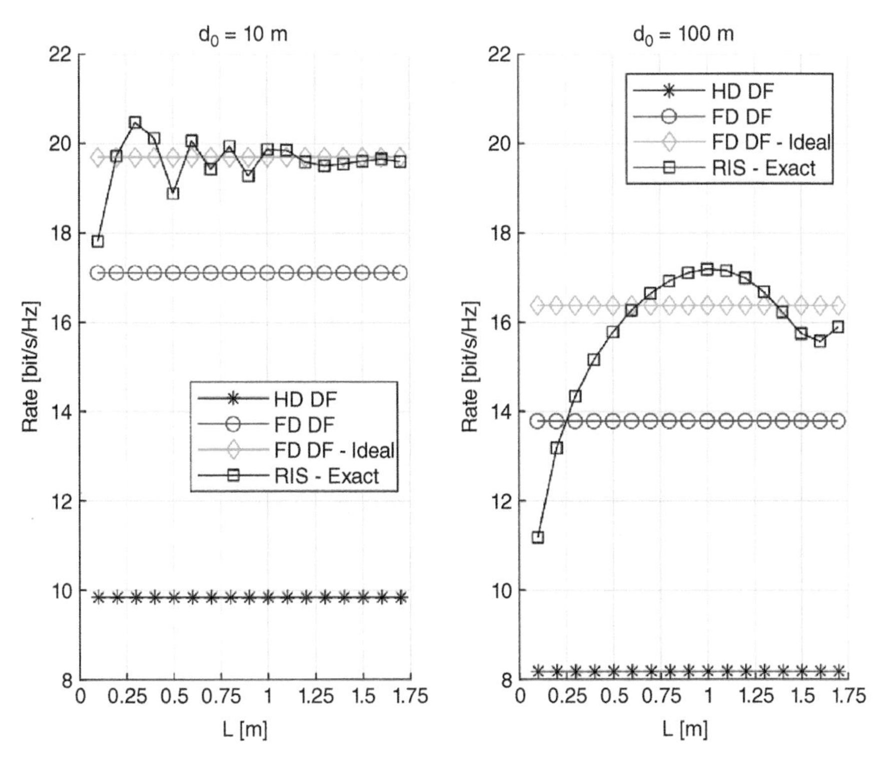

Figure 6.5 Data rate of RISs and relays versus the size of the RIS. *Source:* Di Renzo et al. [10]. Licensed under CC BY 4.0.

same setup described above, at two different transmit distances of 10 and 100 m for different RIS lengths. Once again it is demonstrated that for an electrically large RIS the achieved bitrate is higher than that of the FD and HD relay, which motivates the pervasive use of RISs in the network in order to be employed particularly for small transmit distances.

6.3 Experimental Activities

Having outlined the key technologies employed for the implementation of RISs as well as the main factors influencing their performance as well as their advantages and limitations, the present section will focus on the practical implementations of RISs and the experimental activities available in the literature. In particular, the potential of the RIS technology toward realizing smart and controllable radio environments for the first time in history has already spurred the practical implementation of RIS prototypes both discrete and continuous, corroborating the theoretical results presented above with respect to the enhancement of received power and capacity, and a brief survey of those prototypes and of their characteristics is performed hereafter.

6.3.1 Large Arrays of Inexpensive Antennas

6.3.1.1 RFocus

One of the first prototypes realized based on discrete arrays of inexpensive antennas is the RFocus prototype implemented at the Computer Science & Artificial Intelligence Laboratory (CSAIL lab) of Massachusetts Institute of Technology (MIT) [26]. The RFocus prototype, depicted in Figure 6.6, employs 3200 inexpensive antennas distributed over a surface of 6 square-meters and constitutes (to the best of the authors' knowledge, among all published configurations) the configuration comprising the largest number of antennas

Figure 6.6 The RFocus prototype surface. *Source:* Arun [63].

ever employed for a single communication link. The prototype constitutes, in reality, a software-controlled surface with thousands of plain switching elements. Each of the switching elements is a two-way RF switch, with two distinct states. Any impinging wave is either reflected or traversing through the element. Moreover, a controller is employed to configure all constituent elements in the way that ensures that any signal impinging upon the RFocus surface from the transmitter will be directed toward the receiver. The controller is agnostic to the location of the transmitter or receiver and just switches between different optimized states to maximize the signal power between the different endpoints of the transmitter and receiver. The RFocus redirected signals are not amplified and the controller can choose between a "mirror" or a "lens" operation, where the endpoints can reside either on the same or on different sides of the surface. The measurements at the premises of the MIT CSAIL laboratory report that the employment of RFocus improved the median signal strength by a factor of 9.5× and the median channel capacity by a factor of 2.0× [26]. At this point, however, it should be noted that these measurements and gains correspond only to an indoor environment, whereas similar outdoor measurements are required in order to demonstrate the efficiency of the prototype in less controllable environments.

In practice RFocus serves as a beamformer, but the beamforming function is now shifted from the radio endpoints to the radio environment itself. This shift of the beamforming functions to the environment allows for the deployment of huge beamforming antenna arrays (of a 6 square-meter area or more) that could not be deployed before due to the space limitations typically imposed to infrastructure BSs or access points (which are hard to deploy). Thus, it allows for the deployment of massive and potentially multiple beamforming antenna arrays. Moreover, such a beamforming antenna array is not in need of connecting each antenna element to a full-fledged radio transmit/receive circuitry of increased power consumption and cost. However, the deployment of such massive antenna arrays of an excessively high number of antenna elements incurs an optimization overhead cost. In particular, the optimization of all 3200 antenna elements by the controller incurs a significant latency. After collecting Transmission Control Protocol (TCP) throughput data, the latency of the optimization algorithm is equal to 4000 packets. Hence, an interesting trade-off arises between the number of antenna elements and implicitly the level of reflected power and the incurring latency of the optimization algorithm at the controller. This trade-off needs to be extensively studied, and in this direction smaller discrete arrays of reduced latency have also been studied and implemented in practice, like the one mentioned next.

6.3.1.2 The ScatterMIMO Prototype

The ScatterMIMO prototype is another discrete array implementation of an RIS similar to that of RFocus, implemented by the University of California [27] (Figure 6.7). The ScatterMIMO prototype encompasses 48 antenna elements in the form of three tiles of 4 × 4 MIMO antenna arrays forming a virtual access point. This virtual access point complements the active access point by reflecting its signal in a controlled and an optimized manner either vertically or horizontally toward creating additional MIMO streams to double the throughput. Moreover, the location of this small RIS can itself be appropriately optimized in order to ensure that either an line of sight (LOS) link between the transmitter and the receiver is established or that the RIS will reflect the maximum amount of power toward the receiver. In the latter case, the RIS needs to be positioned close to the receiver or

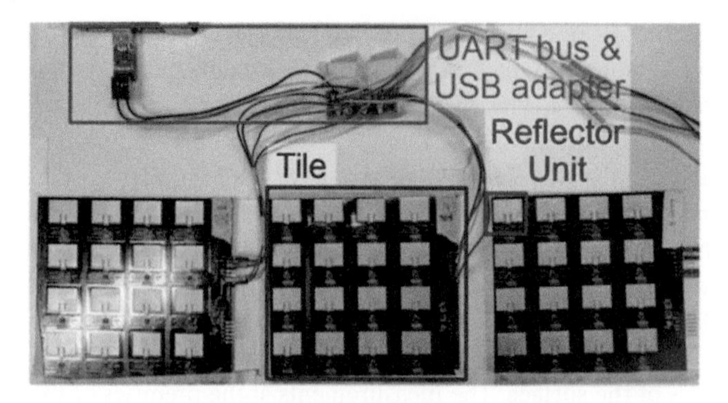

Figure 6.7 The ScatterMIMO hardware prototype. *Source:* [27].

the transmitter, and since the position of the receiver is not fixed, the RIS needs to be positioned in the vicinity of the fixed transmitter. Thus, the distance to the receiver or transmitter arises as an additional degree of freedom. By exploiting this degree of freedom and strategically placing the RIS at the optimal location, the authors of [27] have demonstrated that the same QoS can be achieved as that provided by an RIS encompassing more antenna elements but at a suboptimal location. Moreover, [27] demonstrates that the RIS of 48 antenna element incurs only a minimal optimization-related latency cost, which is equal to 3 packets, as opposed to the 4000 packets of RFocus. Hence, the ScatterMIMO approach demonstrates an interplay between the antenna elements and the latency (already demonstrated by the RFocus approach) but also with respect to the distance between the RIS and the receiver or the transmitter, respectively. Once again all reported measurements of [27] pertain to an indoor environment.

6.3.2 Metasurface Approaches

Further to the discrete array implementation of RISs mentioned above, numerous metamaterial-based implementations of RISs have also been fabricated, with the majority of those approaches also being tested primarily in indoor and controllable environments. As already mentioned the nonlocal design of such metasurfaces (i.e. the coupled optimization of their constituent elements) endows an interesting set of properties such as perfect anomalous reflection at large angles of reflection. One of the first approaches achieving a perfect design of an anomalously reflective surface employing a strongly nonlocal design is that of Aalto University [23] with the fabricated metasurface being depicted in Figure 6.8. The principles of nonlocal design have already been developed in detail in Section 6.2 and are therefore omitted at this point. However, for the sake of completeness, it should be noted that the fabricated metasurface employs 10 unit cells clustered at a super-cell level with each unit cell having a width of 3.5 mm (on the x-axis) and each super-cell having a width of 40 mm (on the x-axis) and a height of 18.75 mm (on the y-axis). The overall dimensions of the metasurface are 440 mm on the x-axis and 262.5 mm on the y-axis, which for an operating frequency of $f = 8$ GHz corresponds to 11.7 λ and 7 λ, respectively. The subwavelength dimensions of the unit cells give rise to a homogenizable metasurface and more importantly pave the way for the advent of homogenizable structures of even smaller

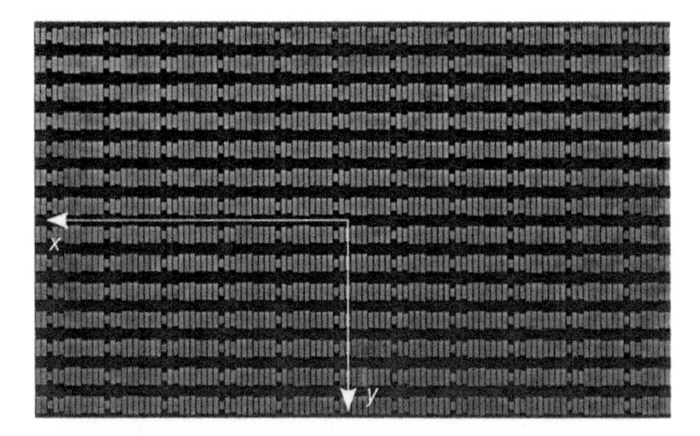

Figure 6.8 Aalto-fabricated metasurface. *Source:* [23].

granularity (i.e. of even smaller unit cells) that can give rise to structures aesthetically unobtrusive like the one presented below.

Such an aesthetically unobtrusive metasurface-based RIS is that fabricated by DOCOMO and depicted in Figure 6.9 [28]. The DOCOMO prototype employs unit cells of extremely small size, smaller than the 3.5 mm of the Aalto-fabricated metasurface or even of the 2 mm size of other contemporary state of the art metasurfaces. These miniscule unit cell sizes and the implicit continuous granularity give rise to a *transparent* RIS that can be deployed on glass building facades, vehicles, and billboards, as previously visualized in Figure 6.2. Such transparent RISs do not interfere aesthetically or physically with the surrounding environment, facilitating the proliferation of RISs in wireless network environments in an unobtrusive manner. Thus allows for the transformation of wireless networks into RIS-empowered smart radio environments, through the employment of large RISs deployed at small transmit distances, residing in the vicinity of the receiver or transmitter. In this course, relevant measurements not only in indoor and controllable environments but also in outdoor environments need to be reported and this constitutes one of the key challenges of future research work in the field.

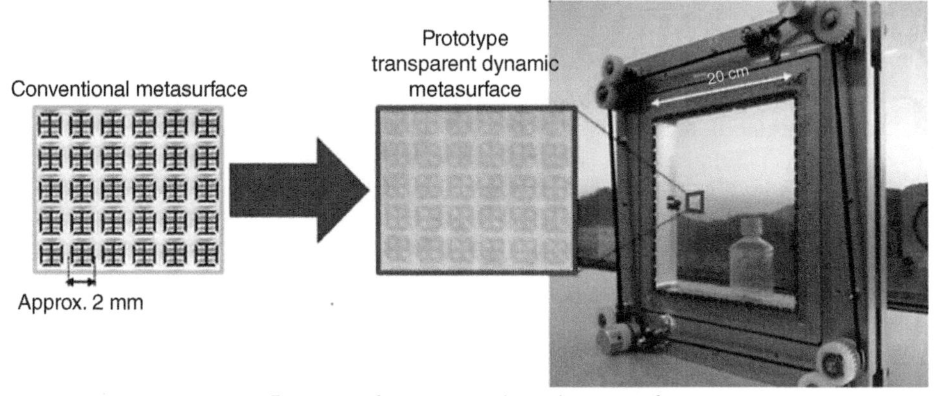

Figure 6.9 DOCOMO prototype of transparent dynamic metasurface. *Source:* [28].

6.4 RIS Research Areas and Challenges in the 6G Ecosystem

The previous sections have epigrammatically outlined the potential of RISs, by investigating their potential benefits and limitations, while also comparing them with similar and well-established technologies such as conventional relays [10], MIMO, and phased arrays [24, 25]. Moreover, since RISs are a novel paradigm in the context of wireless communications, the realization of proof-of-concept platforms and hardware testbeds is outlined in the previous section [23, 26–28] in order to corroborate the theoretical findings and demonstrate the potential benefits arising from the utilization of RISs by 6G systems. In this course, a multitude of additional experimental activities can also be found in the literature, focusing, for instance, on RIS elements of finer (2-bit) phase resolution [29] or RIS testbeds employed for ambient backscatter communications [30].

The employment of RISs and their amalgamation with legacy transmission technologies (such as that of [30]) is also attracting the interest of the wireless community, with RIS being seamlessly conflated with other wireless sections, such as physical layer security [31–34], non-orthogonal multiple access [35–37], Internet of Things and backscattering communications [38], aerial communications [39, 40], wireless power transfer [41], multiple-access edge computing [42], millimeter-wave, terahertz, and optical wireless communications [43, 44], as well as software-defined networking protocols for the control and programmability of the RISs [45]. For a thorough literature review on the application of RISs on the holistic wireless ecosystem, the reader can also refer to a number of available overviews and tutorials on RISs and the references therein [3, 16, 46–48].

Further to the aforementioned cross-sectorial implementation of RISs, a number of research challenges have also arisen from the standalone implementation of RISs. In particular, one of the major research challenges pertains to the development of simple but accurate models to account for the power at the receiver, after a transmitter emits radio waves that are reflected from an RIS. Some early research works have already provided models and measurements for this RIS operation [12, 49–51], but additional work is required in order to analyze the performance limits of RISs and optimize their operation. In this direction, some early approximate and asymptotic analytical frameworks have already been developed, toward quantifying the gains and limitations of RISs in different network scenarios, like in the presence of phase noise, characterizing the distribution and scaling laws of the signal-to-noise ratio (SNR) [52], or by employing point processes, such as stochastic geometry and random spatial processes to compute the probability that an RIS-coated object in the network acts as a reflector [53].

As already stressed, the next step after the network analysis and performance evaluation in the presence of RISs is the optimization of the RIS operation. In fact, the resource optimization problem in the presence of RISs (both model-based and data-driven) is one of the most investigated research topics in the context of RIS-empowered networks. The majority of the research activities focus on the problem of active and passive beamforming in multiple input, single output (MISO) systems, i.e. the optimization of transmit (active) beamforming and (passive) RIS phase shifts, with respect to energy and spectral efficiency as well as system rate and optimization-introduced overhead. In general, both analytical techniques [54–58] such as the alternating maximization (for the non-convex problem) and data-driven, machine learning, and artificial intelligence techniques [59, 60] have been employed. However, the RIS optimization remains an under-explored research challenge, since the problems arising in RIS-enabled MIMO systems, or in systems comprising multiple RISs, remain yet to be explored, with the complexity increasing with the number of antennas and RISs.

Another open research challenge related to RIS is the channel estimation that accounts for the effect of the RIS on the channel between the transmitter and the receiver. Either the RIS's transformation of the channel must be reported by a feedback mechanism to the transmitter or the channel state information (CSI) of the combined channel (transmitter-RSI-receiver) must be estimated by the transmitter as a whole (for all possible RIS states). In general, the nearly passive operation of RISs and their minimal processing capabilities impose the investigation of new algorithms and protocols for the channel estimation while avoiding signal processing onboard the RIS to the extent possible. Some research works in this direction employing neural networks already exist [61]; however, channel estimation of RIS-enabled networks remains a little explored research avenue.

Last but not least, one of the most promising applications of RISs is the modulation and encoding of data directly on the RIS scattering elements and on the reflected radiation pattern. This is in fact an RIS-enabled variation of the spatial modulation and index modulation concepts [13, 16, 62], which, as already explained in detail in Section 6.2.1, provides a low complexity, power-conscious alternative to massive MIMO chains.

References

1 Wong, K.K., Tong, K.F., Chu, Z., and Zhang, Y. (2020). A vision to smart radio environment: surface wave communication superhighways. arXiv preprint arXiv:2005.14082, arxiv.org.

2 https://www.comsoc.org/publications/ctn/what-will-6g-be

3 Di Renzo, M., Debbah, M., Phan-Huy, D. et al. (2019). Smart radio environments empowered by reconfigurable AI meta-surfaces: an idea whose time has come. *Journal on Wireless Communications and Networking* 2019: 129. https://doi.org/10.1186/s13638-019-1438-9.

4 Yu, N., Genevet, P., Kats, M.A. et al. (2011). Light propagation with phase discontinuities: generalized laws of reflection and refraction. *Science* 334 (6054): 333–337.

5 Liu, F., Tsilipakos, O., Pitilakis, A. et al. (2019). Intelligent metasurfaces with continuously tunable local surface impedance for multiple reconfigurable functions. *Physical Review Applied* 11 (4): 044024.

6 Kaina, N., Dupré, M., Lerosey, G., and Fink, M. (2014). Shaping complex microwave fields in reverberating media with binary tunable metasurfaces. *Scientific Reports* 4: 6693. https://doi.org/10.1038/srep06693.

7 Tretyakov, S., Asadchy, V., and Díaz-Rubio, A. (2018). Metasurfaces for general control of reflection and transmission. In: *World Scientific Handbook of Metamaterials and Plasmonics*, World Scientific Series in Nanoscience and Nanotechnology. World Scientific.

8 Tretyakov, S.A., Asadchy, V., and Diaz-Rubio, A. (2018). Metasurfaces for general control of reflection and transmission. In: *World Scientific Hand-Book of Metamaterials and Plasmonics*, 249–293.

9 Tariq, F., Khandaker, M.R.A., Wong, K.K. et al. (2019). A speculative study on 6G. arXiv preprint arXiv:1902.06700v2, arxiv.org.

10 Di Renzo, M., Ntontin, K., Song, J. et al. (2020). Reconfigurable intelligent surfaces vs. relaying: differences, similarities, and performance comparison. *IEEE Open Journal of the Communications Society* 1: 798–807.

11 Qian, X., Di Renzo, M., Liu, J. et al. (2020). Beamforming through reconfigurable intelligent surfaces in single-user MIMO systems: SNR distribution and scaling laws in the presence of channel fading and phase noise. arXiv preprint arXiv:2005.07472, arxiv.org.

12 Di Renzo, M., Danufane, F.H., Xi, X. et al. (2020). Analytical modeling of the path-loss for reconfigurable intelligent surfaces – anomalous mirror or scatterer? 2020 IEEE 21st International Workshop on Signal Processing Advances in Wireless Communications (SPAWC), Atlanta, GA, USA, 1–5.

13 Karasik, R., Simeone, O., Di Renzo, M., and Shamai, S. (2019). Beyond Max-SNR: joint encoding for reconfigurable intelligent surfaces. arXiv preprint arXiv:1911.09443, arxiv.org.

14 Liaskos, C., Nie, S., Tsioliaridou, A. et al. (2018). A new wireless Communication paradigm through software-controlled metasurfaces. *IEEE Communications Magazine* 56 (9): 162–169. https://doi.org/10.1109/MCOM.2018.1700659.

15 H2020 VISORSURF project (2017–2020). A hardware platform for software-driven functional metasurfaces. http://www.visorsurf.eu/

16 Basar, E., Di Renzo, M., De Rosny, J. et al. (2019). Wireless communications through reconfigurable intelligent surfaces. *IEEE Access* 7: 116753–116773.

17 Monti, A., Soric, J., Barbuto, M. et al. (2016). Mantle cloaking for co-site radio-frequency antennas. *Applied Physics Letters* 108: 113502.

18 https://www.metawave.co/

19 Wang, X. (2020). Surface-impedance engineering for advanced wave transformations. Aalto University publication series Doctoral Dissertations, 75/2020.

20 Asadchy, V.S., Albooyeh, M., Tcvetkova, S.N. et al. (2016). Perfect control of reflection and refraction using spatially dispersive metasurfaces. *Physical Review B* 94 (7): 075142.

21 Díaz-Rubio, A., Asadchy, V.S., Elsakka, A., and Tretyakov, S.A. (2017). From the generalized reflection law to the realization of perfect anomalous reflectors. *Science Advances* 3 (8): e1602714.

22 Lavigne, G., Achouri, K., Asadchy, V.S. et al. (2018). Susceptibility derivation and experimental demonstration of refracting meta-surfaces without spurious diffraction. *IEEE Transactions on Antennas and Propagation* 66 (3): 1321–1330.

23 Díaz-Rubio, A., Asadchy, V.S., Elsakka, A., and Tretyakov, S.A. (2017). From the generalized reflection law to the realization of perfect anomalous reflectors. *Science Advances* 3 (8): e1602714.

24 https://pivotalcommware.com/wp-content/uploads/2017/12/Holographic-Beamforming-WP-v.6C-FINAL.pdf

25 https://pivotalcommware.com/wp-content/uploads/2019/10/HBF-vs-APA-White-Paper-2019.pdf

26 Arun, V. and Balakrishnan, H. (2020). RFocus: beamforming using thousands of passive antennas. USENIX NSDI, 1047–1061.

27 Dunna, M., Zhang, C., Sievenpiper, D., and Bharadia, D. (2020). ScatterMIMO: enabling virtual MIMO with smart surfaces. *ACM MobiCom*: 14.

28 https://www.nttdocomo.co.jp/english/info/media_center/pr/2020/0117_00.html

29 Dai, L., Wang, B., Wang, M. et al. (2020). Reconfigurable intelligent surface-based wireless communication: antenna design, prototyping and experimental results. *IEEE Access* 8: 45913–45923.

30 Fara, R., Phan-Huy, D.-T., Ourir, A. et al. (2020). Polarization-based reconfigurable tags for robust ambient backscatter communications. *IEEE Open Journal of the Communications Society* 1: 1140–1152.

31 Yang, L., Yang, J., Xie, W. et al. (2020). Secrecy performance analysis of RIS-aided wireless communication systems. IEEE Transactions on Vehicular Technology. Early access.

32 Yu, X., Xu, D., and Schober, R. (2019). Enabling secure wireless communications via intelligent reflecting surfaces. 2019 IEEE Global Communications Conference (GLOBECOM), Waikoloa, HI, USA, 1–6.

33 Guan, X., Wu, Q., and Zhang, R. (2020). Intelligent reflecting surface assisted secrecy communication: is artificial noise helpful or not? *IEEE Wireless Communications Letters* 9 (6): 778–782.

34 Qiao, J. and Alouini, M.-S. (2020). Secure transmission for intelligent reflecting surface-assisted mmWave and terahertz systems. IEEE Wireless Communications Letters. Early access.

35 Mu, X., Liu, Y., Guo, L. et al. (2020). Exploiting intelligent reflecting surfaces in NOMA networks: joint beamforming optimization. IEEE Transactions on Wireless Communications. Early access.

36 Ding, Z. and Vincent Poor, H. (2020). A simple design of IRS-NOMA transmission. *IEEE Communications Letters* 24 (5): 1119–1123.

37 Liu, X., Liu, Y., Chen, Y., and Poor, H.V. (2020). RIS enhanced massive non-orthogonal multiple access networks: deployment and passive beamforming design. IEEE Journal on Selected Areas in Communications. Early access.

38 Zhang, Q., Liang, Y.-C., and Poor, H.V. (2020). Large intelligent surface/antennas (LISA) assisted symbiotic radio for IoT communications. arXiv [Online].

39 Li, S., Duo, B., Yuan, X. et al. (2020). Reconfigurable intelligent surface assisted UAV communication: joint trajectory design and passive beamforming. *IEEE Wireless Communications Letters* 9 (5): 716–720.

40 Yang, L., Meng, F., Zhang, J. et al. (2020). On the performance of RIS-assisted dual-hop UAV communication systems. IEEE Transactions on Vehicular Technology. Early access.

41 Pan, C., Ren, H., Wang, K. et al. (2020). Intelligent reflecting surface aided MIMO broadcasting for simultaneous wireless information and power transfer. *IEEE Journal on Selected Areas in Communications* 38 (8): 1719–1734.

42 Bai, T., Pan, C., Deng, Y. et al. (2020). Latency minimization for intelligent reflecting surface aided mobile edge computing. IEEE Journal on Selected Areas in Communications. Early access.

43 Zuo, J., Liu, Y., Basar, E., and O. Dobre. (2020). Intelligent reflecting surface enhanced millimeter-wave NOMA systems. IEEE Communication Letters. Early access.

44 Perovic, N.S., Di Renzo, M., and Flanagan, M.F. (2020). Channel capacity optimization using reconfigurable intelligent surfaces in indoor mmwave environments. IEEE International Conference on Communications (ICC), Dublin, Ireland, 1–7.

45 Liaskos, C., Tsioliaridou, A., Nie, S. et al. (2019). On the network-layer modeling and configuration of programmable wireless environments. *IEEE/ACM Transactions on Networking* 27 (4): 1696–1713.

46 Di Renzo, M., Zappone, A., Debbah, M. et al. (2020). Smart radio environments empowered by reconfigurable intelligent surfaces: how it works, state of research, and road ahead. *IEEE Journal on Selected Areas in Communications* 38 (11): 2450–2525. https://doi.org/10.1109/JSAC.2020.3007211.

47 Huang, C., Hu, S., Alexandropoulos, G.C. et al. (2020). Holographic MIMO surfaces for 6G wireless networks: opportunities, challenges, and trends. *IEEE Wireless Communications* 27 (5): 118–125. https://doi.org/10.1109/MWC.001.1900534.

48 Liu, Y., Liu, X., Mu, X. et al. (2020). Reconfigurable intelligent surfaces: principles and opportunities. arXiv. [Online].

49 Tang, W., Chen, M.Z., Chen, X. et al. (2019). Wireless communications with reconfigurable intelligent surface: path-loss modeling and experimental measurement. arXiv [Online].

50 Danufane, F.H., Di Renzo, M., Xi, X. et al. (2020). On the path-loss of reconfigurable intelligent surfaces: an approach based on Green's theorem applied to vector fields. arXiv [Online].

51 Gradoni, G. and Di Renzo, M. (2020). End-to-end mutual-coupling-aware communication model for reconfigurable intelligent surfaces: an electromagnetic-compliant approach based on mutual impedances. arXiv [Online].

52 Qian, X., Di Renzo, M., Liu, J. et al. (2021). Beamforming through reconfigurable intelligent surfaces in single user MIMO systems: SNR distribution and scaling laws in the presence of channel fading and phase noise. *IEEE Wireless Communications Letters* 10 (1): 77–81. https://doi.org/10.1109/LWC.2020.3021058.

53 Di Renzo, M. and Song, J. (2019). Reflection probability in wireless networks with metasurface-coated environmental objects: an approach based on random spatial processes. *EURASIP Journal on Wireless Communications and Networking* 99 https://doi.org/10.1186/s13638-019-1403-7.

54 Zhou, S., Xu, W., Wang, K. et al. (2020). Spectral and energy efficiency of IRS-assisted MISO communication with hardware impairments. *IEEE Wireless Communications Letters* 9 (9): 1366–1369.

55 Han, H., Zhao, J., Niyato, D. et al. (2020). Intelligent reflecting surface aided network: power control for physical-layer broadcasting. IEEE International Conference on Communications (ICC), Dublin, Ireland, 1–7.

56 Zhou, G., Pan, C., Ren, H. et al. (2020). Robust beamforming design for intelligent reflecting surface aided MISO communication systems. IEEE Wireless Communications Letters. Early access.

57 Zappone, A., Di Renzo, M., Shams, F. et al. (2020). Overhead-aware design of reconfigurable intelligent surfaces in smart radio environments. IEEE Transactions on Wireless Communications. Early access.

58 Perović, N.S., Tran, L.N., Di Renzo, M., and Flanagan, M.F. (2020). Achievable rate optimization for MIMO systems with reconfigurable intelligent surfaces. arXiv. [Online].

59 Zappone, A., Di Renzo, M., and Debbah, M. (2019). Wireless networks design in the era of deep learning: model-based, ai-based, or both? *IEEE Transactions on Communications* 67 (10): 7331–7376.

60 Gacanin, H. and Di Renzo, M. (2020). Wireless 2.0: towards an intelligent radio environment empowered by reconfigurable meta-surfaces and artificial intelligence. arXiv [Online].

61 Liu, S., Gao, Z., Zhang, J. et al. (2020). Deep denoising neural network assisted compressive channel estimation for mmWave intelligent reflecting surfaces. *IEEE Transactions on Vehicular Technology* 69 (8): 9223–9228.

62 Li, A., Wen, M., and Di Renzo, M. (2020). Single-RF MIMO: from spatial modulation to metasurface-based modulation. arXiv [Online].

63 Arun, V. and Balakrishnan, H. (2020). RFocus: Practical beamforming for small devices. arXiv [Online].

7

Potential of THz Broadband Systems for Joint Communication, Radar, and Sensing Applications in 6G

Robert Müller and Markus Landmann

Electronic Measurements and Signal Processing (EMS) Department, Fraunhofer Institute for Integrated Circuits IIS, Ilmenau, Germany

5G already provides the vision of high data rates [1], new services [2], and new applications [3–5] utilizing large instantaneous bandwidths (BWs) up to several GHz [6–10] to reach the goal of 100 Gbit/s and higher data rates in mobile networks. Additionally, in the field of Wi-Fi systems (IEEE802.11/802.15) [11, 12], different activities from mmWave to THz and in the light spectrum target broadband communications with several Gbit/s. For the 6th Generation of mobile networks and future wireless technologies, it is envisioned to achieve on long-term, until the year 2030 and beyond, wireless data rates in the range of Tbit/s per wireless link. Analyzing the data rates of wireless systems till now, the significant improvements were based on increased instantaneous BW rather than through sophisticated transmission schemes, such as spatial multiplexing. Tbit/s in wireless networks can be achieved only when BWs of above 10 GHz per link per channel are available. The respective spectrum for such high data rates can be identified at carrier frequencies beyond 100 GHz. These systems are often called sub-THz or THz wireless systems. In this chapter we focus on THz communication in the range from 100 to 400 GHz.

That THz communication attracts increasing attention becomes obvious not only through several demonstrations on component and research level [13–15] but also through a first communication standard for THz wireless communications operating between 252 and 325 GHz [11]. The Institute of Electrical and Electronics Engineering (IEEE) has published the first THz standard (IEEE802.15) already in 2017. Furthermore, standardization organization started the regulation and unification of wireless transmission at THz [16].

Important applications for THz systems range from networks for data centers [17], machine to machine (M2M) communications in industry environments [18], inter-device communication for smart devices [14], Wi-Fi applications [19], THz backhauling for "6G" [20], on body networks [21], wireless chip-to-chip communication [22], and much more (Figure 7.1).

However, Industry 4.0 environments will introduce significant challenges even for 5G specifications (up to Release 16 3GPP) and Wi-Fi 6 systems. These challenges are reflected

Shaping Future 6G Networks: Needs, Impacts, and Technologies, First Edition.
Edited by Emmanuel Bertin, Noel Crespi, and Thomas Magedanz.

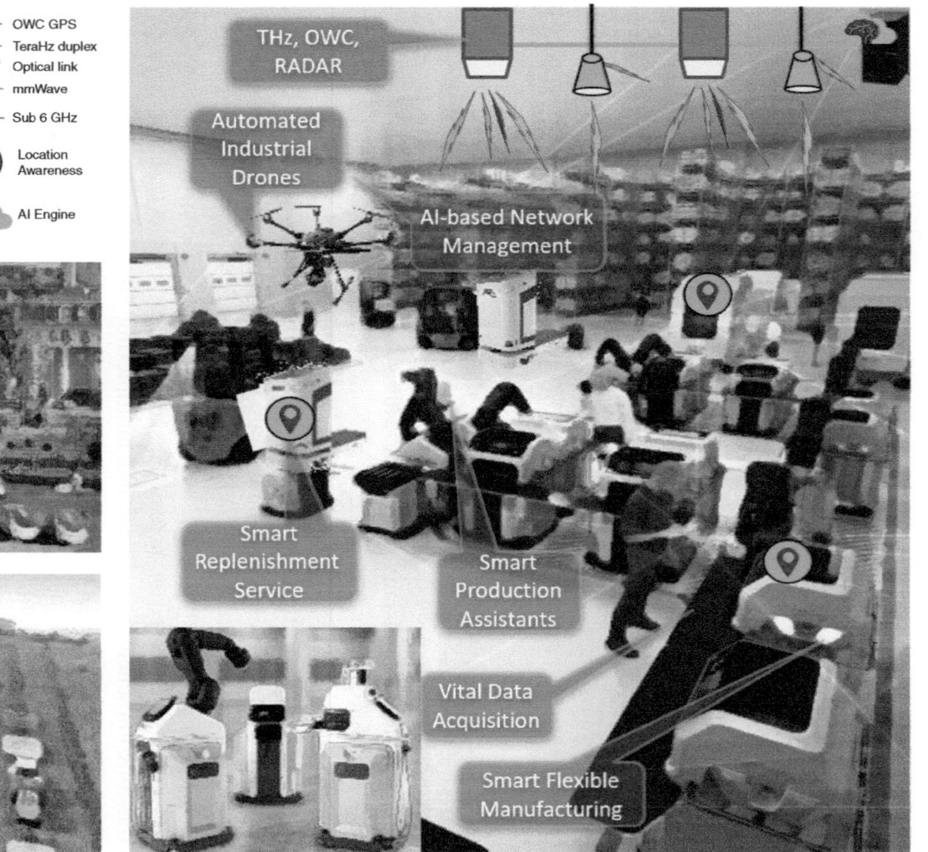

Figure 7.1 Vision of the EU-Project 6G-BRAINS for THz applications in the industrial environment [23].

in many parts of the communication system like spanning congestion, interference, security and safety concerns, high power consumption, restricted propagation, and poor location accuracy within the current radio technology inside buildings. The Industry 4.0 revolution will be accelerated with 5G and introduces new complementary use cases of the radio frequency (RF) interface of communication systems. In this context research and development already started on the idea of the fusion of sensor systems (e.g. radar) and communications systems to reach in long term the mass market by means of low-cost sensor-communication systems applying standardized communication interfaces [24]. The use of broadband communication interfaces as sensors system is applicable not only in industry environments but also in the automobile sector [25]. Hereby, the combination of radar and V2X communication systems is one possible application for Joint Communication Sensor Systems (JCSS). Another application can be the contactless recording of vital data using THz systems [3].

The use of such system can lead to the availability of new features of mass product communication systems in almost all areas of daily life. However, for most applications a high BW is required to obtain the necessary position, location, or structural depth resolution. Depending on the targeted application scenario, the physical required resolution of the sensor is larger than current available BW at 5G mmWave and ISM (Industrial, Scientific, and Medical) frequency bands. A solution can be the lower THz range between 116 and 400 GHz, which provides several free bands with an available BW over 20 GHz for JCSS applications. This allows resolutions in the millimeter range for radar application and can approach the resolution of a lidar [26]. The accuracy of the positioning and localisation applications dependents direct on the bandwidth, the localisation accuracies will be better in THz systems as in current wireless communication or localisation systems [27] with bandwidth less than 1 GHz. The positioning accuracy can be further increased through the cooperative use of the various transmission positions of the wireless devices in a communication system since the devices can coordinate the transmission signals in a time-division-multiplex (TDM) or frequency-division-multiplexing (FDM) to constitute a multi-static localization and RADAR system. The merge of broadband communication technology and sensor functionality in the THz range may lead to inexpensive and very precise hybrid system concepts providing data rates over 1 Tbit/s in combination with positioning resolution in the submillimeter range for sensor applications. Based on these features, the needs of industrial environments can be easily addressed and open the chance for mass market THz technology.

Despite this high potential of THz technologies for communication as well as for JCSS, there are many open questions. The challenges for THz system can be divided into the following categories:

1) Technologies to counteract the high pathloss at THz frequencies.
2) Channel model for medium and large size scenarios for frequencies over 100 GHz.
3) Low-cost and reliable semiconductor-integrated RF-chips.
4) Low-cost basebands with BW over 10 GHz.
5) Optimizing the phase noise in THz systems to increase the modulation depth and improve the beamformer quality.

To overcome the high pathloss at THz frequencies, additional techniques are required to enable the use of THz systems in larger scenarios. One simple solution to handle the pathloss at THz is the use of high-gain antennas. However, high-gain antennas always bring a high directivity without any flexibility to steer into different directions. For some applications like a point-to-point Tbit/s backhaul or fronthaul link, the use of high-gain antennas can be an acceptable solution. Meanwhile, for the most 6G and future wireless communication applications like wireless LAN (Wi-Fi) or M2M communication, it is mandatory to use beamforming solutions to cover scenarios with propagation distances over a couple of meters in a wider angular range. One approach can be analogue beamforming (ABF), which uses the adaptive weighing of amplitude and phase of the individual antenna elements or subgroups of the antenna array to adaptively change the antenna characteristics [28, 29] by means of the beam direction. Different solutions for the implementation of complete THz systems including an ABF antenna array were proposed [28, 30]. These research projects and THz system-on-chip (SoC) solutions show exemplarily that from the technology point of view with respect to manufacturing and integration, THz applications in large scenarios are likely in principle. The challenge with this approach is that many phase shifters as well as variable amplifiers are required for the ABF network. This makes the system more susceptible to technical failures and increases energy losses in the THz system. Furthermore, the increased demand for chip area, which is necessary for implementation of ABF networks, creates additional costs. A further approach for beamforming is the implementation of fully digital beamforming (DBF) [31]. The DBF requires at every antenna element or subgroups of antenna elements of the antenna array an own digital to analogue (D/A) converter and an RF converter into THz range. A similar configuration with analogue to digital (A/D) converters and down converters from the THz band is necessary for the receiver. The beamforming will be applied in the digital domain by adjusting the amplitude and phase shifts directly to multiple copies of the signal to be transmitted. This configuration enables to form multiple independent beams in combination with an antenna array in almost arbitrary directions. Further advantages of the DBF are the improved dynamic range, the possibility to control multiple beams, a more precise control of amplitude and phase, as well as a higher control speed of the beam steering in contrast to ABF. To reach the required gain at THz to compensate the pathloss, the numbers of A/D as well D/A converters are prohibitively large. Additionally, the required numbers of RF THz up and down converters lead to complex and costly systems architecture for a THz system. A hybrid approach as a combination of an ABF and a DBF can be implemented to increase the performance and flexibility of the single beamforming techniques. This hybrid beamforming brings the advantages of the two approaches mentioned. They provide digital flexibility of being able to dynamically form many beams and nulls with no change in hardware and without a full RF chain per antenna element. The hybrid beamforming requires only a full RF chain per sub-array and still offers a high flexibility with respect to the number of beams and nulls.

Another approach is the use of multiple high-gain antennas or fixed antenna arrays, which are aligned in different directions. This sector-based approach with high-gain antennas enables MIMO (Multi-Input Multi-Output) transmissions and a faster detection of the link between the THz users. The required hardware is equivalent to the hybrid beamforming approach where the number of chains with A/D and D/A converters depends on the beam width of the high-gain antennas and can be optimized with respect to the targeted beamwidth based on scenarios and applications. However, output amplifiers with slightly higher outputs are required to get an equivalent communication range as with ABF or DBF

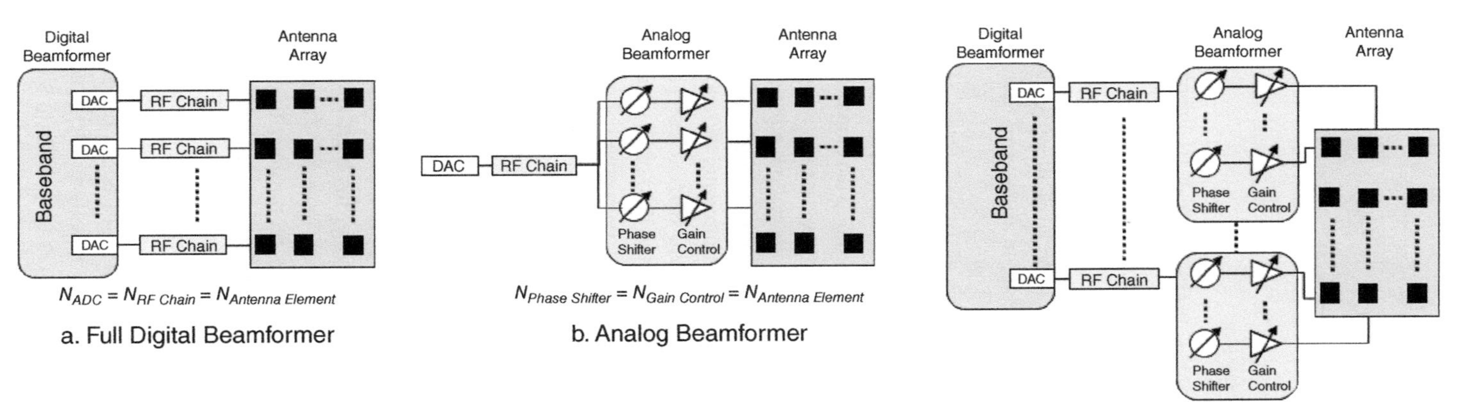

Figure 7.2 Schematic representation of various approaches to handle the pathloss at THz.

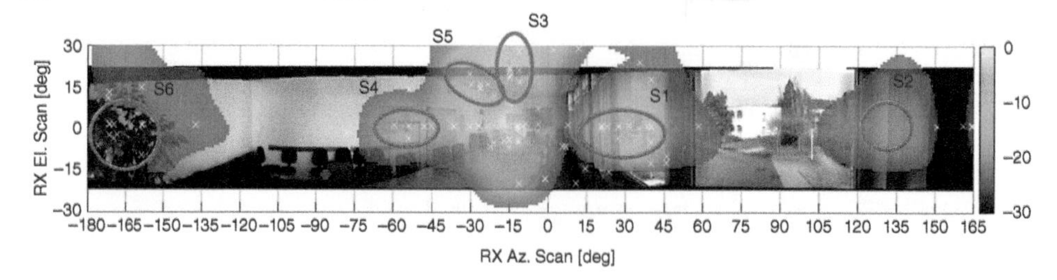

Figure 7.3 A 200 GHz channel measurement result in a meeting room at TU Ilmenau. The colored areas show the powers of the reflected paths of the 200 GHz channel. The white cross shows the results from the ray tracing simulation. More information in [19].

solutions. All the five approaches are illustrated in Figure 7.2. Different combinations of these approaches can also be used depending on the applications and required dynamic ranges (Figure 7.3).

Nevertheless, before THz technology will be a part of standard communication systems, some research questions need to be answered, beginning at the electromagnetic wave propagation. THz communication systems cannot be used and characterized as a scaled-up version of mmWave or microwave systems. To analyze the potential and to develop suitable technologies and standards in the THz range, the electromagnetic wave propagation must be understood, to develop THz systems. With respect to modeling of THz wave propagation, a reasonable amount of research work is already performed. In small and medium size scenarios like offices, data centers, entrance halls, etc., many channel measurements with different channel sounders (CSs) and vector network analyzers (VNA) were carried out [17, 19, 32–36]. Nevertheless, there are still many unanswered questions in the field of channel characterization at THz. Parameters like the delay spread, Doppler spread, respective coherence time, and BW are largely influenced by the spatial characteristic of the wave propagation at THz in the targeted scenarios. Since there are currently no antenna-independent channel measurements and channel models in the THz field available, further research work must be carried out to characterize the propagation channel at different THz frequency bands to develop reliable and antenna-independent channel models at THz. This is one of the important key openers to bring the THz technology into standardization committees like the 3GPP, IEEE, etc.

The characterization of the electromagnetic wave propagation at THz is one fundamental key aspect to enable the potential of THz technology for high-speed data transmissions as well as to develop advanced JCSS. However, without cheap and reliable integrated THz SoCs, it will not be possible for the THz technology to enter the mass market. Additionally, the required energy to generate THz carrier frequencies that is dependent on the Planck's quantum of action and proportional to the applied carrier frequency [28] shows that the energy to generate RF power at THz is higher. Therefore, it is even more important to develop energy-efficient THz semiconductors. The integration of such future wireless compact systems is demonstrated in different research activities and prototypes [37]. Most of the integrated solutions are based on CMOS (Complementary Metal Oxide Semiconductor) [38], GaAs [30, 39], SiGe [40], and InP [41]. All these SoCs show the potential of the use of BWs over several 10 GHz BW at THz. Technology wise one of the biggest challenges is the implementation of control-capable and inexpensive RF hardware. Without the existence of a standard technology that enables the generation of THz radio waves in an energy-efficient manner, the THz technology will only be available for expensive niche applications. A second technological challenge is the access to broadband baseband technology, which can generate and compute BWs over 20 GHz in real-time [42]. An overview of possible energy-efficient receiver structures for 6G application in THz bands can be found in [43]. Another possible solution can be the use of 1-bit A/D converters to increase the SNR (signal-to-noise ratio) by applying higher oversampling rates [46, 47]. Furthermore, solutions have to cope with other hardware impairments such as the much higher phase noise. To minimize phase noise at THz, an approach could be to increase the reference clock signal to couple of GHz. Another solution could be the use of new photonic microwave oscillators [44], which are able to generate very stable reference clocks. A low phase noise is necessary in order to implement higher modulation orders [45] in the THz system, to achieve the targeted data rates in the Tbit/s range. However, if all these hardware challenges are overcome, THz communications and sensor systems have the potential to open the way to several new applications.

In summary to enable the THz technology for future wireless communication and sensor applications, high-gain antennas or phased array beamforming antennas are necessary. The THz system architecture must be adapted to the changed conditions of wave propagation and the hardware limitations at THz. The paradigm shift compared to lower frequency systems comes from the spatial filtering and the huge available BW in the frequency range above 100 GHz. The spatial filtering at very high frequencies is a result of the used high-gain antennas and beamformers. Compared to lower frequency systems, the propagation channel is spatially very limited like a view through a telescope. These circumstances are additionally challenging in terms of beam management between the transmitter and receiver. To overcome these challenges, beam-tracking [46] mechanization also in combination with artificial intelligence (AI) can be applied for example based on position information from microwave and mmWave systems [47–50]. In this field a significant research effort is necessary to determine the optimal approach for THz JCSS. The use of THz technologies can lead to further revolution in the degree of automation, especially in the industrial environment. With AI significant improvements are also expected in sensor applications [51]. Here, intelligent swarm organization of the JCSS can be optimized applying AI, which leads to a further improvement for multi-static sensor/radar and localization application. However, before AI can be used meaningfully for THz applications, the challenges that currently exist for THz systems must be solved. Furthermore, the joint use of the communication system as sensor must be considered in the design and standardization from the beginning. This is the only way to exploit the full potential of THz system as sensor and communication systems.

All of this could start now with the vision of 6G at THz.

References

1 Samsung (2015). Samsung 5G vision. Samsung White Paper.

2 Redana, S., Bulakci, Ö., Zafeiropoulos, A. et al. (eds.) (2019). *5G PPP Architecture Working Group: View on 5G Architecture. Version 3.0.* Belgium: European Commission, 166pp.

3 Gavras, A., Ai Keow Lim Jumelle, Alois Paulin et al. (2015). 5G and e-Health. 5G-PPP White Paper.
4 Haerick, W., Gupta, M., Jean-Sebastien Bedo et al. (2015). 5G and the factories of the future. 5G-PPP White Paper.
5 Thrybom, L., Kapovits, A., Krister Landernäs, et al.. (2015). 5G and energy. 5G-PPP White Paper.
6 Rappaport, T.S., MacCartney, G.R., Samimi, M.K., and Sun, S. (2015). Wideband millimeter-wave propagation measurements and channel models for future wireless communication system design. *IEEE Transactions on Communications* 63 (9): 3029–3056. https://doi.org/10.1109/TCOMM.2015.2434384.
7 Solomitckii, D., Semkin, V., Karttunen, A. et al. (2020). Characterizing radio wave propagation in urban street canyon with vehicular blockage at 20 GHz. *IEEE Transactions on Vehicular Technology* 69 (2): 1227–1236.
8 Müller, R., Häner, S., Dupleich, D. et al. (2016). Simultaneous multi-band channel sounding at mm-Wave frequencies. 2016 10th European Conference on Antennas and Propagation (EuCAP), 1–5.
9 Salous, S., Degli Esposti, V., Fuschini, F. et al. (2016). Millimeter-wave propagation: characterization and modeling toward fifth-generation systems. [wireless corner]. *IEEE Antennas and Propagation Magazine* 58 (6): 115–127.
10 Papazian, P.B., Remley, K.A., Gentile, C., and Golmie, N. (2015). Radio channel sounders for modeling mobile communications at 28 GHz, 60 GHz and 83 GHz. Global Symposium on Millimeter-Waves (GSMM), 1–3.
11 Thomas Kürner (2015). TG3d Applications Requirements Document (ARD)/ IEEE P802.15 /Wireless Personal Area Networks. IEEE P802.15 Working Group for Wireless Personal Area Networks (WPANs).
12 Status of IEEE 802.11 Light Communication TG, March. https://www.ieee802.org/11/Reports/tgbb_update.htm
13 Lee, S., Dong, R., Yoshida, T. et al. (2019). 9.5 an 80Gb/s 300GHz-band single-chip CMOS transceiver. 2019 IEEE International Solid-State Circuits Conference – (ISSCC), 170–172.
14 Sen, P., Pados, D.A., Batalama, S.N. et al. (2020). The TeraNova platform: an integrated testbed for ultra-broadband wireless communications at true Terahertz frequencies. *Computer Networks* 179: 107370.
15 Kuerner, T. (2018). Turning THz communications into reality: status on technology, standardization and regulation. 2018 43rd International Conference on Infrared, Millimeter, and Terahertz Waves (IRMMW-THz), 1–3.
16 Kuerner, T. and Hirata, A. (2020). On the impact of the results of WRC 2019 on THz communications. 2020 Third International Workshop on Mobile Terahertz Systems (IWMTS), 1–3.
17 Cheng, C. and Zajic, A. (2020). Characterization of propagation phenomena relevant for 300 GHz wireless data center links. *IEEE Transactions on Antennas and Propagation* 68 (2): 1074–1087.
18 Chowdhury, M.Z., Shahjalal, M., Ahmed, S., and Jang, Y.M. (2020). 6g wireless communication systems: applications, requirements, technologies, challenges, and research directions. *IEEE Open Journal of the Communications Society* 1: 957–975.

19 Dupleich, D., Müller, R., Skoblikov, S. et al. (2020). Characterization of the propagation channel in conference room scenario at 190 GHz. 2020 14th European Conference on Antennas and Propagation (EuCAP), 1–5.

20 Paoloni, C., Alexiou, A., Bouchet, O. et al. (2019). ICT beyond 5G cluster: seven H2020 for future 5G. European Conference on Networks and Communications (EuCNC), Valencia (18–21 June 2019).

21 Yang, K., Abbasi, Q.H., Qaraqe, K. et al. (2014). Body-centric nano-networks: em channel characterisation in water at the terahertz band. 2014 Asia-Pacific Microwave Conference, 531–533.

22 Fu, J., Juyal, P., and Zajic, A. (2019). THz channel characterization of chip-to-chip communication in desktop size metal enclosure. *IEEE Transactions on Antennas and Propagation* 67 (12): 7550–7560.

23 Horizon 2020 call (2021). H2020-ICT-2020- 2, Call ICT-52-2020: 5G PPP – Smart Connectivity beyond 5G, 6G BRAINS: Bringing Reinforcement learning Into Radio Light Network for Massive Connections.

24 Feng, Z., Fang, Z., Wei, Z. et al. (2020). Joint radar and communication: a survey. *China Communications* 17 (1): 1–27.

25 Petrov, V., Fodor, G., Kokkoniemi, J. et al. (2019). On unified vehicular communications and radar sensing in millimeter-wave and low terahertz bands. *IEEE Wireless Communications* 26: 146–153.

26 Daniel, L., Phippen, D., Hoare, E. et al. (2017). Low-THz radar, lidar and optical imaging through artificially generated fog. International Conference on Radar Systems (Radar 2017), 1–4.

27 El-Absi, M., Abbas, A.A., Abuelhaija, A. et al. (2018). High-accuracy indoor localization based on chipless RFID systems at THz band. IEEE Access, 1–1.

28 Aminu, M.U., Tervo, O., Lehtomki, J., and Juntti, M. (2018). Beamforming and transceiver HW design for THz band. 2018 52nd Asilomar Conference on Signals, Systems, and Computers, 1547–1551.

29 Lin, C. and Li, G.Y. (2015). Adaptive beamforming with resource allocation for distance-aware multi-user indoor terahertz communications. *IEEE Transactions on Communications* 63 (8): 2985–2995.

30 Alibakhshikenari, M., Virdee, B.S., See, C. et al. (2019). A novel 0.3-0.31 THz GaAs-based transceiver with on-chip slotted metamaterial antenna based on SIW technology, *2019 IEEE Asia-Pacific Microwave Conference (APMC)*, 2019, pp. 69–71, doi: 10.1109/APMC46564.2019.9038371.

31 Ntouni, G.D., Merkle, T., Loghis, E.K. et al. (2020). Real-time experimental wireless testbed with digital beamforming at 300 GHz. 2020 European Conference on Networks and Communications (EuCNC), 271–275.

32 Kleine-Ostmann, T., Jastrow, C., Priebe, S. et al. (2012). Measurement of channel and propagation properties at 300 GHz. 2012 Conference on Precision electromagnetic Measurements, 258–259.

33 Priebe, S., Jastrow, C., Jacob, M. et al. (2011). Channel and propagation measurements at 300 GHz. *IEEE Transactions on Antennas and Propagation* 59 (5): 1688–1698.

34 Zantah, Y., Sheikh, F., Abbas, A.A. et al. (2019). Channel measurements in lecture room environment at 300 GHz. 2019 Second International Workshop on Mobile Terahertz Systems (IWMTS), 1–5.

35 Guan, K., Peng, B., He, D. et al. (2019). Channel sounding and ray tracing for train-to-train communications at the THz band. 2019 13th European Conference on Antennas and Propagation (EuCAP), 1–5.

36 Schmieder, M., Keusgen, W., Peter, M. et al. (2020). THz channel sounding: design and validation of a high performance channel sounder at 300 GHz. *2020 IEEE Wireless Communications and Networking Conference Workshops (WCNCW)*, 2020, pp. 1–6, doi: 10.1109/WCNCW48565.2020.9124887.

37 Nakasha, Y., Shiba, S., Kawano, Y., and Takahashi, T. (2017). Compact terahertz receiver for short-range wireless communications of tens of gbps. *Fujitsu Scientific & Technical Journal* 53 (02): 9–14.

38 Laskin, E., Khanpour, M., Nicolson, S.T. et al. (2009). Nanoscale CMOS transceiver design in the 90–170-GHz range. *IEEE Transactions on Microwave Theory and Techniques* 57 (12): 3477–3490.

39 Liu, Y., Zhang, B., Feng, Y. et al. (2020). Development of 340-GHz transceiver front end based on GaAs monolithic integration technology for THz active imaging array. *Applied Sciences* 10 (11): 7924.

40 Pfeiffer, U.R. (2012). Silicon CMOS/SiGe transceiver circuits for THz applications. 2012 IEEE 12th Topical Meeting on Silicon Monolithic Integrated Circuits in RF Systems, 159–162.

41 Hamada, H., Fujimura, T., Abdo, I. et al. (2018). 300-GHz. 100-Gb/s InP-HEMT wireless transceiver using a 300-GHz fundamental mixer. 2018 IEEE/MTT-S International Microwave Symposium – IMS, 1480–1483.

42 Fatemi, A., Kahmen, G., and Malignaggi, A. (2021). A 96-Gb/s PAM-4 receiver using time-interleaved converters in 130-nm SiGe BiCMOS. *IEEE Solid-State Circuits Letters* 4: 60–63.

43 Skrimponis, P., Hosseinzadeh, N., Khalili, A. et al. (2021). Towards energy efficient mobile wireless receivers above 100 GHz. *IEEE Access* 9: 20704–20716.

44 Giunta, M., Lessing, M., Yu, J. et al. (2021). Photonic microwave oscillator based on an ultra-stable-laser and an optical frequency comb. 2020 50th European Microwave Conference (EuMC), 591–594.

45 Iqbal, M., Lee, J., and Kim, K. (2000). Performance comparison of digital modulation schemes with respect to phase noise spectral shape. 2000 Canadian Conference on Electrical and Computer Engineering. Conference Proceedings. Navigating to a New Era (Cat. No.00TH8492), vol. 2, 856–860.

46 Jeong, J., Lim, S.H., Song, Y., and Jeon, S.W. (2020). Online learning for joint beam tracking and pattern optimization in massive MIMO systems. IEEE INFOCOM 2020 – IEEE Conference on Computer Communications, 764–773.

47 Li, H., Li, C., Gao, H. et al. (2021). Study of moving targets tracking methods for a multi-beam tracking system in terahertz band. *IEEE Sensors Journal* 21 (5): 6520–6529.

48 Tan, J. and Dai, L. (2020). Wideband beam tracking based on beam zooming for THz massive MIMO. GLOBECOM 2020 – 2020 IEEE Global Communications Conference, 1–6.

49 Gao, X., Dai, L., Zhang, Y. et al. (2017). Fast channel tracking for terahertz beamspace massive MIMO systems. *IEEE Transactions on Vehicular Technology* 66 (7): 5689–5696.

50 Stratidakis, G., Ntouni, G., Boulogeorgos, A.-A. et al. (2020). A low-overhead hierarchical beam-tracking algorithm for THz wireless systems. *2020 European Conference on Networks and Communications (EuCNC)*, 2020, pp. 74–78, doi: 10.1109/EuCNC48522.2020.9200946.

51 Deng, H., Geng, Z., and Himed, B. (2018). Radar target detection using target features and artificial intelligence. 2018 International Conference on Radar (RADAR), 1–4.

8

Non-Terrestrial Networks in 6G

Thomas Heyn[1], *Alexander Hofmann*[2], *Sahana Raghunandan*[2], *and Leszek Raschkowski*[3]

[1] *Broadband and Broadcast Department, Head of Mobile Communications Group, Fraunhofer IIS, Erlangen, Germany*
[2] *Department RF SatCom Systems, Fraunhofer Institute for Integrated Circuits, Erlangen, Germany*
[3] *Wireless Communications and Networks Department, Fraunhofer Heinrich Hertz Institute HHI, Berlin, Germany*

8.1 Introduction

In recent years, a high connectivity demand started being experienced in wireless communication. Practically, everyone and everything needs to be connected because of the huge variety of applications existing today. This is a challenging situation for terrestrial telecommunications infrastructure that they cannot address on their own. Therefore, the 3rd Generation Partnership Project (3GPP) started in 2017 to study the integration of satellites as a part of the 5G ecosystem involving both cellular and satellite stakeholders. The substantial value added by satellites as part of the access technology mix for 5G is now becoming clear, especially for mission critical and other applications where ubiquitous coverage is crucial. For example, non-terrestrial networks (NTN) can broaden service delivery to unserved or underserved areas, by complementing and extending terrestrial networks.

In this chapter, we will elaborate on the current standardization of NTN in 5G and further detail the architecture and research challenges toward 6G-NTN.

8.2 Non-Terrestrial Networks in 5G

The use of satellite-based networks to provide connections to different user equipment (UE) is also referred to as 5G-NTN in the 3GPP community. These so-called NTN cover satellites as well as airborne vehicles, also called "high-altitude platforms (HAPs)." The satellite either employs a transparent (bent pipe) payload or in future 3GPP releases a regenerative payload and can be placed into geostationary Earth orbit (GEO), medium-Earth orbit (MEO), or low-Earth orbit (LEO). The airborne vehicles are unmanned aerial vehicles (UAVs), operating at a height between 8 and 50 km.

Shaping Future 6G Networks: Needs, Impacts, and Technologies, First Edition.
Edited by Emmanuel Bertin, Noel Crespi, and Thomas Magedanz.

A transparent satellite works as a relay between the UEs and the base station, also known as gNB (next-generation NodeB), implemented on the gateway side on ground. In contrast, a regenerative satellite acts as a flying gNB, with a backhaul link to the 5G core network on ground.

In the 3GPP Service and Systems Aspect (SA) specification group, a study on using satellite access in 5G is summarizing the use cases including satellite in [1].

Another study report [2] "Study on architecture aspects for using satellite access in 5G" details the role of satellite links in 5G networks. From an architectural point of view, satellites either act as backhaul for gNBs on ground or provide a direct access with 5G New Radio (NR) to UEs.

The satellite direct access with 5G NR has been accepted to the roadmap in the cellular standardization organization 3GPP in the Radio Access Networks (RANs) specification groups. After finalizing two study items in Release 15 [3] and Release 16 [4], the 3GPP RAN working groups currently specify the extension of 5G NR to support NTN as part of the Release 17 [5]. So, for the first time, satellite communication with direct access will be supported by the 3GPP standards, which were formerly limited to terrestrial cellular networks. The RAN work item [5] covers a frequency range from 2 to 30 GHz and GEO, MEO, and LEO satellite constellations and states that these extensions are implicitly compatible with HAP stations. Different terminal types are considered, either smartphone type with the regular transmit power of 200 mW (UE Power Class 3) and omnidirectional antennas or very small aperture terminals (VSAT) with directional antennas (cf. Figure 8.1).

A detailed link budget analysis for various system constellations as a combination of GEO and LEO satellites, VSAT and handheld terminals, and frequency bands is included in [4]. The book [6] includes a comprehensive description of the adaptations in 5G RAN to support NTN.

To complement the upcoming 5G NR broadband standard for satellites, another study item is carried out in 3GPP Release 17 on the adaptation of the LTE-based technologies Narrowband Internet of Things (NB-IoT) and enhanced Machine Type Communication (eMTC) to support low data rate use cases with satellites [7].

The current 3GPP Release 17 is a crucial working point in the 5G standardization groups in order to develop and to approve the technical specifications to enable a direct access technology via satellite links. The goal is to deploy NTN as part of 5G by approximately 2025 in order to meet the challenges of mobile network operators in terms of reachability, availability, and resiliency. Satellite industry is gaining more and more interest in this

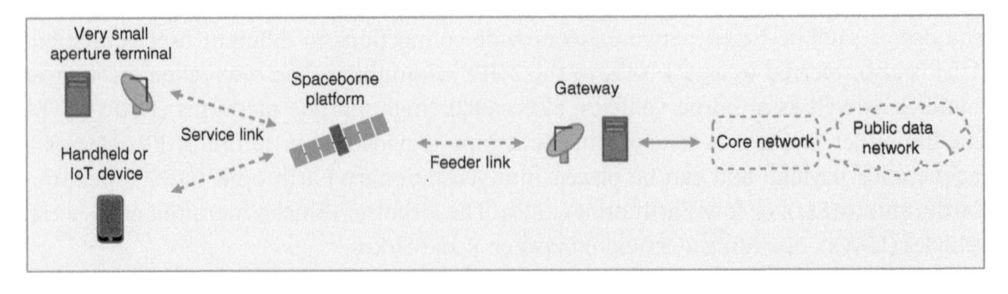

Figure 8.1 Satellite access network with 5G NR direct access to VSAT terminals or handheld devices. *Source:* Fraunhofer IIS.

emerging topic, and several companies even participate actively in the traditional terrestrial standards organization. Pre-standard trials of Release 17 5G-NTN technology elements are already reported in [8].

8.3 Innovations in Telecom Satellites

Telecom satellites have constantly evolved over the years, and further improvements can be expected, which are highly beneficial for future networks, e.g. 6G.

For GEO satellites there was a huge improvement in the total system throughput over the past years with the introduction of high throughput satellites (HTS) and even very high throughput satellites (VHTS). This new class of satellites could achieve total throughputs of more than 100 Gbit/s in case of HTS and more than 1 Tbit/s in case of VHTS per satellite, thanks to transparent digital processing payloads and flexible antenna technology.

Another innovation is seen in the upcoming LEO constellations within the past years. LEO constellations have been already deployed in the 1990s with Iridium and Globalstar as one of the first companies fulfilling the vision of a worldwide coverage that (in case of Iridium) also covers the polar regions. On the latest upgrade of the Iridium constellation called "Iridium Next," a more advanced regenerative payload that is fully reconfigurable has been introduced thanks to field-programmable gate array (FPGA) technology.

The Starlink constellation by SpaceX currently already has a total number of more than 1000 satellites in place and is planning to increase the number to up to 12 000 satellites [9]. Given such high numbers of satellites, technologies, e.g. Application-Specific Integrated Circuits (ASICs), are also getting more and more cost efficient due the high numbers of necessary chips, where in the past due to the limited number of satellites the development of new ASICs for one application were too costly. Thanks to LEO constellations, the latency of satellite systems could be improved dramatically compared to a GEO solution. In the example of Starlink latencies from 40 ms down to 20 ms can be reached. Unfortunately, Starlink is not designed for 5G direct access, but we see the need for new constellations to be ready for 5G and beyond and 6G direct access.

Independent of GEO or LEO, most of the innovations are based on the evolution of the payload in combination with the antenna. The payloads of the satellites are more and more equipped with flexible on-board processors (OBPs). For GEO satellites, due to the high bandwidth and throughput, the processing is mostly based on a transparent basis, called digital transparent OBP without re-encoding of the entire data-stream onboard the satellite. In this case the OBP performs channelization only, where only parts of data are decoded directly on the satellite, e.g. for other features (beam adjustments) or also routing decisions if necessary.

For LEO, the payload could be assumed to be an integral part of the network. This may offer the processing of lower layer (similar to a 5G gNB distributed unit, gNB-DU), higher layer (e.g. entire gNB), and network functions (e.g. edge computing) between the user on ground and the core network. Therefore, a more complex type of OBP with more processing capability is needed, which performs signal regeneration in addition to routing decisions. Especially for LEO the problem of optimized routing between different satellites using the inter-satellite links (ISLs) needs to be solved. OBPs that perform decoding and encoding are called "regenerative OBPs" as the signal between users on ground and the gateway is partially or totally regenerated. Not only could this ability improve the overall signal-to-noise ratio (SNR) of the signal, but it also offers the ability to optimize the bandwidth utilization in general as completely different protocols could be used for the feeder link and the user links. Also, the resource handling, interference management, etc. could be performed directly onboard the satellite, which is beneficial in case of feeder link switch. In future networks more core network functions will be moved closer to the RAN where more processing is needed inside the satellite. In addition to more processing power directly available within the satellite a flexibility is needed to support these features. This will require a more software defined payload for future networks.

On the antenna side, a lot of improvements have been made. At the current stage of development, beamforming, either analog or digital (or hybrid), e.g. with a phased array antenna on the satellite, is not new and has been already successfully deployed in orbit. The principal architecture of a satellite with beamforming is depicted in Figure 8.2, enabling a satellite coverage planning as done for cellular networks.

Beside the technology and beam steering or moving capability, which is obviously necessary for future networks (e.g. in LEO) also, other types of technologies have experienced a constant evolution, e.g. beam switching/beam hopping. Especially, beam hopping is necessary in today's satellites in order to solve the limitation of the available power per beam and be able to scale capacity to different beams/regions. In this domain we also expect a merge between the two worlds: satellite and terrestrial networks, which might bring other enhanced technologies to solve the technical challenges of future satellites in general.

In both domains more flexibility thanks to more available processing in space is the key, thus we expect this evolution to continue.

In addition to the GEO and LEO satellite improvements, other types of orbits might be envisioned for future networks, such as MEO, highly elliptical orbit (HEO), and very low earth orbit (VLEO). Especially, the VLEO (<500 km altitude) seems to be attractive to reach lower latency requirements below 20 ms. Nevertheless, this orbit suffers from higher atmospheric pressure due to a higher number of particles, which lead to velocity

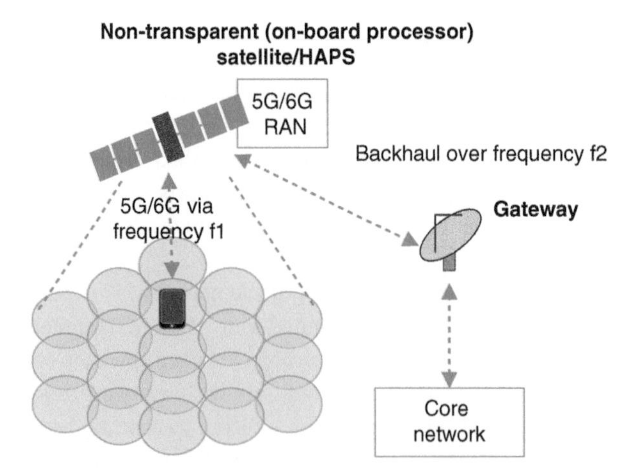

Figure 8.2 Satellite with on-board processor, supporting beamforming. *Source:* Fraunhofer IIS.

degradation and in consequence altitude degradation if no thrusters are used to maintain the orbit.

CubeSat satellites represent a special new class within nanosatellites, whose mass is specified between 1 and 10 kg. The peculiarity of the CubeSat satellites is the standardized dimensions, which have the shape of a cube and in the smallest version 1U of 11.35 cm × 10 cm × 10 cm. Hence 3U (3× 1U, i.e. about 33 cm × 10 cm × 10 cm) corresponds to a mass of about 4 kg. The power classes of nanosatellites are currently in the range of a few multiples of 10 W for the whole satellite. However, this power is not completely available for the payload, since power is needed for the satellite itself and its communication and control. Therefore, the link budget is difficult to close at all or only with very low spectral efficiency, which is not suitable for enhanced mobile broadband (eMBB) type of applications with 5G NR but potentially for massive machine type communications (mMTC) applications.

Currently, there are only LTE-based specifications available for Internet of Things (IoT), such as LTE Cat-NB (NB-IoT) and LTE Cat-M, which are intended for much lower bandwidths (down to 180 kHz) or data rates. Nevertheless, satellites represent a high potential for IoT applications, and 6G-NTN should address these types of applications as well. Section 8.5.5 will further elaborate on the need for the scalability of 6G-NTN carrier bandwidths.

8.4 Extended Non-Terrestrial Networks in 6G

8.4.1 Motivation

In general, 6G will be influenced by two different major aspects.

The first aspect is that 6G will cover certain solutions that are not included in 5G due to several reasons. Related to NTN, one of the major influencing factors is the timeline of the implementation of such a satellite system. Especially for satellites the development and deployment time either for GEO satellites or for LEO satellite constellations might be relatively long compared to terrestrial solutions. Also, the ability of upgrading such a system is limited compared to terrestrial solutions. Given this boundary, not every evolution of technical features or enhancements for NTN could be adopted in a short time frame. In addition to that, it has already been mentioned that with 5G it was the first time to introduce such technologies. So, there is a high chance that some challenges will be visible after first operation. So not every feature to support NTN in the most efficient way might be fully integrated in 5G and consequently will be left for 6G.

The second major aspect is that 6G introduces new requirements for the network, which also may need an extension of the features and capabilities of 5G-NTN. There are many different visions of additional features and requirements from various public and research organizations as well as companies involved in the 3GPP standardization. In order to provide at least an overview of the envisaged features and requirements of 6G, we will focus on the current vision of the International Telecommunication Union (ITU). In the past, the ITU Focus Group on network aspects of International Mobile Telecommunication (IMT)-2020 was established in May 2015 to analyze how emerging 5G technologies will interact in future networks. After a first meeting in Geneva in July 2018, the ITU-T Focus Group Technologies for Network 2030 (FG NET-2030) was established, which concluded its activity on July 2020 and intends to study the capabilities of networks for the year 2030 and beyond. The outcome is directing and shaping the way toward 6G. In their white paper [10], the vision for network 2030 is given in the picture in Figure 8.3.

The given vision of the ITU illustrates three major directions for future networks: new verticals, new communication services, and new infrastructures. 5G already extended the classical eMBB from 4G by mMTC and ultrareliable low-latency communication (URLLC). This graphic shows that 6G extends this concept from 5G to an entire network in terms of verticals and services along new infrastructures. In order to provide this, new regions, e.g. with currently no coverage of terrestrial networks, need to be connected as well. We believe that new verticals and services require a global coverage maybe in combination with seamless connectivity in order to support new types of applications and business models where NTN plays a major role to achieve these goals. Obviously, NTN will add new verticals and communication services and is by itself a new type of infrastructure.

This will introduce multiple access among different types of infrastructures. In order to achieve this goal, the flexibility and the dynamics of these different architectures need to be addressed within 6G (cf. Section 8.5)

Besides the aforementioned aspects of what will be part of 6G in the domain of NTN, the evolution of technical solutions in the field of NTN itself will influence the contribution to 6G. As new technologies for NTN as described in Section 8.3 are also available and enhanced in the future, new solutions will arise and benefit 6G.

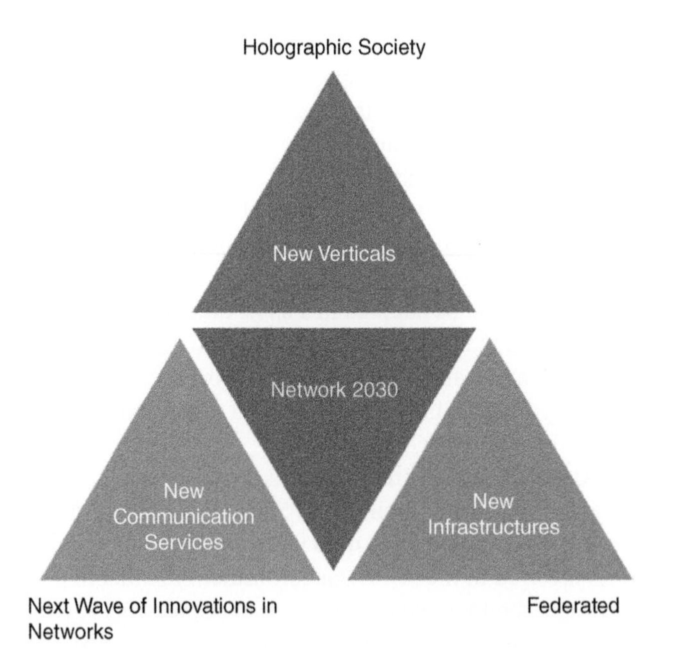

Figure 8.3 ITU-T Focus Group Technologies for Network 2030 (FG NET-2030) vision. *Source:* Fraunhofer IIS.

It is important to note that no one can predict exactly what type of new features and technologies will be part of 6G, but it is possible to identify possible challenges of NTN within 6G in order to push the state of the art and encourage collaborative solutions.

8.4.2 Heterogeneous and Dynamic Networks in 6G

In 6G, an evolution of the network infrastructure can be expected with an increasing integration of various layers, located at air and space level. This results in a more heterogeneous network compared to 5G. This evolution is also expected for broadband connectivity by the 6G flagship white paper from the University of Oulu [11]. In their 6G research vision, the major increase in integration is seen between terrestrial and NTN with UAVs and LEO satellites playing a major role in extending coverage and offloading traffic. In addition, there is an evolution in terms of microsatellites and nano-satellites beneficial for massive IoT use cases. NTN can easily solve coverage problems and thus can contribute to the extension of 6G in general. In 6G, we will see a more seamless integration of the different assets compared to 5G, where only the first steps for integrating such new elements will be realized. With this level of integration, a truly global connectivity can be realized.

In general, three different layers of network platforms (in terms of altitude) can be assumed: Ground-based platforms, airborne platforms, and spaceborne platforms as shown in Figure 8.4.

The ground-based platforms will be also extended by moving base stations, which could be realized by buses, trucks, trains, etc. that provide connectivity either direct or, e.g. via integrated access and backhaul or relaying. In the airborne layer, we expect to see a huge number of different platforms, e.g. drones for delivery services, HAPs, airplanes, air taxis and also autonomous UAVs. At the space level a huge variety of different platforms could be envisioned, starting with the more familiar GEO and LEO satellites, and also not so common types of satellites, e.g. MEO or HEO and VLEO. In addition to the heterogeneity, higher dynamics will be added to the network, introduced by the motion of the different network elements/platforms. The dynamics are introduced by the ad hoc integration of new platforms, e.g. for a limited time only.

This flexible, heterogeneous, and dynamic 6G network architecture including airborne and spaceborne platforms imposes several research challenges like spectrum and interference coordination and flexibility of radio access technologies, which are described in the next section.

8.5 Research Challenges Toward 6G-NTN

In the section above, we introduced possible complex scenarios of future networks and highlighted the benefits by including NTN. But given all this flexibility and dynamics, various challenges will arise that are not yet solved or even touched today in 5G. Therefore, further research is necessary to solve these challenges, which are described in the following.

There is limited literature available so far especially on NTN in 6G. The white paper on broadcast connectivity in 6G [11] identifies the need for a so-called integrated space and

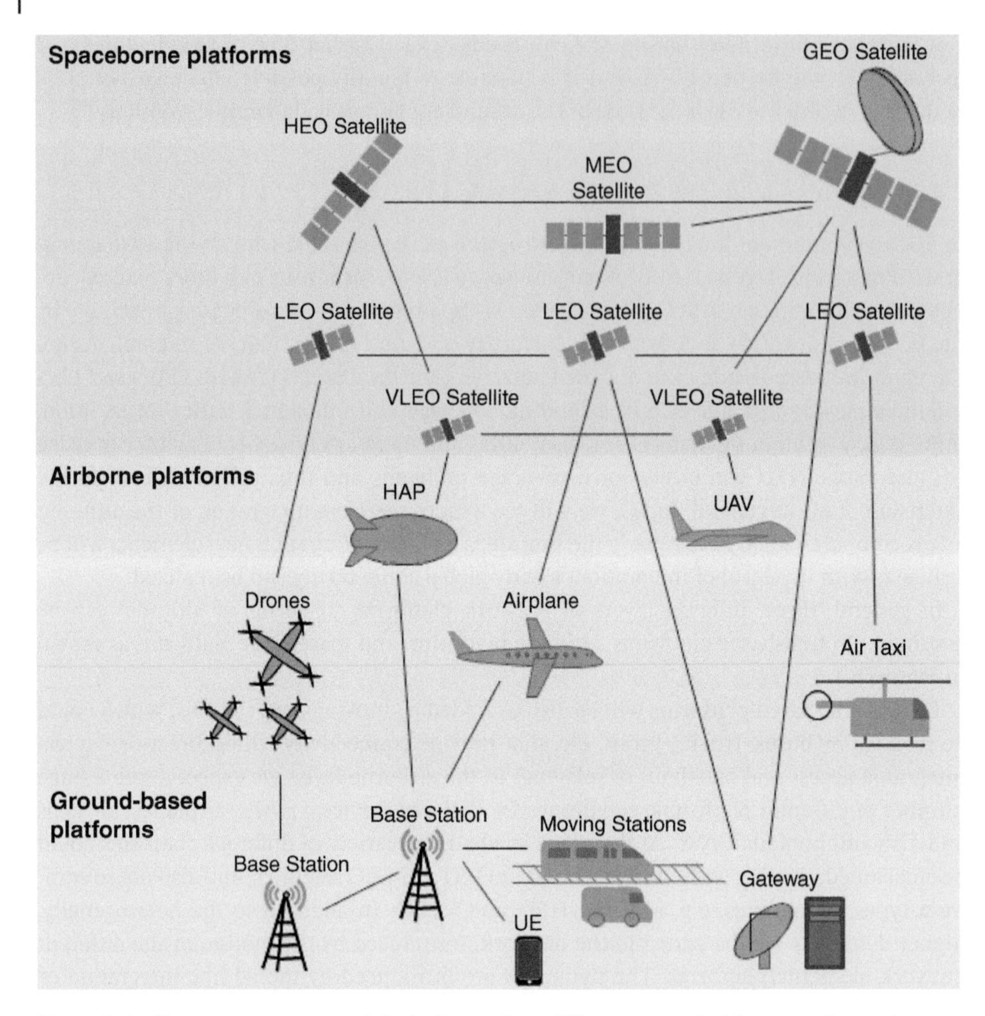

Figure 8.4 Heterogeneous network including various different ground, airborne, and spaceborne platforms. *Source:* Fraunhofer IIS.

terrestrial network (ISTN) and introduces a high-level architecture of a ground network, which is extended by an airborne and a spaceborne network layer.

Another white paper from the 6G flagship project in Finland [12] identifies in agreement to the 3GPP study already performed in 5G [1] the need for NTN in 6G due to, e.g. communication resilience and communication on the move. The reported research targeting 6G-NTN in [12] covers enhancements of the physical layer to cope (like 5G-NTN already) with long delays experienced in satellite links. Another research item is to enable a trade-off between coverage and data rate, and a third one is to introduce a satellite-friendly signal for reduced output back-off (OBO) at the amplifier onboard of a satellite. This last item is important because it is directly related to the achievable throughput of a power-limited satellite system and is explained in Section 8.5.4. In the access and network layers, the challenge is to optimize the resource utilization especially in hybrid systems, even if other non-3GPP radio access technologies are included.

Consistently with [11], in the paper on 5G and beyond 5G (B5G) NTN [13], the authors describe a B5G architecture including a terrestrial network and a multilayered non-terrestrial segment with multiple satellite constellations. The identified high-level challenges in [13] comprise intelligent resource optimization of the flexible architecture, dynamic spectrum management, and flexibility of radio access technologies. Artificial intelligence (AI) is highlighted to address the complex B5G architecture. New antenna designs, e.g. for distributed beamforming, may support the user-centric coverage. A selection of these research challenges is further addressed in the following sections.

8.5.1 Heterogeneous Non-Terrestrial 6G Networks

The complexity of the network will introduce a couple of different challenges, which need to be analyzed and solved to be able to deploy such systems. On the network layer, it is necessary to be able to add and remove these new network nodes on a dynamic basis (e.g. LEO satellites, UAVs). Also, the inter-satellite routing between all these platforms will be a challenge. In addition to that, the delay between the platforms introduced by their motion will vary and the need for an accurate position of all elements is mandatory. The 6G project [14] reports on ongoing research for two aspects to support flexible networks with NTN. One is the channel modeling and simulation for NTN (mobile communications networks based on satellites and flying platforms). This research includes system-level simulations, which enable comprehensive testing of the integration of heterogeneous radio technologies. The other research topic is the implementation of flexible core network software modules for multi-access, dynamic, and AI-based network management; mesh networks; heterogeneous backhauls; and campus networks.

For the transmission over large distances between the platforms, enough power onboard the satellites to close the links is required. As power on mobile platforms and on satellites is limited, new techniques need to be developed to overcome this problem. The power limitation of the satellites is introduced by the maximum area of the solar arrays and more importantly by the maximum heat flux of the satellite radiators. Inside the satellite, it is a huge challenge to get rid of the heat of the electronics using a thermal control system, because in space no convection is available in the vacuum. Given the limited power, new methods for efficient power distribution to specific beams, e.g. beam hopping, needs to be introduced. Another challenge on the physical layer is the interference caused by the different connections between all platforms. Here the coexistence is one of the challenges, which need to be solved.

The propagation channel is predominantly line-of-sight (LOS) in most of the cases, which seems to be not a challenge. Nevertheless, there are some other challenges on the propagation channel. First, due to the lack of a sufficient amount of spectrum, the so-called feeder links between the gateways and the satellites are located at the mmWave band (state of the art) and might be extended to THz in 6G. At the current stage of the development, there are no channel models available for frequencies in the THz bands, which also cover high speed motion.

8.5.2 Required RAN Architecture in 6G to Support NTN

At the current stage of the standardization for 5G-NTN, various types of architecture elements in case of satellites have been introduced. In general, there are two different types of payload architectures available: transparent and regenerative as described in Section 8.3. A transparent payload uses a so-called bent-pipe architecture, which performs only physical layer filtering, frequency conversion, and power amplification without "touching" the content of the signal of the transmission scheme. Therefore, it is called "transparent" to the network, as there is no processing of the data in between two architecture elements. In this case, the satellite is just forwarding the signal between a base station and the user on ground.

A regenerative payload performs a complete encoding of the signal and therefore can perform also various data processing and can be considered as a real network element. This capability allows implementing features of base stations directly inside the satellite, which offers several different opportunities depending on the split option (e.g. lower layer split or higher layer split). These different options vary from implementing a remote radio head inside the satellite to implementing an entire base station inside the satellite.

Due to the complexity and the high amount of effort, analyzing all different types of options for a regenerative payload and for the sake of simplicity, the transparent option has been prioritized in 5G to achieve progress and to focus on a first simple approach, even if the efficiency might be higher for a regenerative payload.

This means for 6G that all different options regarding the satellite payload by introducing a more capable but more complex regenerative payload enabling intelligent routing and switching as well as edge computing on the satellite are to be addressed. Regenerative and reconfigurable payloads are beneficial for the implementation of standard improvements along the 3GPP releases. Despite designing the optimal onboard processing architecture to support 6G, research is required for the different interfaces between the network nodes for higher and variable latency.

8.5.3 Coexistence and Spectrum Sharing

The academic community has investigated the topic of coexistence or spectrum sharing for over two decades. Most of the initial research involved cognitive radios with spectrum sensing capabilities. In recent times, owing to scarcity of spectrum with the introduction of larger sets of use cases and terminal types, sharing has become essential for different services and device types regardless of cognitive capabilities. The coexistence of LTE/NR networks or NB-IoT/eMBB devices are two such examples of scenarios where spectrum-sharing techniques have been addressed by 3GPP. Nonetheless, this has not yet included NTN. Since the LEO mega-constellations are poised to operate in the same band as incumbent GEO operators, the topic of spectrum reuse and interference avoidance has been explored within the satellite community. But the solutions so far have been restricted to more traditional methods of issuing primary and secondary licenses and do not incorporate dynamic spectrum-sharing methodologies. With an increasing number of broadband terrestrial networks expected to be deployed in the same frequency bands as traditional mobile satellite service (MSS) and fixed satellite service (FSS) bands both below and above 6 GHz, dynamic spectrum-sharing techniques involving NTN will be an integral component of enabling coexistence in 6G networks.

8.5.3.1 Regulatory Aspects

The key criteria to enable spectrum sharing is to ensure that the interference experienced by all users in the network is below a certain threshold. This interference apportionment is normally predetermined based on the type of communication service such as MSS or FSS. The ITU and other national regulatory bodies play a key role in determining and enforcing the acceptable interference for different operators and service providers. As can be seen from the agenda of the next ITU World Radiocommunication Conference (WRC) scheduled to be held in 2023 [15], there are several frequency bands that have been identified for compatibility studies and sharing on a co-primary or secondary basis. Some of these include NTN as discussed in Section 8.4.2 such as HAPs, satellite, and aeronautical services envisioned to operate in the same band as terrestrial users.

To provide capacity extension to carriers on a co-primary basis, enable geographical area licenses or license-by-rule usage of spectrum, innovative shared access models have been proposed. The European Telecommunications Standards Institute (ETSI)/European Conference of Postal and Telecommunications Administrations (CEPT) in Europe and Citizen Broadband Radio System (CBRS) alliance in the United States have introduced use of shared repositories to enable shared access using evolved License Shared Access (eLSA) and CBRS frameworks, respectively. These frameworks provide spectrum regulators and service providers with the ability to honor access priorities, while at the same time ensuring quality of service (QoS) for end users. Heterogeneous systems including NTN could be easily incorporated in such frameworks as well.

On the other hand, several unlicensed frequency bands allow different users to access spectrum on an opportunistic basis. Even though these bands are suitable for deployment of low-cost services such as IoT, it may not always be possible to guarantee fair access. Additionally, unlicensed operations must cope with nonuniform regulatory constraints across different geographical regions. Nonetheless, for several CubeSat missions and IoT accesses in NTN, these unlicensed spectrum bands can be beneficial. Further analysis of efficient and fair spectrum sharing in unlicensed bands is necessary specifically for 6G NTN waveforms.

8.5.3.2 Techniques for Coexistence

There are several factors to be considered while investigating techniques that enable coexistence. Some of these include the frequency band of operation that influences the channel propagation characteristics, type of waveforms and channel access schemes, device transmission power limitations, and type of array antennas. Coexistence can further be enhanced in the presence of a central database as is the case with eLSA. A combination of these influencing factors can be exploited to enable different users to coexist in frequency, time, power, and spatial or code domains. Each of these needs to be evaluated for different service types, direction of communication (i.e. uplink (UL) or downlink (DL)), duplex modes, and traffic characteristics. The presence of a connected sky implies larger coverage area, complex geometry, and power constraints. If a receiver can incorporate in-band interference cancellation or excision techniques like in non-orthogonal multiple access (NOMA), sharing of the same time-frequency slots is possible. In other cases where a waveform can support beam hopping or beamforming such as in DVB-S2X or 5G NR, sharing is feasible by duty-cycling and coordination in time and spatial domains. If joint localization and data transmission is enabled as part of mmWave beamforming, spectrum reuse and sharing in spatial domain can be established.

The level of dynamicity in spectrum-sharing approaches is largely dependent on the channel coherence properties and traffic load both at the user level and network level. A combination of short-term and long-term adaptivity may be essential to make use of all available degrees of freedom to achieve optimal spectrum reuse and channel occupancy. Use of AI and ML algorithms can be very effective in implementing dynamic spectrum sharing by enabling signal classification, interference identification, anomaly detection, and policy enforcement. Such algorithms can be integrated into the non-real-time (NRT) and real-time (RT) RAN intelligent controllers of the open radio access network (O-RAN) standard [16], as a tool to enable coexistence of different NTN and terrestrial air interfaces.

8.5.4 Energy-Efficient Waveforms

Energy efficiency as described in Chapter 4 will play an important role in 6G. Despite the benefits of the overall power consumption of a cellular network and power-efficient end devices like IoT, energy-efficient waveforms are relevant for satellites as well and not only for terrestrial transmissions in higher frequency bands (mmWave, THz). The 3GPP study item "5G beyond 52.6 GHz" within Release 17 [17] identified that power amplifier efficiency decreases with increasing frequency toward 100 GHz. 3GPP, however, does not yet include the topic of peak-to-average power ratio (PAPR) reduction in the Release 17 specifications and continues to use orthogonal frequency-division multiple access (OFDMA) in downlink, which results in an even lower amplifier efficiency at frequencies above 52.6 GHz, compared with frequency range 2 (FR2).

Satellite systems are sensitive to the PAPR of a transmission waveform, because they are power limited and need to operate as close as possible to the maximum output power with small OBO. Since 5G, especially the NR downlink is based on cyclic prefix orthogonal frequency division multiplex (CP-OFDM), the PAPR is in the order of 10 dB or higher [18], requiring a higher OBO than traditional single-carrier waveforms like DVB-S2X used in satellite applications. Simulation results of the 5G NR DL and assuming a single 5G NR carrier per satellite transponder are reported in [19]. The results showed that the total degradation is dependent on the selected modulation (quadrature phase shift keying (OPSK), 16-quadrature amplitude modulation (16-QAM), 64-QAM) and amplifier type (either solid-state power amplifier (SSPA) or traveling-wave tube amplifier (TWTA)), leading to a decreased data throughput over the satellite link. The effective loss is called total degradation and is in the range between 2.5 and 6.5 dB (see Figure 8.5).

Consequently, to support an efficient transmission of a 6G signal over satellites, an optimized waveform with lower PAPR would be beneficial. This need is already identified in [20] for higher spectral efficiencies than 2 bit/s/Hz, but so far is not included in the work plan of 3GPP for 5G.

For multi-carrier operation over a satellite transponder, the difference between single-carrier waveforms like DVB-S2X and CP-OFDM decreases [18]. In case of satellites operating with multiple carriers per transponder (e.g. with beamforming capability), the implication of the waveform is reduced. Different variants of CP-OFDM including discrete Fourier

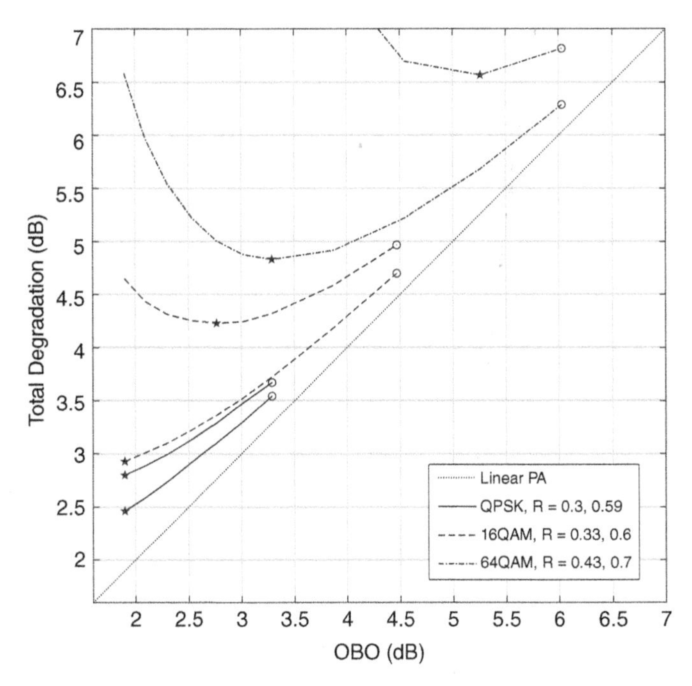

Figure 8.5 Total degradation of a single 5G NR downlink carrier over a satellite transponder with satellite with solid-state power amplifier. *Source:* 3GPP R1-1908996 [19].

transform (DFT)-based precoding, spreading, filtering, and windowing can be evaluated for transmission over different satellite links with specific power amplifier models and frequency-dependent channel characteristics. Preliminary analysis of these variants for different numerologies is presented in [21]. Further investigations of performance of waveform variants including the use of single carrier in the return link and further PAPR reduction mechanisms are necessary to identify the best option for transmission and reception in NTN.

8.5.5 Scalable RF Carrier Bandwidth

Scalability of the radio frequency (RF) carrier bandwidth is another crucial point to enable various deployments of the wireless access technology in non-terrestrial systems. 3GPP specifies LTE for a carrier bandwidth from 1.3 to 20 MHz, 5G NR for a bandwidth from 5 to 100 MHz in frequency range 1 (FR1), and from 50 to 400 MHz in FR2. For the new frequency range from 52.6 to 71 GHz, even a bandwidth up to 2.16 GHz is proposed in [22]. The general problem for the application to NTN scenarios is that the available frequency bands and especially the available transmission power are limited for satellites. So, the increasing bandwidth over the cellular standard generations limits the usability of 5G-NTN for certain frequency bands and types of satellites. As an example, the applicability of 5G-NTN with small CubeSats is described to highlight the need for flexible 6G waveforms with lower minimum RF carrier bandwidth than in 5G.

One of the first performance characteristics for usability is the link budget. In order to estimate this performance characteristic, two parameter sets have been defined within

Table 8.1 Satellite parameters for system-level calibration [4].

Satellite orbit		**LEO-1200**	**LEO-600**
Satellite altitude		1200 km	600 km
Satellite antenna pattern		Section 6.4.1 in [3]	Section 6.4.1 in [3]
Downlink			
Equivalent satellite antenna aperture (Note 1)	S-band (i.e. 2 GHz)	1 m	1 m
Satellite equivalent isotropically radiated power (EIRP) density		34 dBW/MHz	28 dBW/MHz
Satellite transmitter (Tx) max gain		24 dBi	24 dBi
3dB beamwidth		8.8320 deg	8.8320 deg
Satellite beam diameter (Note 2)		190 km	90 km
Uplink			
Equivalent satellite antenna aperture (Note1)	S-band (i.e. 2 GHz)	1 m	1 m
Gain-to-noise-temperature (G/T)		-4.9 dB K^{-1}	-4.9 dB K^{-1}
Satellite receiver (Rx) max gain		24 dBi	24 dBi

Source: 3GPP TR 38.821 [4].

3GPP for 5G system-level simulation. Satellite parameter sets with the lower requirements are defined in [4] for system-level simulator calibration and are presented in Table 8.1.

In system configuration SC25 in [4], 10 MHz bandwidth is provided for the 5G downlink from the satellite to the UE, which requires an EIRP of 68 dBm to just close the link budget (carrier-to-noise ratio [CNR]: 0.6 dB). This means that a resulting power of at least 40 dBm (= 10 Watts) is required at the satellite per UE with a broadband satellite antenna gain of 28 dBi. Assuming smaller antennas for CubeSats with a gain of 6–12 dBi, the required electrical transmit power is 16–22 dB higher, i.e. 400–1500 W, which significantly exceeds the available power of a CubeSat. Even with 5G NR's minimum bandwidth of 5 MHz, which still results in a required transmit power of 200–750 W.

As a conclusion, the minimum bandwidth in 6G should be significantly lower than 5 MHz to enable a wide range of 6G-NTN deployment scenarios, ranging from narrowband IoT with small satellites to broadband eMBB applications with high power satellites.

8.6 Conclusion

NTN will be an integral part of 6G to provide global connectivity with seamless coverage. The initial introduction of NTN in 5G is a significant step to establish a global standard for hybrid scenarios with terrestrial and non-terrestrial networks. However, a much more flexible approach to integrate dynamic network elements such as UAVs, LEO/VLEO satellites, and small satellites is required compared to the initial introduction of NTN in 5G. Innovations in satellite technologies like onboard processing will enable an improved integration in 6G and must be considered in the initial phase of research activities toward 6G. Introducing flexibility in the 6G core network and in the RAN is important to support various NTN deployment scenarios, NTN elements, as well as coexistence scenarios with spectrum sharing.

References

1 3GPP TR 22.822 (2020). Study on using Satellite Access in 5G; Stage 1 (Release 16). V16.0.0 (2018-06). *Tech. Rep.*

2 3GPP TR 23.737 (2020). Study on Architecture Aspects for Using Satellite Access in 5G (Release 17). V17.1.0 (2020-07). *Tech. Rep.*

3 3GPP TR 38.811 (2020). Study on New Radio (NR) to Support Non-Terrestrial Networks (Release 15). V15.3.0 (2020-07). *Tech. Rep.*

4 3GPP TR 38.821 (2019). Solutions for NR to Support Non-Terrestrial Networks (NTN) (Release 16). V16.0.0 (2019-12). *Tech. Rep.*

5 3GPP RP-202908 (2020). Solutions for NR to Support Non-Terrestrial Networks (NTN). RAN#90e meeting, Thales.

6 Sirotkin, S. (ed.) (2020). *5G Radio Access Network Architecture.* Wiley. ISBN: ISBN 978-1-119-55088-4.

7 3GPP RP-202689 (2020). Study on NB-Io/eMTC Support for Non-Terrestrial Network. RAN#90e Meeting, MediaTek.

8 Fraunhofer IIS (2021). Press release on pre standard trials of 5G-NTN. https://www.iis.fraunhofer.de/en/pr/2021/20210312_5G_new_radio.html

9 Spacenews (2019). https://spacenews.com/spacex-submits-paperwork-for-30000-more-starlink-satellites/

10 Network 2030 FG-NET-2030 (2019). A blueprint of technology, applications and market drivers towards the year 2030 and beyond. Network 2030 – A Blueprint of Technology, Applications and Market Drivers Towards the Year 2030 and Beyond.

11 Rajatheva, N., Atzeni, I., Björnson, E. et al. (2020). 6G research visions, No. 10, white paper on broadband connectivity in 6G. University of Oulu, ISSN 2669-963X (online), ISBN 978-952-62-2679-8 (online).

12 Saarnisaari, H. (ed.) (2020). 6G research visions, No. 5, 6G white paper onconnectivity for remote areas. ISSN 2669-963X (online), ISBN 978-952-62-2675-0 (online).

13 Vanelli-Coralli, A., Guidotti, A., Foggiy, T. et al. (2020). 5G and beyond 5G non-terrestrial networks: trends and research challenges. 2020 IEEE 3rd 5G World Forum (5GWF), Bangalore, India, 163–169. https://doi.org/10.1109/5GWF49715.2020.9221119.

14 6G-SENTINEL project. https://www.iis.fraunhofer.de/en/ff/kom/mobile-kom/6g-sentinel.html

15. ITU-R Preparatory Studies for WRC-23. htttps://www.itu.int/en/ITU-R/study-groups/rcpm/Pages/wrc-23-studies.aspx

16 O-RAN Specifications (2021). https://www.o-ran.org/specifications (accessed 25 June 2021).

17 3GPP TR 38.808 (2021). Study on Supporting NR from 52.6 GHz to 71 GHz (Release 17). V1.1.0.

18 3GPP RP-192233 (2019). PAPR required for OFDM and TDM in satellite downlink, Thales, Newport Beach (Tdoc withdrawn, but available).

19 3GPP R1-1908996 (2019). Downlink performance evaluation in NTN. Fraunhofer IIS, Fraunhofer HHI, Prague.

20 3GPP RP-192237 (2019). OFDM operation over satellite, Inmarsat, Intelsat, HNS, Thales, ESA, Fraunhofer IIS, Fraunhofer HHI, Newport Beach, US.

21 Jayaprakash, A., Evans, B., Xiao, P. et al. (2020). New radio numerology and waveform evaluation for satellite integration into 5G terrestrial network. Communications (ICC) ICC 2020-2020 IEEE International Conference on, 1–7.

22 3GPP RP-202926 (2020). Summary of [90E][08][52.6-71GHz_WI_scoping] second intermediate summary. Moderator (CMCC), RAN#90e plenary.

9

Rethinking the IP Framework

David Zhe Lou[1] *and Noel Crespi*[2]

[1] *Huawei Technologies Düsseldorf GmbH, Munich, Germany*
[2] *IMT, Telecom SudParis, Institut Polytechnique de Paris, Paris, France*

9.1 Introduction

In the 6G era, the societal and industrial digitalization is shifting the communication focus from human-to-human toward machine-to-machine. High-volume data transmission is no longer the top requirement, and the assumption of "large bandwidth is equal to high quality" is not universally applicable any more [1]. Instead, determinism becomes an important and indispensable network feature to guarantee the in-time and on-time packet delivery required by many emerging applications. In addition, security and privacy are expected as basic network functionalities to protect devices and applications from malicious attacks and to safeguard against thefts of personal data, respectively.

With trillions of heterogeneous and often resource-constrained devices being connected to 6G networks, the network needs to cope with highly dynamic and complex heterogeneous topologies, while also supporting multichannel concurrency and collaboration. In the light of industrial and societal digitalization, emerging applications expect more demanding Service-Level Agreements (SLAs) that can hardly be supported by the existing Internet Protocols (IPs) and technologies. For instance, host-oriented communication using fixed-length address and predetermined packet semantics might not be an ideal design for the interconnection of highly dynamic and resource-constrained networks. Statistical multiplexing and best effort oriented network transmission cannot guarantee bandwidth, latency, jitter, and loss ratio, which come as strict requirements in machine-to-machine communication.

Although there are many initiatives trying to tackle the aforementioned challenges, they tend to be either too disruptive without backward compatibility, hence difficult to be adopted by the market, or too sporadic to create a holistic and an impactful solution. In this chapter, we aim to identify key requirements, challenges, and technologies to support the emerging trends of digitalization by revisiting the network layer, which is the

Shaping Future 6G Networks: Needs, Impacts, and Technologies, First Edition.
Edited by Emmanuel Bertin, Noel Crespi, and Thomas Magedanz.

foundation of the current Internet but also a bottleneck in enabling demanding applications.

9.2 Emerging Applications and Network Requirements

Toward the year of 2030 and beyond, many novel applications are expected to emerge as others mature, leading to increasingly intertwined human and machine communications [2]. These applications can be classified in two broad categories: those that target the consumer society and those for industrial communications systems.

Among the consumer-oriented applications, multimedia has been witnessing significant advances that are expected to enable even richer and more immersive and interactive communication experiences in the future, as depicted in Figure 9.1. The COVID-19 pandemic accelerates, to some extent, this media evolution since the need for virtual interaction has increased substantially, e.g. for remote education, meetings, conferences, etc. Besides the usual audiovisual senses, some virtual reality systems could convey the sense of touch and holographic communication may even enable taste and smell, which demand multi-capturing synchronization.

Holographic communication, which is shown at the upper part of Figure 9.1, is expected to digitally deliver 3D images from one or multiple sources to one or multiple destination nodes in an interactive manner. It is foreseen that fully immersive 3D imaging will impose great challenges on future networks. First of all, it requires a huge bandwidth. The 5.9 inch hologram display plays video at about 12.6 Gbps, assuming a compression ratio of 1000:1. The 70 inch hologram display will ask for 1.9 Tbps bandwidth support [3, 4]. Second, ultralow latency is crucial to alleviate simulator sickness, especially when head-mounted displays are used. Last but not the least, in order to support multiparty holographic

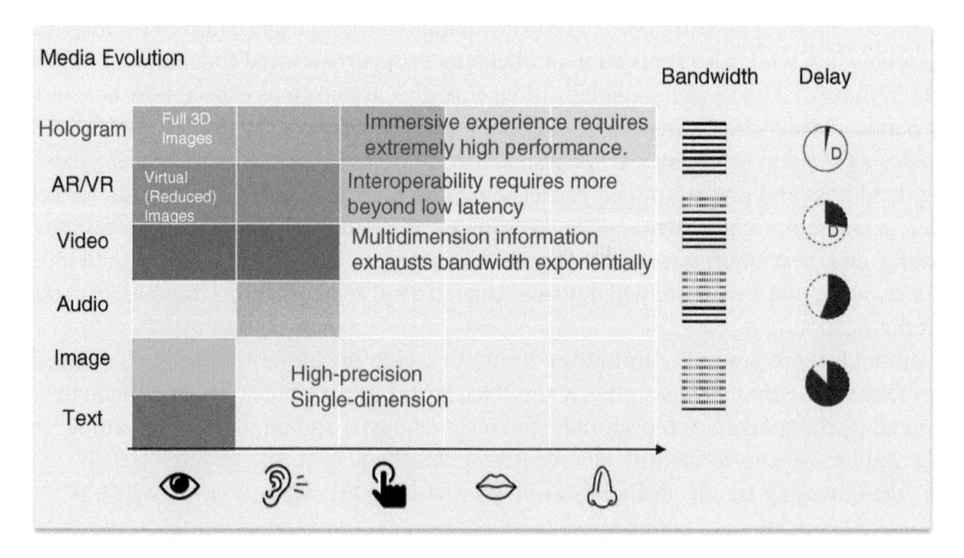

Figure 9.1 The media evolution.

communications or multi-master and single-slave control [5], multiple transmission paths or data streams with diverse geo-locations are expected to be synchronized appropriately with limited arrival time differences, usually at the level of millisecond intervals.

Meanwhile, in the manufacturing domain we are witnessing the fourth industrial revolution where information technology (IT) converges with operational technology (OT) to transform the traditional product lines toward more intelligent and flexible manufacturing systems. Machine-to-machine communication in large-scale deployments is an important capability of such systems, which are expected to provide increased automation, self-monitoring, and self-diagnosis. A key communication requirement for fulfilling these expectations is determinism, for example, on loss rates, jitter, and latency, appropriate for real-time applications. An equally challenging task is securing the industrial network against cyberattacks, which is a likely scenario if multiple sites are interconnected through the public Internet or in case traffic is sent to a public cloud infrastructure to leverage compute resources for big data analysis. Effectively mitigating attacks will prevent damages to the manufacturing lines and significant economic losses. Protecting communications from malicious attacks not only is essential for the industrial domain but also becomes a basic requirement for almost all applications running in the Internet.

To meet the needs of Industry 4.0, cloud and edge computing technologies have been applied to the production line not only to reduce the installation and maintenance costs but also to increase the flexibility of a manufacturing system. Automation companies such as Siemens, Schneider, and Bosch have been promoting their own cloud/edge platforms. As shown in Figure 9.2, the edge computing architecture evolved from a simple edge gateway to real-time edge compute where programmable logic control (PLC) functions could be virtualized and executed on commodity hardware in a data center.

Traditional industrial Ethernet technologies like EtherCAT, Profinet, Powerlink, Modbus, etc. [6] are capable of supporting small-scale manufacturing cells with deterministic capabilities. Typically, such a small-scale manufacturing network contains no more than 255 nodes (including end devices). To control the manufacturing devices from the edge/cloud, however, the instructions will need to pass through a large-scale network with thousands of connected devices. Existing layer-2 industrial Ethernet technologies cannot

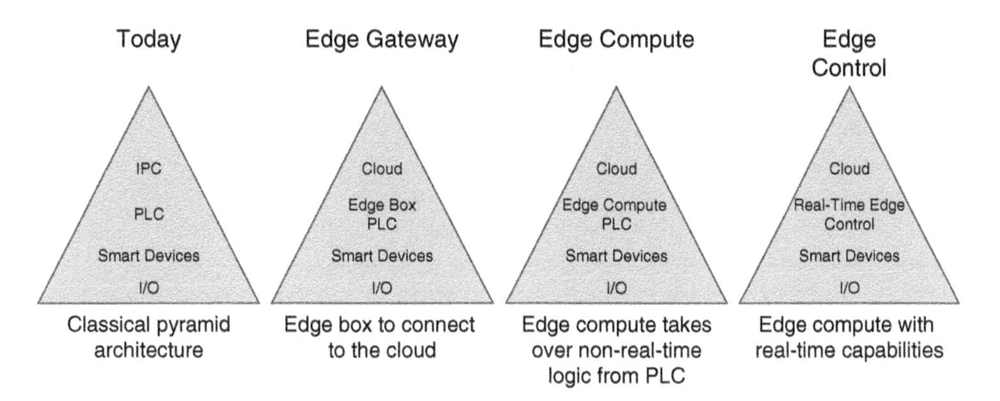

Figure 9.2 Evolution of the edge computing architecture.

cope in these settings and network layer innovation is required to enable deterministic communication in large-scale networks.

In addition to the strict performance and scalability requirements of future applications, security and privacy are key functionalities that should be provided by design. Statistics have shown that more than 87% of service providers have experienced distributed denial-of-service (DDoS) attacks, leading to millions of dollars loss per attack. The most visible series of DDoS attacks occurred as early as February 2000 when Internet giants Yahoo, Amazon, eBay, E-trade, and others were attacked intermittently over a period of several days, which had a cumulative lost of about US $1.2 billion [7]. Besides the downtime of the services and sites, 26% of the attacks led to additional damage including loss or theft of sensitive/private data. In addition, traffic surges due to attacks generate extremely high load in the network devices, which results to unnecessary energy consumption. In order to protect the infrastructure from DDoS attacks, it is necessary to combine authentication and accountability to build source verification schemes. At the same time, privacy-oriented intrinsic security schemes should be considered, which strike the right balance between security and privacy protection. This is especially important in the light of the European Union General Data Protection Regulation (EU GDPR) [8], which seems to be gaining traction even outside Europe.

9.3 State of the Art

The IP invented in 1969 as the foundation of the Internet has achieved unprecedented success for the past 50 years, thanks to its simple and open design. As depicted in Figure 9.3, the IP waist of the open systems interconnection (OSI) model enables diversified services toward upper layers and adapts to heterogeneous lower layers. Although statistical multiplexing and best effort forwarding in current IP networks can achieve high bandwidth usage efficiency, they cannot fulfill the requirements of machine-to-machine communication. Also, the host-oriented communication paradigm using a fixed-length address and predetermined packet semantics is not suitable for highly dynamic networks (e.g. vehicle to vehicle [V2V] and unmanned aerial vehicle [UAV] networks) and resource-constrained networks (e.g. Internet of Things [IoT] networks). Furthermore, security, a key functionality of any networking solution, does not fall in the core of the original IP design principles. Therefore, the foundation of the Internet needs to be revisited in order to look for a holistic solution that meets the requirements of emerging applications. It is time to overcome the IP ossification and to enhance the network layer with new capabilities that can unleash the true power of communications.

Studies to evolve the internetworking capabilities of networks have a long tradition and have been an integral part of the evolution of digital communication networks, such as the Internet, mobile networks, and other vertical communication networks. As early as 2000 [9], the "Next Generation Internet Architecture" efforts pointed to already well-recognized problems to justify the needed architectural evolution of the Internet. In this study, the authors advocate a holistic view on use cases, requirements, capabilities, architecture, and protocols, which in turn will have to be progressed in the context of a

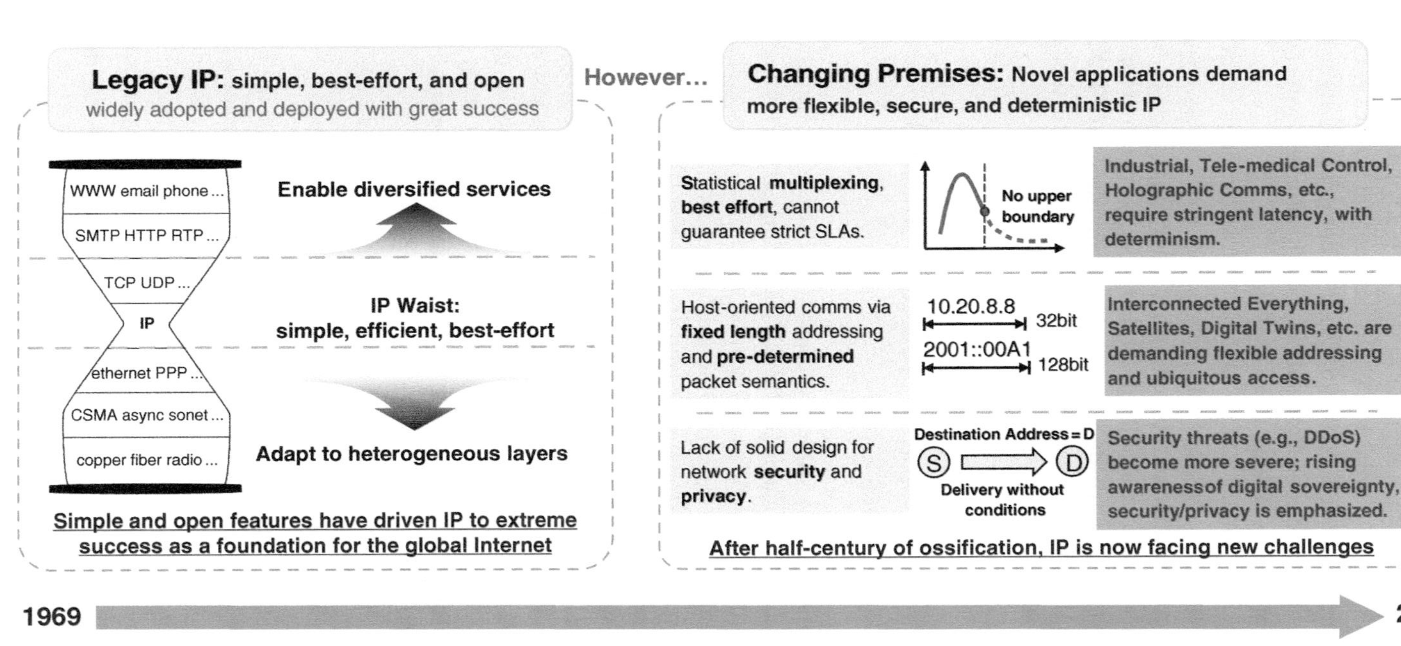

Figure 9.3 New challenges for the IP network.

standards developing organization (SDO) with comprehensive objectives. In addition, dedicated initiatives spawned during the past two decades alone, advocating the rethinking of aspects related to the network layer as well as its key protocols and technologies. Efforts such as Internet2 [10] in the late 1990s, the National Science Foundation (NSF)-funded Future INternet Design (FIND) initiative [11], European efforts on the clean-slate Internet research in Framework Programme 7 [12], and the Asia Future Internet efforts [13] in the early 2010s are only a few examples.

All those past activities emphasized the holistic view on use cases, requirements, capabilities, architecture(s), protocols, and enabling technologies, clearly recognizing that a concerted effort is needed that goes beyond what other SDOs provide. Yet, even SDO-driven efforts, such as those in ITU-T IMT2020 [14], or European Telecommunications Standards Institute (ETSI) efforts on next-generation protocols (NGP) [3], or on non-IP networking (NIN) [15], take a similar holistic view on the evolution of digital communication networks beyond the currently deployed technologies.

9.4 Next-Generation Internet Protocol Framework: Features and Capabilities

The next-generation IP should provide new capabilities like large-scale high-precision and deterministic transmission, semantic and flexible addressing, service routing, high throughput, user-defined network operations, and security and privacy, in order to support futuristic industrial and societal applications. Different from clean-slate network solutions, the network layer of new IP networks should ensure the necessary flexibility to enable interoperability and compatibility with existing IP networks, in order to protect the historical investment. The new network layer protocols are not meant to replace the existing IPs. Instead, the goal is to support the needs of upcoming novel networked applications using existing protocols where possible and enhancing them through extensions and updates where necessary. New protocols, or their components, should only be developed when shortcomings of existing protocols prevent application needs from being addressed. Key features and capabilities the new IP network and protocols should offer, but not be limited to, are described below.

9.4.1 High-Precision and Deterministic Services

Different from best effort services, the current Internet provides that the new IP network should provide high-precision and deterministic services that guarantee low upper bound of end-to-end latency, jitter, and loss ratio for flows in order to satisfy the increased quality of service (QoS) of emerging and future applications mainly coming from the fourth industrial revolution. To realize reliable and deterministic communication in the connectionless IP/Ethernet-based networks, many complex technologies such as resource reservation, scheduling, classification, shaping, and time synchronization need to be in place. These can be implemented in different layers ranging from layer-1 (physical layer) to layer-3 (network layer), as shown in Figure 9.4.

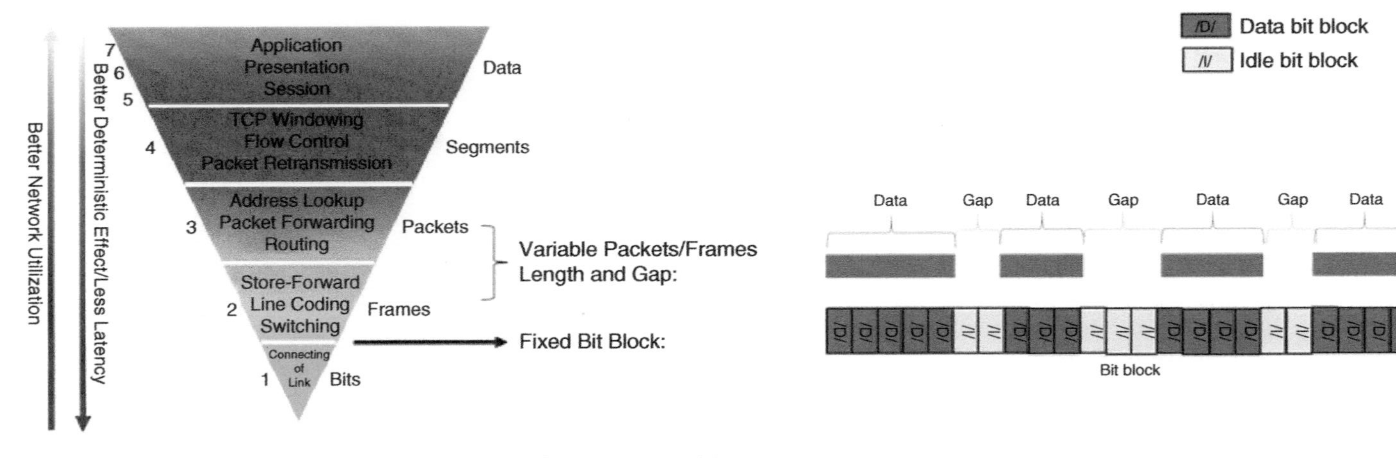

Figure 9.4 Technique to guarantee deterministic performance at each layer.

Typically, techniques implemented in lower layers can achieve better latency and smaller jitter at the expense of less dynamic in-network configurations. For instance, optical transport network (OTN) technologies can achieve 1 microsecond latency and 10 nanoseconds jitter per hop and Time-Sensitive Networking (TSN) [16] from IEEE802.1 can achieve the similar performance. However, OTN and TSN only solve determinism in small-scale layer-2 networks. Implementing deterministic packet delivery in layer-3 will not be able to reach the same performance. However, it can be applied onto large-scale networks to support cross-domain interactive applications. As one of the key techniques for determinism, time synchronization among a large amount of network nodes is not an easy task. The propagation delay between two adjacent nodes cannot be omitted. Therefore, suitable data plane extensions with accompanying control plane solutions need to be explored to provide centralized, distributed, and hybrid control signaling. Frequency synchronization may be utilized to remove the need for tight time synchronization between IP routers.

9.4.2 Semantic and Flexible Addressing

Semantic addressing caters to an increasing number of services that utilize the data plane of the Internet, recognizing existing trends toward improved content delivery through replication in specially deployed content delivery network (CDN) systems as well as from previous initiatives such as information-centric networking [17] and in-network computing [18]. The objective of semantic addressing is to incorporate the semantics (e.g. ID, service, location, and entity) into the address field so that the router can interpret the destination of the packet without consulting another dedicated translation service (e.g. domain name system [DNS]). It actually brings the routing to key services much "closer" to the basic data plane operations in order to reduce the routing latency. Many of the arguments presented in [14] hold in terms of offering an internetworking capability that goes beyond locator semantic only but instead provides an addressing scheme that is more directly aligned with the services offered over the internetworked communication, while the underlying routing natively supports the multi-semantic addressing. The driver behind supporting such varying semantics is that networks are not limited to those operating by the locator-based addressing and routing solutions of the Internet but instead use sector-specific packet forwarding solutions that may use service or content identifiers, sensor/host identifiers, path identifiers, or others, reconciling into a hierarchical addressing scheme over which a unified routing infrastructure will allow for true interconnection across all those networks.

Flexible addressing in length is the key to enable the aforementioned integration of multi-semantic addresses, which caters to an increasing number of specialized network deployments. This feature is driven by the long-standing recognition of the IP header overhead, moving instead to a variable length addressing approach [19] that can be efficiently supported alongside the global reachability, while ensuring the future extensibility of the provided hierarchical addressing used by the routing system. Solutions here may devise a flexible, free-choice addressing scheme for the routing in the network layer, enabling backward compatibility with IPv6 but also allowing for intra-domain short length addresses (e.g. for IoT networks) and the integration of vertical-specific name resolution/mapping systems, such as the aforementioned semantic addressing.

Protocol solutions could utilize, for example, embedding service information into the addressing scheme used for packet delivery [19], allowing for fast redirection to the "optimal" endpoint without requiring DNS-level operations as it is the case today. For this, service name information is encoded into the addressing at the network layer. Anycast routing might be utilized for directing the service request to the most suitable service instance with the given semantic address and for realizing in-network multicast delivery of information requests to the same service or content.

9.4.3 ManyNets Support

The current Internet infrastructure, originally built to provide global connectivity to all services through a common network, is suffering from ossification, which, along with continuing demands, is increasing the fragmentation of the infrastructure and makes the handling of "ManyNets" (many networks) increasingly more difficult [20]. There is ample evidence of this fragmentation with some representative examples being: (i) the "flattening" of the Internet, and a more recent step forward toward zero-hop networking architectures where servers of large content providers are placed within access networks; and (ii) global, publicly accessible, "purpose-built" networks that have been built recently, constituting true bypass networks (one of the various examples being the recent deployments of low-power wide area networks [LPWANs]).

While the current Internet design shares the same vision of ManyNets support ("IP runs over anything"), many of its solutions are skewed toward network technologies with rather static capabilities, as expressed in the use of fiber optical and fixed-network technologies. Dynamic systems, such as satellite, vehicular, and cellular systems, are well supported at the level of access technologies. However, layer-3 networks formed of those more dynamic transport technologies are not well supported due to a range of issues:

- Link state management is significantly more fluid and dynamic in a network where nodes move relative to each other, forming new relations (with other network nodes) on a frequent basis. Unlike satellite networks, in which such dynamic relations may be predictable (from orbital information), other networks such as those formed from fast moving vehicles may be significantly less predictable. Flooding approaches, common in protocols such as Interior Gateway Protocol (IGP), are not suitable for such networks (of moving objects) due to reasons of incurred overhead but also convergence times.
- Addressing may need to expand from a network to a geographical location, where such an extended address semantic can aid the link state but also the data forwarding mechanism.
- Routing may need to move from a distributed state to a centralized solution, e.g. utilizing path-based source routing approaches instead, enabling the coordination of link and topology information in a centralized controller for optimized resource allocation and scheduling.
- With the move to zero-hop architectures, the realization of such latency-optimized deployments is an open question in a satellite network that includes caching and other in-network services for improved service completion times.

Protocols are required here for routing state management in highly dynamic network topologies, extending existing flooding-based protocols like IGP. Furthermore, routing solutions need development that utilize geographical instead of network locator addressing to improve on data plane operations, particularly solutions for in-network support for zero-hop operations (to facilitate reduced latency).

9.4.4 Intrinsic Security and Privacy

The security and privacy aspects were not considered as one of the original "seven design principles" [21], which makes the current IP-based networks vulnerable due to source address spoofing, privacy leaks, trust model weaknesses, and DDoS attacks. According to the STRIDE security model [22], a network architecture with intrinsic security and privacy should maximally protect user privacy, consolidate distributed trust basis, and build secure and trustable networks, which would meet strict privacy protection requirements and the security and trustworthiness requirements of industry-wide interconnection.

The need for better privacy practices is further reinforced by the EU GDPR [8], which requires data protection by design and by default. The regulation has been driving significant changes in the ways in which personal data is presented and handled, and has classified IP addresses as such data. Pseudonymization techniques [23] have been investigated by the EU's security agency ENISA (the European Union Agency for Cybersecurity) as a prominent method for protecting personal data. In addition, GAIA-X [24] is an important European initiative that is somehow reminiscent of the Schengen routing approach [25] but also extends in the compute domain. GAIA-X, with the support of key European players and beyond, aims to develop a federated data infrastructure for Europe, which will meet the highest standards in terms of digital sovereignty, and where data and services can be made available, collected, and shared in an environment of trust.

Privacy-preserving protocol solutions may decouple the locator addressing from the user identification at the routing level, creating an ephemeral routing identifier that can only be traced to its originating autonomous system (AS) while long-term tracking of users becomes impossible. Denial-of-service prevention is enabled through allowing for removing the coupling of user identity to the ephemeral locator based on signaling from the attached AS toward the originating AS.

9.4.5 High Throughput

High throughput aims at evolving current transport layer technologies toward the need for new communication scenarios. These do not only include the evolution of video-based communication (with expected Tbps needs for new use cases such as holographic communication [3]) but also the optimal scheduling of resources across multiple flows of communication through an increased use of multipath and flow coordination in the network through, for example, interleaving bursts of flows, where the overall utilization of both compute (e.g. cloud) resources and communication resources is taken into account [15]. High throughput over a number of concurrent access technologies ensures not only the needed quality of the envisioned new services (with video technologies being core to driving the requirements for bandwidth) but also catering to the emergence of many networks at the access level, including satellite, 5G (and evolutions), and many others. Protocols for multipath delivery [19] of network-coded information (to reduce ACK feedback and therefore reduce overall flow completion latency) can be devised, while exposing network capabilities through an extended transport application programming interface (API) for an improved matching against application requirements.

9.4.6 User-Defined Network Operations

User-defined network operations enable application constraint-based routing as well as optimal scheduling of multi-flow forwarding relations [15]. This evolves the separation of control and data planes, expressed and realized in software-defined networking by enabling end users (and their applications) to have a greater control over the end-to-end path of packets, as well as the utilization of communication resources, including the cross-flow treatment of packets for scenarios of video communication with embedded sensor information (e.g. for interactive control). It is being seen as central to utilizing in-network capabilities of the forwarding plane to improve on service experience. Contrasting the flow-based approach of contemporary solutions, such as software-defined network (SDN) [26] and P4 [27], are approaches that embed such operations into the packet delivery for consideration in the intermediary forwarding element, e.g. for multi-flow treatment, support for sync operations across flows and other use cases. Solutions here may extend packets at the network layer with metadata to define in-network processing options for handling at intermediary components [19]. The trade-off between overhead, complexity, and service quality is a key focus of investigation in this area.

Furthermore, new approaches will be needed to largely simplify existing practices that define how packets are treated in the network. Such approaches can, for example, build upon the principles of ongoing efforts on intent-based networking [28] and allow users to (re-)configure packet headers with appropriate operations by defining intended outcomes as opposed to using rule-based or algorithmic logic. An advanced level of programmability will enable adaptive behavior in the forwarding plane with enhanced flexibility in the packet treatment, which form key features in meeting the requirements of demanding services and applications. Important challenges to address in this domain include the definition of appropriate interfaces to describe intents, mechanisms to automatically decompose abstract intents and select concrete operations to encode in the packet header, and feedback mechanisms to ensure that intents are being met by the network.

9.5 Flexible Addressing System Example

This section focuses on a specific feature of the next-generation IP framework by providing a representative example. The flexible addressing system was chosen for this purpose, as the dramatic impact it will have on the protocol format will lead to a fundamental change in the IP framework altogether.

The scope of the Internet has continuously expanded from an experimental and local area network (Advanced Research Projects Agency Network [ARPANET]) to the all-industry Internet, which consists of IoT, satellite networks, cellular networks, industrial manufacturing networks, smart campus networks, etc. However, the fixed-length IP address is not applicable to all kinds of scenarios, especially for resource-constrained

IoT. According to statistics, the average payload length of IoT packets is 25 bytes, which is much less than the overhead of IPv6. Moreover, the Internet is expected to connect every person and everything, and IP with topology-oriented semantic cannot meet the requirements of emerging communication scenarios. For example, the MobilityFirst, a novel non-IP network architecture, is proposed to support mobile communication scenarios. Nevertheless, communication among heterogeneous network entities will break the original address allocation and the routing rules of each network, and a costly network address translator or gateway is needed.

To determine the root cause, we propose an addressing model and we analyze the nature of the current IP address system. In our model, the IP address (can be regarded as a name) is defined as a label of the object it represents, and the corresponding namespace is a set of object labels. The namespace has two key elements: the name assignment (including its space, length, semantic, and scope) and the binding between name and communication endpoint. Currently, the IP address system uses a namespace at the network layer, so all communication entities must be bound to that namespace. The length of the name (address) is fixed (e.g. 128 bits for IPv6), and each name is bound to a unique node in the topology (semantic). If a heterogeneous network needs to communicate through the Internet, it must be bound to the IP namespace only, even if the heterogeneous network has been bound to another namespace or a more suitable namespace (ZigBee for IoT). Thus, the topology-oriented semantic of the current IP namespace with a fixed length constrains the flexibility and scalability of the IP address system and makes it difficult to interconnect heterogeneous networks.

According to the above analysis, we propose an Open Generalized Address system, where the length of the name and the semantics can be decided based on the requirements. The main design principle of this system is to allow the coexistence of multiple namespaces, as shown in Figure 9.5. Names of a same object can be expressed as a hierarchical structure. Different names in a different namespace can identify the same object or a group of objects (virtual object). Because objects are addressed in an independent namespace, the address of the namespace is not restricted by the upper-layer namespace (there is no dependency

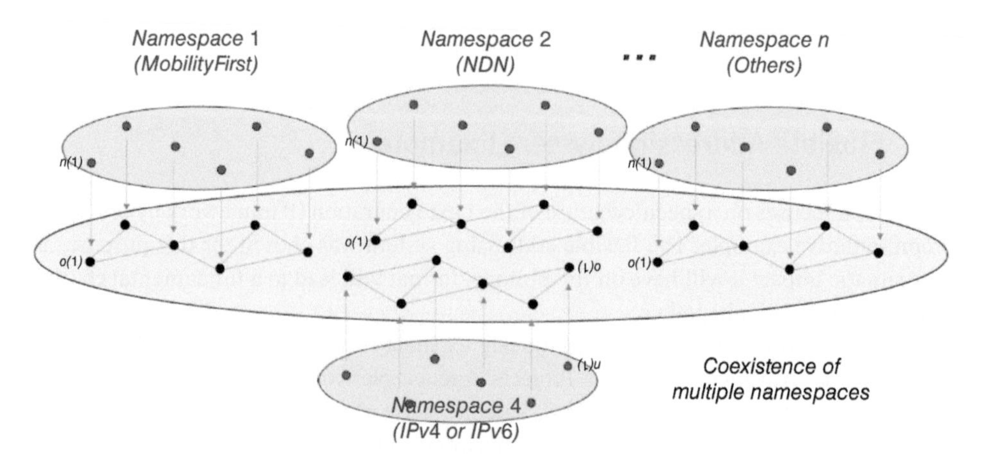

Figure 9.5 Coexistence of multiple namespaces in the Open Generalized Address system.

relationship) and can be named as required. The address system can be iterated down infinitely until there is only a single object in a namespace and no virtual object.

To achieve the interconnection of heterogeneous networks, the Open Generalized Address system defines the Network Index Address (NIA) as the top-level namespace to identify networks, which is allocated and advertised in a decentralized way. Since the namespace of each network (or subnetwork) is independent, a hierarchical address structure is used for routing among different namespaces. For example, as illustrated in Figure 9.6, Network 1 obtains NIA = 100 and advertises it to other networks in the same top-level namespace via a modified Border Gateway Protocol (BGP). The hierarchical global address of a communication entity in Network 1 can be expressed as <100, 10.32>, where 10.32 is the local address. Therefore, the communication entity in other networks, such as Network 2, can find the routes to Network 1 via NIA = 100. Moreover, the hierarchical global address can be extended downward, for instance, a n-level hierarchical address can be constructed as <NIA, 1st level network address, 2nd level network address, . . ., nth level network address>.

9.6 Conclusion

Since it was invented in 1969, the IP has achieved unprecedented success for the past 50 years. Meanwhile, the ossification of IP networks has made it harder and harder to meet the networking requirements of emerging applications due to the paradigm shift from human-oriented to machine oriented communications. Instead of sporadic optimizations as was the case for the past 20 years, it is time to reconsider upgrading the foundation of the Internet beyond statistical multiplexing and best effort oriented services. The high-precision

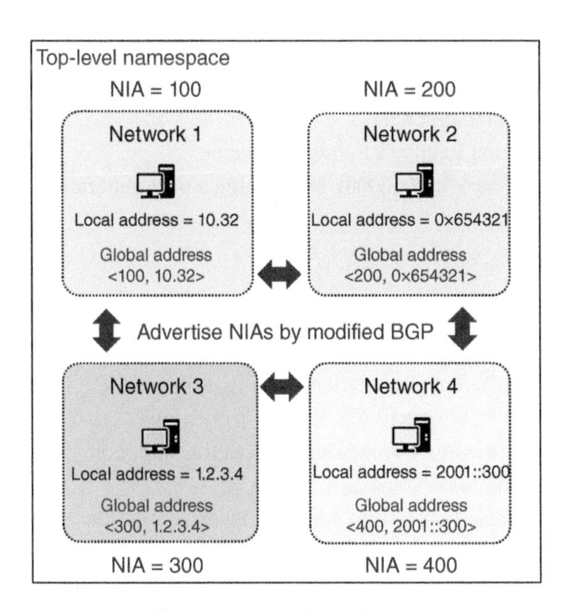

Figure 9.6 The interconnection of heterogeneous networks with hierarchical address.

and deterministic service, flexible and semantic addressing, and well-balanced security and privacy that are required by the future machine-to-machine communication will become the key features and capabilities of the new IP networks. We believe that not only the radio access network (RAN) and core networks but also the evolution of the IP network in between plays an essential role in the research and development of 6G networks.

References

1 A brief introduction about new IP research initiative. https://www.huawei.com/en/technology-insights/industry-insights/innovation/new-ip.
2 Da, B. and Carugi, M. (2020). Representative use cases and key network requirements for Network 2030. ITU-T Technical Report (16 January 2020). https://www.itu.int/dms_pub/itu-t/opb/fg/T-FG-NET2030-2020-SUB.G1-PDF-E.pdf.
3 Xu, X., Pan, Y., Lwin, P., and Liang, X. (2011). 3D holographic display and its data transmission requirement. *International Conference on Information Photonics and Optical Communications* (21–23 October 2011). https://ieeexplore.ieee.org/abstract/document/6122872.
4 Oh, K., Choo, H., and Kim, J. (2016). Analysis on digital holographic data representation and compression. *Asia-Pacific Signal and Information Processing Association Annual Summit and Conference (APSIPA)* (13–16 December 2016). https://ieeexplore.ieee.org/document/7820789.
5 Shahbazi, M., Atashzar, S.F., and Patel, R.V. (2018). A systematic review of multilateral teleoperation systems. *IEEE Transactions on Haptics* 11 (3).
6 Lachello, L., Wratil, P., Meindl, A. et al. (2013). Industrial Ethernet Facts, issue 2 (February 2013). https://www.ethernet-powerlink.org/fileadmin/user_upload/Dokumente/Dokumente/EPSG_IEF2ndEdition_en_WEB.pdf.
7 Distributed denial-of-service (DDoS) attacks: an economic perspective. https://nsfocusglobal.com/wp-content/uploads/2017/01/Distributed_Denial_of_Service_Attacks__An_Economic_Perspective__Whitepaper.pdf.
8 The European Union General Data Protection Regulation. https://gdpr.eu/
9 Braden, R., Clark, D., Shenker, S., and Wroclawski, J. (2000). Developing a next-generation internet architecture. White paper, DARPA.
10 Internet2. https://www.internet2.edu/.
11 NSF FIND. http://www.nets-find.net/.
12 Future Internet Assembly in EU FP7 Programme. http://futureinternetassembly.eu/
13 Asia FI Forum. http://www.asiafi.net/.
14 Trossen, D., Sarela, M., and Sollins, K. (2010). Arguments for an information-centric internetworking architecture. ACM CCR.
15 Zhang, J. and Clayman, S. (2019). EBS: electric burst scheduling system that supports future large bandwidth applications in scale. ACM Conext.
16 IEEE 802.1. Time-Sensitive Networking (TSN) Task Group. https://1.ieee802.org/tsn/.
17 Awais, M. and Shah, M. (2017). Information-centric networking: a review on futuristic networks. *23rd International Conference on Automation and Computing (ICAC)* (7–8 September 2017). https://ieeexplore.ieee.org/document/8082033.

18 Zilberman, N. In-network computing. https://www.sigarch.org/in-network-computing-draft/.

19 Chen, Z., Wang, C., Li, G. et al. NEW IP framework and protocol for future applications. *Proceedings of 2020 IEEE/IFIP Network Operations and Management Symposium (NOMS).*

20 Ammar, M.H. (2017). The service-infrastructure cycle, ossification, and the fragmentation of the internet. Georgia Institute of Technology. https://arxiv.org/pdf/1712.04379.pdf.

21 Clark, D. (1988). The design philosophy of the DARPA internet protocols. *SIGCOMM Computer Communication Review* 18 (4): 106–114. https://doi.org/10.1145/52325.52336.

22 Microsoft. The STRIDE Threat Model. https://msdn.microsoft.com/en-us/library/ee823878(v=cs.20).aspx.

23 Pseudonymisation techniques and best practices. ENISA report (3 December 2019). https://www.enisa.europa.eu/publications/pseudonymisation-techniques-and-best-practices.

24 GAIA-X project. https://www.data-infrastructure.eu/GAIAX/Navigation/EN/Home/home.html.

25 Donni, D., Machado, G., Tsiaras, C., and Stiller, B. (2015). Schengen routing: a compliance analysis. *Proceedings of IFIP International Conference on Autonomous Infrastructure, Management and Security (AIMS)*, Ghent, Belgium.

26 https://www.opennetworking.org/sdn-definition/.

27 P4 Open Source Programming Language. https://p4.org/.

28 Clemm, A., Ciavaglia, L., Granville, L., and Tantsura, J. (2019). Intent-based networking – concepts and over- view. Draft-clemm-nmrg-dist-intent-03, IETF.

10

Computing in the Network: The Core-Edge Continuum in 6G Network

Marie-José Montpetit[1] *and Noel Crespi*[2]

[1] *Concordia University, Montreal, Canada*
[2] *IMT, Telecom SudParis, Institut Polytechnique de Paris, Paris, France*

10.1 Introduction

6G offers the opportunity for leveraging the most recent networking innovations to improve the telecommunication infrastructure. This will not only result in improved overall performance and resource utilization but also allow the introduction of new players, improved economics, and new services with applications in media, Internet of Things (IoT), manufacturing, and autonomous systems to name a few.

While there has been few major changes in the basic architecture of the Internet since its inception, the networking community has witnessed the clouderization, the softwarization (software-defined networking or SDN), and the virtualization of network functions (network function virtualization or NFV) over the last two decades. Novel concepts of data and connectivity management to ensure robustness and evolvability have also been developed and deployed and are still the topic of academic research. This includes the introduction of white boxes as network nodes, end-to-end encryption, and now the deployment of new transports such as the QUIC (Quick UDP Internet Connections) protocol [1].

At the same time, data centers (DCs) and edge networks have gained prominence in a network that is more and more heterogeneous and application centric. There has been active research in next-generation architectures, that question the basic tenets of the Internet itself, to respond to the move to data-centric and mobile networking. In particular, the use of data-driven approaches and recent advances in the fields of optimization theory, game theory, machine learning (ML), and artificial intelligence (AI) derived from big data analytics are now applied improving the performance of networks and networked systems. These not only are topics of innovative research and development but also are now being integrated to operational platforms [2]. New approaches to network management execute different decision mechanisms in the network nodes themselves, and self-orchestration and planning guarantee predictable performance. Taken collectively, all of those developments share a common trait: that computing and networking are becoming inseparable for

Shaping Future 6G Networks: Needs, Impacts, and Technologies, First Edition.
Edited by Emmanuel Bertin, Noel Crespi, and Thomas Magedanz.

innovation to happen. "Computing in the network," the melding of computing and networking everywhere in the network is the topic of this chapter.

Traditionally, there was little integration of computing tasks into networks because of the notion that they operated in two distinct domains: the relatively dumb networking nodes implement protocols that enable forwarding between more intelligent computation entities. The intelligence remains in the end nodes, and the network fabric is optimized for speed and security. These concepts were of course inherited from the early days of networking when adding computations to network nodes was incompatible with the goals of keeping the network node simple and consequently very fast. Hardware has evolved of course, and the recent developments in virtualization and data processing have shown that there may be new ways to design in-network services based on local processing of the data (summarized in [3]). Some of these innovations will be introduced in this chapter.

While 5G defined the concept of slices [4] to allow the colocation of many subnetworks on the same infrastructure and their isolation, this is still an end-to-end solution and does not define a new level of distribution and layering of network functionality and data management. But the heritage of alternative Internet architectures like information-centric networking (ICN) [5] combined with the emergence of new programmable data planes and of novel data-centric services is creating a renewed interest in adding computing inside the network node. The edge positioning of network resources closest to the consumption of services and the use of local computing and switching capacities in DCs are providing new frameworks and platforms for network design and operations. Hence, the idea of merging computing and networking is getting traction and, considering the level of activity that is dedicated to it, will most likely be central to the networks in the 6G time frame. In 6G, it is now possible to reimagine the network where routing, computing, and storage blend based on service and application requirements. The impacts of locating programmability inside the network will be introduced in this chapter with some of its enabled use cases.

10.2 A Few Stops on the Road to Programmable Networks

At Sigcomm 2018, Professor Jennifer Rexford's keynote [6] presented an evolution of computing in the network: it is important to present some of the milestones in this history before going further. Because, in fact, computing in the network is in itself not totally new: original network nodes were computers used to implement connectivity protocols and determining routes by executing local algorithms. Even when the need for speed, reliability, and security optimized those nodes by minimizing generic computations, in the past 25 years many attempts were made to improve network services with programmability and to change the Internet to reflect the data needs of the web-centric world. This section is a short summary of some of salient contributors to the merging of computing and networking that we are witnessing today.

10.2.1 Active Networks

Many of the concepts related to functional execution in network nodes derive from "active networking" [7–9]. Active networking was a popular research topic from about 1996 to 2001 and remains a pioneer of the current move to programmable network nodes. The goal of active networking was to adapt the network to dynamic requirements by adding executables on demand in nodes traversed by packets. Custom functionality inside the network could take two forms: application inside packets (capsules) that could be executed along a path and code in routers [9]. Active networking introduced the concept that transit packets may define states for future packet, something still used in SDN today [6]. Sadly, active networking did not find much deployment outside research testbeds due to the inherent security issues of executable in packets and its implementation complexity. However, the idea of in-network computing lives on.

10.2.2 Information-centric Networking

From the early 2000s, the global deployment of Internet services especially the World Wide Web (WWW) justified the research on alternative Internet architectures, both evolutive and clean slate, supported by major grants from the likes of the National Science Foundation (NSF) in the United States and the European Union. Among those, content-centric networking (CCNx) [10] and named data networking (NDN) [11] addressed jointly under the ICN [5] umbrella reimagined an Internet not based on addresses, the legacy of the telephony infrastructure, but on content naming, more in line with the type of searches on the WWW. Indeed, the predominance of web-based services, wordy universal resource locators, and the subsequent introduction of powerful search engines to navigate the Internet was making it obvious that application developers and end users were not looking only for a site but for the content in that site. The means to discover and deliver that content needed some fundamental rethinking. With basic interest and data messaging (to request data and respond), ICN has shown to be particularly well-suited for the multisource multidestination architecture of media experience and IoT (see the many contributions in [5]). It provides a natural functional decomposition between the requests for information and its delivery. The combination of interest messages to signal what content is needed combined with the data responses help to coordinate the different streams and multiple users. In ICN, access to web content has evolved from single, context-independent streams to context-dependent information components. Those can adapt dynamically to the changes in the network and to the immersive nature of some data-centric services and can be delivered securely and efficiently.

ICN requires in-network caching to provide its services efficiently. Thus, it introduced major concepts necessary to computing in the network including storage (the network as a distributed database as one researcher calls it), namespaces, and content discovery [12], all heirs to ICN research. Recent ICN research in function (named function networking or NFN [13] or the ICE-AR [14] ICN-Enabled Secure Edge Networking with Augmented Reality (project for advanced augmented reality (AR)) are moving closer to computing in the network concepts. ICE-AR introduces "acceleration as a service" for the exploration of the design and usage of computation at the edge, including the wireless edge, to improve the performance of delay-sensitive AR applications.

10.2.3 Compute-first Networking

ICN has evolved over the years with increased core and the edge capabilities. It has influenced or helped create many related architectures that take advantage of its traffic engineering capabilities for optimized data discovery and caching capabilities such as NFN mentioned precedingly. One of its most notable recent avatars and a major in-network computing architecture is compute-first networking (CFN)[15–19]. CFN is based on the joint optimization of resources that is necessary for melding computing and networking. CFN also use distributed networking concepts, as applications can be split into components to be pipelined and executed at run time in network nodes. With node discovery capabilities and timely execution, CFN functions comply to operator policies and application specifications but also user requirements and end-device preferences [18].

CFN is an example of a computing-in-the-network system that is based on distributed programming to provide a general-purpose framework. As stated in [20]:

> In CFN, compute nodes that can execute functions within a given program instance are called workers. The allocation of functions and actors to workers happens in a distributed fashion. A CFN system knows the current utilization of available resources and the least cost paths to copies of needed input data. It can dynamically decide which worker to use, performing optimizations such as instantiating functions close to big data inputs. The bindings that control which execution platforms host which program interfaces (or individual functions/actors) is maintained through a computation graph. To realize this distributed scheduling, workers in each resource pool advertise their available resources. This information is shared among all workers in the pool. A worker execution environment can decide, without a centralized scheduler, which set of workers to prefer to invoke a function or to instantiate an actor. In order to direct function invocation requests to specific worker nodes, CFN utilizes the underlying ICN network's forwarding capabilities – the network performs late binding through name-based forwarding and workers can provide forwarding hints to steer the flow of work.

In the remainder of the chapter, some of the CFN basic ideas will be included for the more generic concept computing in the network.

10.2.4 Software-defined Networking

The move to cloud-based computing and the rise of DCs has been significant in the evolution of networks, and for computing in the network in particular, and will be dealt as a major use case 10.4.1. However, and in addition, in-network computing is also the natural heir of SDN [7]. The Internet, the ubiquitous and global network based on a set of protocols that supports most of the data networking today, has profited in recent years from advances in cloud computing and SDN (and its associated NFV) that have moved traditional embedded operational functions to the "softwarized" network.

SDN has significantly impacted network design and management in the recent years. Like active networking, SDN is promoting pushing some programming into the network under centralized control. It clearly creates a boundary between the control plane, where network policies and handling are decided, and the data plane that forwards the traffic according to the rules from the control plane. The innovation here is that a single control plane element can control multiple data plane elements (routers, switches, etc.) via well-defined application programming interfaces (APIs) such as OpenFlow, that define how that element will handle traffic. SDN enables network applications from access control to load balancing [7, 21]. Network virtualization, NFV, is in turn, made possible by SDN and OpenFlow. Network virtualization creates "containers" that implement a whole application-centric stack [21]. In terms of "computation," NFV instantiates functions on a software platform, some of which can process packets, making it a precursor to more advanced computation in the network.

With SDN and the softwarization of networking in general, the software implementing network functions, protocols, and services are separated from the hardware running them; they can evolve independently. This can lead to more rapid and flexible service creation when added to functional distribution across nodes. SDN has demonstrated that networks can be operated and managed top-down. As will be seen in the use case section of this chapter, these features make it easier to add intelligence to network nodes and to support the distribution of the functions that not only manage and automate networks but also of the services that use them.

10.3 Beyond Softwarization and Clouderization: The Computerization of Networks

Traditional networks operated with dedicated hardware, and the Internet was not different until recently. Large routers were designed for speed and efficiency and to ensure packets went to their destination rapidly and securely. While the basic Internet protocols were open, their implementation was in the realm of closed platforms. With the evolution of computing elements following Moore's law, it was realized about 10 years ago that more generic hardware, the white boxes, could be programmed to support network functionalities with more flexibility and capabilities, the softwarization of networking. At the same time, the "new operators," GAFA[1] as well as Akamai, accelerated the clouderization of networks with the DCs and edge networks needed to support the "as a service" era. And now, as will be seen below while basic role of networking remains to connect things and people to one another and to the services they need, more and more of this role includes processing the packets in network flows, the computerization of networking.

10.3.1 A New End-to-End Paradigm

Most of the requirements for the network services today include mobility, scalability, security, and availability, often in real-time or in a timely manner, anywhere and over a wide variety of end devices. It is now widely accepted that traditional approaches like cloud and client-server architectures lead to complexity and scalability issues especially for automation and decision systems that increasingly depend on fast reactions but as well as on ML and decision algorithms. While it too breaks the "end-to-end" principle of the Internet, there are advantages to be able to add network function execution as a means to improve

1 Google, Amazon, Facebook, and Apple

system performance but also to support a growing number of networked data-driven applications.

The architects and designers of new emerging services agree that innovative solutions are necessary to address the emergence and growth of next-generation distributed intelligent systems. These require self-management, orchestration across components and federation across network nodes [22]. And more and more they involve distributed computing technologies for in-network service and the use of programmable switches, network interface controllers (NICs) along with white boxes.

Beyond management and orchestration, the addition of the capability to transmit fragmented datasets of heterogeneous sizes that are characteristics of data-driven applications needs addressing: those datasets create variable and often heavy and dynamic traffic loads. Facilitating the deployment of data-driven service involves combining:

- Edge resources for data reduction and local rendering and decisions as required.
- Cloud resources for advanced computing and further distribution.

In that context, there could be multiple cloud and edge instances needing interworking among themselves and high-level abstractions for data modeling and a focus on primitives and functionality reuse for classes of applications (not single ones)[23]. While this is an emerging field that is still in the realm of academic research [24], some salient use cases (see Section 10.4) are emerging to:

- reduce the load on DC and cloud infrastructure.
- provide quick reaction to delay sensitive applications everywhere.

Application developers and more and more cloud providers alike have incentives to move computing closer to users: every operation that is done in the network frees central processing unit (CPU) cycles in the cloud and in the user device, providing better throughput and signification latency reduction [25].

10.3.2 Computing in the Network Basic Concepts

Until recently networking innovation has focused mainly on:

- Over the network layer: the cloud and the vertical top-down SND architecture over virtualized nodes.
- Under the network layer in particular in the mobile and wireless infrastructure.
- At the edge of the network the applications and the devices to support them.

Now, the "in the network" or "on the network" innovation is rising, to enable the deployment of the next-generation service: a horizontal innovation, node to node, within the network structure.

In this chapter, computing in the network encompasses and is defined as the execution of native host applications within the network nodes. In an operational sense, this requires resources to be available and programmable and the definition of who can make changes and where [6]. Those resources can be physical or virtualized distributed among many locations [20]. As will be seen, adding computation to the network enables the emergence of a common programmable infrastructure that can be distributed everywhere inside the network fabric. It also means to adding functions inside the routing infrastructure (directly in the routers, in NICS) but also into the white boxes of SDN and edge networking, inside the switches, in microcontrollers or even in sensors and cameras especially for next-generation IoT. Computing becomes as ubiquitous along any network path to support:

- DCs with the goal of combining components into a managed entity, with a network-fabric and "peripherals" devices at the core optimized of for data rate [26].
- Performance improvement mechanisms and telemetry at aggregation points inside the network.
- Edge and fog combinations of applications and devices for critical decision-making and distributed intelligence.

In this, the concepts in-network (directly on the nodes) and on-network (in associated peripherals such as storage node) computing are combined. The scope of the emerging melding of computing and networking is expanded to encompass each touchpoint in the path of a packet. The distributed nature of this computing and its associated, service (and micro-service) architecture defines new paradigms for interconnection, orchestration, and collaboration between node along with discovery of computing and storage resources. The computerized network expands the capabilities of [20]:

- Edge and DC computing with platform-as-a-service and computing-as-a-service.
- SDN to include data plane programming of switches and other hardware.
- Applications to include specific processing frameworks and platforms.

The edge in particular, for computing and networking, is central to these concepts after becoming essential to the success of many autonomous systems and increasingly supporting cloud operations especially in the automotive sector [27] but moving everywhere (see Section 10.4.2) [28, 29].

The ability to deploy, manage, and evolve a system with DC, aggregators, and edge components brings new challenges to how complex applications are orchestrated, both individually and in mutual competition for resources. The orchestration of end-to-end resources between the DC network and the edge is another key topic in computing on the network with increasingly heterogeneous distributed components. Current approaches (e.g. Kubernetes and its distributed versions) are likely to need updating, extending, and/or simplifying in multi-domain network environments. To improve Internet performance overall, the network, the computers, and the storage resources must work jointly in close partnership to provide service to data-intensive or delay-sensitive distributed applications. Embedding the application-specific computing and storage in the network nodes ties them to the application life management cycle, and in consequence updates or changes in the application results in updates to the network and vice versa.

There are now many of distributed networking and computing systems that are the basis of adding computation both in the DC and the edge. They include distributed file systems such as the Hadoop Distributed File System (HDFS) [30] or the Interplanetary File System (IPFS) [31], distributed memory database such as MemCached [32], distributed computing such as MapReduce [33](also from Hadoop), TensorFlow [34], and Spark GraphX [35]. Other frameworks especially for DCs include Pregel from Google [36] and DryadLinq from Microsoft [37]. To ensure security, distributed trust systems on the blockchain, such as hyperledgers and smart contracts are now widely used (shown by many presentations as the Distributed Networking Research Group [38]). The means by which code can be provisioned to the processing node depend on the provider, the type of device, or eventually on the application itself depending on the dynamicity targeted and the level of trust associated with the code source. This relates to the overall security impacts of in-network computing (see Section 10.3.3.4).

Hence, the network of the 6G era will be much more like a multiprocessor computer board and peripherals than the "telephone system" analog of today with routers forwarding not processing packets to destination nodes. The melding of networking and computing is becoming a prevalent research concern as was shown by a large number of related presentations and keynotes at the recent Sigcomm 2020 [39]. The emergence of many commercial implementations and the creation of new companies (see [40] for an example) is also the proof of the viability of the concept.

10.3.3 Related Impacts

With packet processing and associated function beyond forwarding, network nodes now have functionalities that can be turned on and off inside the network with the ability to set the parameters and the performance objective for these functions. Thus, the deployment of novel services as well as the optimization of some traditional ones is facilitated for better operations, management, and the introduction of concepts like ML in networking. However, there are impacts on existing networks that need to be evaluated. A sample of those impacts is briefly introduced next; they are all currently research topics and should evolve in the near future.

10.3.3.1 The Need for Resource Discovery

Discovery is a feature that is essential for adding computing to the network everywhere: discovery of computing resources as well as of associated storage, functionality, and data and of the interface to allow their (dynamic) composition and orchestration [41]. While a lot of effort in discover protocols have been concentrated at the edge [33, 42], their conceptual nature points to the conjecture that developed approaches should be extendable to discover and then allocate computing/storage resources, onboarding and running a program, and defining the best provider for a specific functionality. Adding performance metrics to the discovery as well as device access will also become a part of discovery and allow to dynamically schedule computing tasks based on these specifications and enable more efficient service placement [43]. Computing and networking resource discovery should also support in-time computing allocation to perform telemetry for traffic load inference, as well as resource advertising to facilitate inter-node collaboration and autonomic management based on ML and other AI concepts. This discovery can be in fact integrated to scheduling of the joint networking/computing resources and address equally traditional data and stream requirements with required CPU or even security options in a "computing as a service concept" [44]. This helps define a common data layer architecture that includes discovery as one of its main building blocks (see Section 10.3.1).

10.3.3.2 Power Savings for Eco-conscious Networking

Transporting and storing large amounts of data results in large energy consumption. Comparing computing, versus transporting that data, shows that 70–90% of the power consumed in a data system is consumed to transport that information [25]. An added advantage of distributing the computing inside the network especially on existing devices is energy saving. A single network switch can "replace" multiple server-racks saving in the range of 1 GWh/year [25] with almost zero cost as the devices are already within the network and constantly powered. Of course, latency, bandwidth, and reach must be traded off and deciding whether computing locally is feasible, but gone are the days when CPUs used all the power in a system. The interconnections between processors appear to be the new energy bottleneck, and local processing in network nodes may become key for eco-conscious connectivity.

10.3.3.3 Transport is Still Needed!

Executing programs in network nodes may impact or be impacted by transport layer protocols. In particular, crypto/security features and information that could (or could not) be exposed across network layers, including parameters that might manage congestion and enable quality of service (QoS), will need to be accommodated withing any computing framework. These issues can potentially impair inter-node collaboration and prevent service-specific functions such as discovery and composition to be performed without added overhead and participating nodes authentication. In particular, the friction between end-to-end principles including privacy and security of traditional and new transport protocols alike and network computations needs to be addressed for the latter to be successful and remain functional. Authentication of collaborating nodes under the same domain or operator could facilitate the inclusion of "trusted entity" to the computing community. This in not outside the capabilities of current transport and higher layer protocols. As stated in the goals of the Computing in the Network Research Group (COINRG) [45] and Internet Research Task Force (IRTF) research group, adding computing supports the transport layer will:

- improve multicast and peer-to-peer distribution in bandwidth-constrained environments because of dynamic route selection based on match-actions.
- reduce network loads with local caching, data reduction, compressed sensing, and pre-rendering.
- alleviate congestion with local execution capabilities to manage QoS.
- optimize higher layer protocols to reduce latency.
- enable trust protocols, including blockchains and smart contracts, for secure community building across disconnected nodes and domains [46] (see Section 10.3.3.4 next).
- improve support for nomadicity and mobility because of better traffic planning.
- optimize performance by deduplication, tunneling, improved virtualization, and packet loss protection (see Section 10.5.5.1).

10.3.3.4 How About Security?

For in-network computing to be successful, the suitability of privacy and security preserving algorithms and their per node execution needs to be evaluated for their robustness to attacks and leakages [47]. For example, traditional centralized mechanisms to discover and admit nodes to the network, to provide access right and name resolution, need to be updated to be used in a distributed computerized network. This is critical to support industrial, healthcare, and agricultural applications and services among others. The heterogeneous nature of edge-cloud computing with processors at the DC, the aggregation and the IoT gateways and microcontroller, can be vulnerable to threats of cyber physical and human nature alike. In this integrated system, the authenticity of all computing elements becomes essential [48].

A gateway that is connected to a cluster of micro-devices or intelligent sensors of the next generation needs to authenticate its attached systems. Since an IoT gateway should be under the control of a single operator alleviates some of these issues but not all of them. Various techniques like fully homomorphic encryption [49], third-party audition, as well as cyber physical resource access control and privacy preserving mechanisms such as differentiated privacy noise [50] are already being used across many distributed networks. They can be used to safeguard the data security and privacy locally and require some level of autonomy that can be provided by local processing and supported in the cloud.

However, the large volume of data being generated at the edge devices and sent to DCs for computation and storage may lead to question the safety of the data itself not only with the identity of the node that generated them. One solution is emerging: blockchain technology, with operations performed in a decentralized way. Blockchains are becoming a major scalable means of providing trust and validating provenance in a large number of distributed systems [51]. Smart contracts (on the blockchain) and hyperledgers supply a mechanism to provide the trust and validation for autonomous edge nodes like those connected, for example, by the emerging IPFS [31].

In blockchains, a new participant node is admitted after it has committed to a smart contract that contains the rules and mechanisms to distribute content via this node in a trusted and secure way. This constitutes its proof of validity. After a node is admitted, it will be then provisioned with the appropriate software to become fully operational. Newly admitted nodes will be inserted in the general ledger on the blockchain enabling other nodes to discover them and hence, to form a trusted network. A name resolution authority can also be provided by the blockchain to manage and validate the origin of the content, the proof of origin, and to provide the ability to search such content. The proof of origin can also be used to prevent some content from reaching one or more nodes and implement content filtering based on trusted authorities. This is useful not only for content but also for packets containing functional information that impact a node's operations. Finally, when some content reaches a specific destination, it can be verified against the content rules of the reached node even before it is sent to the application; this allows to provide a proof of delivery for the content, to generate statistics and performance metrics, and to enable the nodes to adapt service requirements. All of the above assumes that the nodes can implement the functions needed by the blockchain and hence once again infers that there is enough computing power in the nodes to perform these operations at an acceptable speed, which is another area for development. But as a result, the computation nodes could create "privacy and data control as a service" [44] and improve not reduce privacy and security in networks.

10.4 Computing Everywhere: The Core-Edge Continuum

We are witnesses to the birth of the "cloud-edge continuum" of computing resources across the network and the birth of the ubiquitous in-network computing. So what is this the "core-edge"? To answer this question, it is possible to both build on and extend the requirements of what Peterson calls the "access-edge" [52], integrating the move to white boxes and commodity hardware to the virtualization and the softwarization of network nodes and distributing functionality across the network landscape.

The resulting core-edge features include:

- Disaggregation
  - Breaking vertically integrated systems into independent components with elements across the network with open interfaces that is essential to next-generation IoT among others.
- Multitenancy
  - In the core-edge, different stakeholders (operators, service providers, application developers, enterprises) can be responsible for managing different components. This emphasizes the need to address isolation between tenants and the extension of the 5G slice concepts [4] since it will be often impossible to run the entire computerized network in a single trust domain (to allow for wider orchestration).
- Data reduction
  - Devices across the network and especially end user devices, the edge's edge, generate a large amount of data, and data reduction will be critical for the 6G networks. This means the need for substantial compute and the refactoring and potential redesign of services and applications to take advantage of computing and storage resources.
- (Near-)real-time
  - The new network landscape is dynamic, with functionality constantly adapting in response to mobility, workload, and application requirements. Supporting such an environment requires tight adaptable control loops for control and management. It also drives the development of new architectures to provide these low delay services [22].

There are different scales of computing resources in the core-edge continuum, from large to small (edge) DCs to end user devices and intelligent sensing devices with limited computing resources. Distributed, decentralized networks and resources required in the edge-cloud continuum are thus impacted by the increased heterogeneity of networking devices. This results in both advantages and limitations that require computing disintermediation and functional decomposition across common data layers, programmable nodes, open APIs, and interchangeable functionality.

10.4.1 A Common Data Layer

Previous sections have described how the melding of computing and networking follows the softwarization of networking functions and alternative Internet architectures: beyond packet forwarding, processing and caching can now be distributed physically over the network. Much has been said in the recent past about the move to edge computing and networking especially for autonomous systems: but more and more it encompasses variety of other data-driven services for networks and commercial applications to perform as specified. However, too often, these service deployments over the Internet and in the IoT are limited by the lack of cloud and local subsystems interoperability or cloud intermittent reachability. Hence, network functionalities are migrating closer to the service location.

While in-network processing may violate end-to-end Internet rules, it does conform to layering. In fact, layering concepts enable the core-edge continuum. For example, as will be seen in Section 10.4, next-generation IoT will profit from a dynamic infrastructure providing self-identifiable elements, adaptive devices, and increasingly autonomous intelligent decision-making. This has a particular impact on the joint use of edge and cloud computing/networking especially for emerging networked applications. Joint optimization of computing and networking resource usage is needed to achieve low latency for certain functions, services, and applications and pipelining of functions in real-time to meet a throughput goal and performing network traffic management (load, congestion) dynamically [53]. SDN has already introduced the concepts of top-down programmability: now horizontal programmability within the network layer is considered. And the network layer in turn is becoming the data layer.

The first interaction of IoT sensors with decision systems are in the microcontrollers that are placed very close to in the sensing environment and connected to gateways for access to the cloud. In general, these controllers have limited roles: forwarding information to the upper layers for processing and executing simple commands in response. But IoT nodes need to recognize critical events locally and react accordingly. However, even simple IoT devices generate massive volumes of data that need to be interpreted to provide the appropriate feedback. Since most IoT devices have limited on-board processing resources, cloud servers have been relied on to enable simple local gateways to aggregate and send data to be analyzed off-premises. But cloud services do not support real-time and time-critical applications adequately and require constant connectivity, an issue for mobile and remote locations with variable or unreliable access. In order to reduce response time and increase reliability, local processing in the path from IoT device to the cloud can be added to the system based on principles of distributed and edge computing, a data layer. The processing capabilities of this data layer facilitate the integration of the automated mechanical and software worlds and embedding intelligence everywhere.

In addition, the fragmentation of the datasets at the edge, both in size and in timeliness, increasingly require that local nodes become autonomous and "intelligent." New services for mixed reality deployment and decision-making require in-network optimization, advanced data structures and interactivity, security, and resiliency. These are important for advanced media systems and are critical for industrial networking. Hence, another function of the common data layer is to extend the current match-action switches (see next Section 10.4.2) based on packet header information to include metadata. The (meta) data generated by local measurements of network and services alike should be filtered and analyzed and their result identified as significant or critical; those will be used to determine a course of local retroaction and fed back to centralized decision-making when appropriate. They require bringing more network, computing, or production capacity online for reducing loads or distributing load across more geographical locations to meet demand. In this context, the common data layer could potentially become many layers of more and more complex processing and reaching even to the physical layer components [54]. With the availability of very inexpensive hardware, microcontrollers and the gateway that have more processing power and storage, can process sensor information to provide the services beyond alarms and including data reduction ahead of offloading. They are also where some publish subscribe messages are interpreted locally to enable decision using mechanisms such as CoAP (Constrained Application Protocol) [55] MQTT (Message Queuing Telemetry Transport) [56] for IoT systems or based on ICN principles (with many contributions in the ICN repository [5]).

Finally, discovery of the resources between an edge and the cloud is a key topic to address in the data layer design (as was introduced in Section 10.2.3). In addition to discover storage and computing capabilities, discovery enables the orchestration of increasingly heterogeneous distributed components and that draw inspiration from current approaches such as Kubernetes [17] or Storm [57] for self-organizing and distributed systems, unikernels [58, 59], as well as small footprint operating systems edge networks [60] or for IoT such as RIOT (Real Time Operating System for the Internet of Things) [61]. It is likely that distributed environments will need updating, extending, and/or simplifying to address the multi-domain production environments that are characteristic of the emerging IoT landscape potentially based on dynamic resource discovery. In particular, ML is a very active research area for decision-making at the edge because of the large datasets needed to identify and tag especially when imaging is added. For example, in TensorFlow parameter updates are small deltas that only change a subset of the overall tensor and can be aggregated locally by a vector addition operation [34]. ML is also more and more used to detect local and network-wide faults and failures and allow rapid responses as well as implement network control and analytics. Hence "intelligence discovery" could soon become another feature of the data layer.

Revisiting the traditional device's stack and introducing a common data layer to provide computing services to the overlying application layer are important for next-generation cloud and edge services. The ensuing interoperability and the enabled introduction of novel concepts of intelligence and decision-making inside the network is an integral part of what 6G could become.

10.4.2 The New Programmable Data Plane

Network algorithms are building blocks of network applications and until recently, most network software was embedded in proprietary hardware. But this is changing due to the softwarization of networking and the rise of virtualization architectures independent of the underlying infrastructure. SDN has stimulated the innovation in domain-specific languages and flexible switch architectures. Hence, the introduction of dedicated hardware for computing in the network has greatly contributed to its adoption in the industry and consequently has promoted the development of a services marketplace for as second-generation SDN.

The latest programmable switches make the concept of the totally programmable and dynamically reconfigurable network closer to reality since they operate at the line speed compatible with operational networks in and out of DCs. And, as distributed systems are increasingly based on memory instead of hard disks, distributed system-dependent application performance is increasingly enhanced by computing. A programmable data plane provides abstractions of different types of network hardware and while limited by the capabilities of the hardware and the related programming language(s) and packets/flow abstractions, match-action and filtering can provide solutions for a growing set of network function. The match-action features enabled by languages like P4 are essential to monitor and measure network performance as well as enabling fast reaction to critical events in the combined data-driven systems and networks [62]. These actions that are applied to packets include rewriting some headers, forwarding, and dropping.

In the data plane programming evolution, the protocol-independent switch architecture (PISA) has played a major role. PISA implements the flexible match-action pipeline that maintains comparable performance to fixed-function switches. While one of the most known is the Barefoot Tofino (now Intel)[63], but Cavium's XPliant [64], Broadcom Trident 4 [65], and Cisco's Nexus [66] follow the PISA architecture as well. In the rest of the chapter, "programmable switch" refers to a PISA switch. In all these architectures, Reconfigurable Match Table, or RMT, [67] models are generally accepted as generic enough to represent limited but essential switch architecture components and functionalities [68].

A typical PISA switch includes a programmable parser, ingress, queue, egress, and de-parser. The parser and de-parser are support user-defined packet header formats and thus enable efficient packet filtering functions that are the basis of, for example, in-network telemetry (see Section 10.5.4.1). The ingress and egress pipelines process packets through match-action tables that are arranged in stages and filled with predefined rules to match the header and perform the corresponding action on the packet. This is essential for enabling diagnostics and advanced caching and processing when combined with a general processing unit (GPU) [69]. Actions also allow to use primitives to modify the nonpersistent resources such as headers (and maybe in the future some metadata). It important to stress the fact that accurate data filtering and match/action functionality are essential to identify critical events or important information to feed decision algorithms. There are needs for mechanisms to distinguish between different data types, for example, those that need immediate action for control and automation and others for training and ML and thus ensure the creation of consistent inputs for all critical and noncritical operations. Beyond header filtering, the data flows also need to be analyzed and characterized to ensure reliable data transfers to GPUs for further analysis such as in the NetCache [69] architecture.

In addition to the emerging commodity programmable switches, the Programming Protocol-Independent Packet Processors (P4) language [70] for target-independent programming is spurring a renewed interest for implementing algorithms directly into the network fabric. P4 does not aim to be a general-purpose language such as C or Python but is a domain-specific language with a number of constructs optimized for network data forwarding. P4 enables to program the data plane behavior of the programmable switches by primitives and reconfigurable match-action tables. While it is outside the scope of this document to provide an in-depth description of PISA programming, it is worth introducing some elements of the P4 programming language. In short, P4 is a data plane programming language for packet processing, an instruction set for packet processing pipelines. P4 describes the data plane of programmable switches and the interface by which the control plane and the data plane communicate; it provides a set of basic constructs to describe a packet processing pipeline and allows to define headers, parser, action, and table in a packet pipeline. For example, P4 can implement a parser to extract information from the header and make a decision on the next parsing steps. Tables can be defined with the key that the table will match on and the action to be performed on a packet if a match is true (for example, dropping a packet). An action can also take arguments, which is from an entry in the table. The P4 compiler generates the target-specific data plane configurations and P4Runtime APIs from the P4 programs. P4Runtime APIs are target-independent compiler outputs that define interfaces for controlling the data plane elements and enable control plane functions. In turn, the control plane executes the functions that include port management, table entry modification, and register operation [70]. P4 is still evolving with optimization modularization and reuse remaining active fields of development [71].

For the moment, programmable data planes have been mainly deployed in DCs for very targeted network applications (see Section 10.4.1) and limited network management [72, 73] where instructions need to be performed at line speed, and hence the functionality has to be kept minimal. With P4 tightly coupled to the chip-specific architecture, it is difficult to develop higher level data plane programs for more advanced functionality. This enables network performance tools like telemetry or minimal forward error correction but limits more advanced applications such as ML for congestion control. For example, instruction sets are limited, and the operations are mainly simple arithmetic, data (packet) manipulation, and hash operation. New research initiatives such as Lyra [74] enable to add functionalities to create more advanced pipelining to the data plane. This is necessary to push in-network programming to the edge where speed is not as important, but flexibility is key. Interoperability between edge and DC based on similar paradigms is also desirable. Some use cases (see Section 10.4 next) also would need filtering to include more information such as metadata, and there is still development in the field to render advanced filtering possible as per recent conversations in the COINRG [45]. In addition, in order to maintain line speed, only a few operations can be performed on each packet, and while looping could allow a processed packet to reenter the ingress queue, it increases latency and reduces forwarding capabilities; speed requirements are key for defining behavior. As for the devices directly located on distributed network nodes, the processing can reduce the network traffic/congestion and increase the throughput; the level of these improvements is under evaluation.

The disintermediation of data processing away from a centralized approach via the computing node is essential to relieve the pressure on operational bottlenecks in the cloud that can undermine the scalability of critical applications. Scalability is where the programmable data planes will gain its most rewards. While programmable network hardware is still evolving, it should have improved in performance and functionality by the time 6G is deployed.

10.4.3 Novel Architectures Using Computing in the Network

Novel service architectures that will become available in the 6G time frame can also take advantage of the computerization of networks. Two examples are introduced next: a radically new approach to both computing and networking provided by quantum mechanics and a more incremental approach to respond to the short-term service and applications.

10.4.3.1 The Newest and Boldest: Quantum Networking

Quantum networking represents the "next frontier" for true in-network computing, and the field of quantum communication has been a subject of active research for many years. In essence, quantum networks use basic quantum mechanics principles to create distributed systems of quantum devices to provide communications. The physics of these systems involve superposition, entanglement, and quantum measurement and capabilities to provide services and achieve performances that are beyond what is provided by current non-quantum networks [75].

Quantum devices ranges from simple photonic devices capable of preparing and measuring only one quantum bit (qubit) at a time have been introduced leading to large-scale quantum computers of the future that are still in laboratories. These quantum computers target ultra-large data systems such as molecular and chemical reaction simulations but also large-scale optimization in manufacturing and supply chains, real-time financial modeling, ML, and enhanced security [75, 76].

The quantum Internet aims to interconnect these quantum devices via a "quantum network" capable of sharing quantum states among remote nodes. Entanglement[76] is at the basis of that communication capability [75] and provides the means to transmit qubits while keeping the quantum physics rules of that relationship [76]. Using local operations and an entangled pair of qubits, it is possible to use "quantum teleportation" to "transmit" an unknown quantum state between two remote quantum devices. Hybrid quantum/classical networks are now being envisaged with the core quantum network providing services such secure communications, distributed quantum computation, and quantum-enhanced measurement networks. As such they truly represent a novel approach to computing in the network. A review of quantum networking and the requirements and architecture of the quantum Internet are available in [75, 77].

10.4.3.2 Creating the Tactile and the Automated Internet: FlexNGIA

The FlexNGIA (Flexible Next-Generation Internet Architecture) is one example of an architecture to target the tactile Internet and delay sensitive applications that require edge processing with cloud support [22] as well as advanced security and automation [78]. FlexNGIA wants to leverage in-network computing and the recent advances in virtualization technology, SDN, and data plane programmability. The goal is to provide a fully flexible architecture from the edge to the core of the network.

FlexNGIA allows to design and develop network services, define network functions, or discover the ones offered by the network to optimize application performance. Layer-3 and above communication protocols are not specified and could be customized to the application. Because of its focus on programmability and software implementation, FlexNGIA wants to deploy customized network functions that are not limited to traditional packet forwarding but could include or interface with in-network match-action devices, ICN overlays and nodes, and deterministic networking (DetNet) functions implemented in both a distributed or centralized manner [22].

10.5 Making it Real: Use Cases

This section presents some of these use cases gathered from next-generation applications and services such as interactive extended reality (XR), industrial IoT, and autonomous decision systems beyond what is currently supported by 5G: not just requiring large bandwidth and low delay but computing, storage, function execution, as well as distributed orchestration. They all rely on the functional decomposition that is made possible by the codesign of layered approaches with computing (and related caching) positioned at selected nodes inside the network fabric. These use cases are inspired by work in the IETF (Internet Engineering Task Force) and IRTF beyond the group on computing in the network (COINRG) [45] such as DetNet [38], decentralized networks (DINRG [Decentralized Internet Infrastructure Research Group]) [51], and IoT (T2TRG) [79]. The new network paradigms such as ICN defined in Section 10.2.2 are also essential for the creation of these use cases and their implementation. Clearly, in-network computing profits from all trends in the future of networking.

The salient present below offers a sampling of what made possible when computing and networking mix, as the breadth of the field is still expanding both in academic research and in the innovation economy. They, however, illustrate how the location of computing directly inside the network will have major impacts in the way 6G services will be offered to applications and how "computing service providers" [44] and "computing as a service" could offer applications for the level of processing they need to be successful as their data transits through the network. Integrating computing with networking is considered promising if not essential for delivering adequate performance and efficiency for these exciting new use cases and illustrate the advantage of terminating data at the edge before it gets to the network and the cloud. This inherent data reduction increases scalability of application on existing networks, hence the life of the networking infrastructure [80].

The use cases presented in this section are fairly holistic: enabled by softwarization and virtualization added programmability with the goal of providing optimal performance such as low latency in compute-intensive interactive applications such networked XR to maintain the immersive nature of the experience [81]. Many of the use cases, beyond delay minimization, require the additional privacy or critical decision-making provided by processing data locally or by distributed computations within the network fabric itself. For example, IoT and intelligent industrial controllers all have requirements that require some level of local processing to support existing cloud capabilities to manage data loads, provide local real-time decisions, and increase reliability and data security. Low-latency requirements and event criticality demand that functions be located close to where the services are needed to help far-away cloud management: the rise of the DCs and cloud/edge functional decomposition for data-intensive and delay-sensitive applications has helped define the functionality required for many current and future services. As will be seen in the next section, DCs were the first target for in-network computing and many instances of the use of programmable data planes exist in the confines of DCs. The use cases will illustrate that there is indeed an advantage to be able to perform in-network packet processing including filtering, monitoring, and performance enhancement even if it breaks the end-to-end principles of the Internet: the advantages are worth the disruption.

10.5.1 Computing in the Data Center

In the past few years, virtualization and cloud computing have redefined the DC boundaries [82]. As a consequence, more and more servers are combined with storage, and network monitoring and connectivity are becoming one. In addition, highly performing DC hardware components enable implementing critical functions to handle more dynamic, service-centric, and rules-based configurations. In the subsections below, three in-network scenarios that have demonstrated value for DCs are presented: aggregation for on-path computing, key-value (K-V) cache maintenance, and consistency management [83].

10.5.1.1 Data and Flow Aggregation

The goals on on-path computing in DC is to reduce delay and/or increase throughput for improved performance with packet processing, to better balance traffic and manage network loads and to alleviate congestion [82].

The performance of data-intensive applications such as big data analysis, data reduction, deduplication and ML, graph processing, and stream processing in workload mode is improved by partition and aggregation. In such large-scale applications, scalability is achieved by distributing data and computing to many servers: each process a part of the data and then collectively update the shared state with the final results. Data communication cost and availability of bottleneck resources remain one of the main challenges for big data applications as large amounts of data need to be transmitted frequently in many-to-many mode (multisource and multi-destination). Adding local processing and filtering reduces the network load by pre-analyzing the data, aggregating all single messages together, and consequently reducing the task execution time at the different processing nodes as well as achieving traffic reduction; this can range from 48% up to 93% [82]. Scalability and multitenancy are also improved by aggregating path capacity to individual flows, which in turn improves congestion control by predicting these aggregate flow loads.

The aggregation functions needed to reduce data loads can be done via:

- Galois field arithmetic such as addition.
- logical evaluation such minima/maxima detection.
- comparisons for a certain filtering criterion.

Hence, they are well suited for the current capabilities of P4. They can also be parallelized. Map-reduce experiments show that after aggregation computing, the number of packets decreases by 88–90% with UDP (User Datagram Protocol) and by 40% with TCP (Transmission Control Protocol) [82].

Data reduction using exhaustive decision-making for network management and fault detection involving learning or AI can be added to these basic operations. Performing these functions in the DC at the ingress of the network can be beneficial to reduce the total network traffic, congestion, costs, or as was mentioned previously (see Section 10.3.3.3) energy consumption. There is thus great promise for optimizing computing and storage concurrently. Storage and caching can ensure that no critical data will be lost and that statistics can still be used offline while allowing critical data or computation results to be transmitted more efficiently.

10.5.1.2 Key-value Storage and In-network Caching

Key-value (K-V) stores are ubiquitous features of DCs with associated dynamic and data-intensive workloads. As is the case in any caching systems, popular items in K-V stores are significantly more often queried than other less trendy entries. However, the level of popularity can change or fluctuate rapidly with the occurrence of viral posts, limited time offers, and other trending events. This dynamic behavior may lead to load imbalance across caches and as a result in potentially significant performance deterioration: as a server can be overused in an area or underused in another, the overall system throughput will decrease with response time latency increasing. When the storage server uses per core sharding or partitioning to process high concurrency, this degradation will be further amplified [69]. Unbalanced loads are especially acute for high performance in memory-based K-V store.

A traditional solution to this problem is to selectively replicate popular items across many locations with the caveat of more hardware and energy resource consumption added to the overhead incurred for reliable mechanisms to support data mobility, data consistency, and query routing. In-network caching has shown a performance improvement of 3–10 times over traditional K-V storage: a small frontend cache can provide load balancing for N back-end nodes by caching only O(N logN) entries, even under worst-case request patterns [84] and is sufficient to balance the load for N servers. The pioneering NetCache system [69] uses rack-scale K-V stores with a programmable switch that detects, sorts, and caches popular K-V pairs and thus enables to rapidly process load balancing between storage nodes and enables higher layer functionality. This design guarantees billions of queries per second (QPS) with bounded latencies even under highly skewed and rapidly changing workloads.

10.5.1.3 Consensus

Coordination between nodes in a distributed network is needed to maintain overall system consistency and maintain consensus. Maintaining consistency may require multiple communication rounds in order to reach agreement creating potential messaging bottlenecks in large systems. Even without congestion, failure, or lost messages, a decision can only be reached as fast as the network round trip time (RTT) permits. Added to the verification of instances, this takes away processing capabilities and resources from other potentially critical tasks. This makes it essential to find highly efficient mechanisms for the agreement protocols: one idea is to use the computing capabilities of the network devices themselves. Consequently, in-network computing research and development has particularly targeted consistency and consensus [85].

Mechanisms for ensuring consistency and their associated overhead are some of the most expensive operations in managing large amounts of data [26]. There is a trade-off between reducing the coordination overhead and accepting possible data losses due to inconsistencies. As the demand for more efficient DCs increases, consensus has been removed from the (critical) data path by moving it to hardware [26] and push operation deeper into the network by extending the functionality of middle boxes or adding consensus to intermediate network nodes.

The Paxos improvements provide a case in point [86]. Paxos is a widely deployed DC consensus protocol for fault-tolerant systems. It serializes transaction requests from different clients, ensuring that each message replicator in the distributed system is implemented with the same order [87]. Each request can be an atomic operation from an inseparable operation set, and the specific content of the proposal is not considered. It has been shown that moving Paxos operations to network nodes can yield significant performance benefits for distributed applications [87]. Paxos can be implemented on PISA switches using the P4 language. The network switches take the role of both coordinators (request managers) and acceptors (managed storage nodes). Overall, messages travel over fewer hops in the network reducing the latency for the replicated system to reach consensus and prevent coordinators and acceptors from becoming bottlenecks when they aggregate or multiplex multiple messages. Moving consensus logic into network devices and using multi-ordered, multicast, and multi-sequencing can dramatically improve the performance of replicated systems by close to an order of magnitude [87–89].

10.5.2 Next-generation IoT and Intelligence Everywhere

The current trend in IoT is to develop a common, "thin waist" of protocols to enable a horizontally unified architecture, the data layer (Section 10.3.1), or even the "intelligent layer" [90]. Next-generation IoT, including controls and automation, will profit from recent developments in distributed systems and ML [91].

10.5.2.1 The Internet of Intelligent Things

Already, IoT sensors, microcontrollers, and cameras can be accessed remotely and managed through SDN [92]. However, many of them also fall under the "constrained" definition: no or little CPU and limited storage and batteries or power. IoT gateways that use multiple commercially available devices and systems linked by commercial data busses have traditionally allowed sensor data to be uploaded to the cloud for storage and analysis and to return feedback information to the individual elements [93]. In this configuration, each subsystem communicates with the gateway with low-power wireless or very simple networking.

Yet even simple sensors can generate a large amount of data with one installation often numbering hundreds of devices. In many instances of these data-driven systems, the gateway becomes the bottleneck and includes delays in what should be low-latency closed-loop data management. Streamlining this process is the requirement for large-scale deterministic networks and applications [38] as well as new architectures (see Section 10.3.3.2 and [22]). In addition, the reliance on the cloud for all advanced processing also implies continuous connectivity, something that cannot be ensured as IoT and automation move to rural or remote areas. The location of computational resources locally in the automation network to implement critical functions becomes essential; the on-device capabilities are rapidly increasing over time and supporting and even replacing cloud-based analysis [93]. The gateway is the ideal place to colocate local processing algorithms and schedule processing dynamically and deal with the scatted landscape of IoT devices [90]. As mentioned in Section 10.3.1, this is the objective of the data layer architecture: make resource objects securely accessible to applications across organizations and domains via joint local and cloud processing and composition [94]. And while this is beyond the scope of this chapter, the topic of the sematic web-architectures, efficient and secure data formatting and tagging are essential for operations and as inputs to automation and decision-making [95] local and remote function execution.

10.5.2.2 Industrial Automation: From Factories to Farms

Industrial automation has recently made significant progress because of the use of deep learning, image-based detection, and feature classification. Data-driven systems in industry often suffer from insufficient or sparse data and the training dataset needed to create a deep learning model will achieve the desired diagnostic accuracy. Even by using powerful computing resources, training algorithms can be very expensive due to the need to train for hours or days depending on the tasks or the level of accuracy needed. For this reason, the combination of cloud (for training and network planning) and (mobile) edge (for capture, execution, and data reduction) offers the best solution for advanced industrial intelligence as functionalities can be decomposed and can be developed independently locally and in the cloud [96].

Factory automation and machine control applications typically demand low end-to-end latency and small delay variation to meet critical closed loop control requirements and critical event responses as any break or suspension leads to major monetary or production losses as well as security breaches or emergencies; these are all reasons to process information locally [97, 98]. Vertical agriculture may not necessitate such closed-loop integration, but it needs to coordinate multiple sensors and cameras to maintain yields and production of "living things." The convergence of computing, operational technologies, and manufacturing networking is creating a new landscape in industry, and the softwarization has also reached the factory floor. Because of the rise of data-driven approaches, control functions traditionally carried out by fairly simple either proprietary or customized hardware and programmable logic controllers (PLCs) have been virtualized and moved to edge gateways or into the cloud, the "industrial cloudification" [27]. Control subsystems also run different cycle times, and applications need the result of these controls to perform efficiently. In turn, these different systems need to be integrated into internal system communications but also to other local or remote systems to enable the creation of combined or pipelined applications [95]. This means that pushing the computation and associated tasks (service description, discovery, and storage, for example) to the gateway is important to meet scheduling and execution of critical functions. This is a fundamental requirement to permit cooperation between various devices locally and remotely and keep synchronization between systems and cloud controllers alike [99]. Security and reliability are also increased by pushing some cybersecurity functions inside the local production facilities. In addition to improved and responsive data processing capabilities, distributed intelligence and flexible implementation can lead to a reduction in both capital expenditure (CAPEX) and operational expenditure (OPEX) as it increases the level of autonomous operation of the systems [95].

New dedicated processing include the family of algorithms known as "convolutional neural networks" [100–103] that can now be added to edge nodes software are spurring the development of the intelligent computerized factory. The implementation of these algorithms in a distributed fashion [90] in "intelligent gateways" and the availability of open source software [104] make it possible to combine multiple sensor data (data fusion) and linear feature extraction, main discriminant analysis, feature detection (via the CNN

[Convolutional Neural Network]), or data transformation in local CPU to rapidly and locally select and combine into the robust decision-making most likely to lead to a correct diagnostic or data interpretation. The final decision via ensemble classifiers and entropy-minimizing approaches can be left to cloud operations with ensemble views and other inputs related to economics or operator specifications. Predictive maintenance in smart manufacturing and crop maintenance in agriculture are to profit greatly from these approaches as large data volumes prevent real-time transmission to cloud-based analytics. Instead, with computing embedded in the local network, the data can be stored and processed locally, close to the capture site, to identify and respond to critical events and interface to higher layer analytics. The use of 6G as a next-generation wireless network to move the data in between production and decision-making locations is going to also become essential and the availability of these networks become part of the system specifications.

10.5.3 Computing Support for Networked Multimedia

Networked XR and advanced multimedia require that support functions assist user application by performing certain tasks (such as detailed object or event recognition) as well as analytics that are too compute intensive for some mobile devices or too data heavy for low-capacity connections. Hence, computing in the network can enhance next-generation multimedia as will be seen next.

10.5.3.1 Video Analytics

Cameras are everywhere, and with the recent pandemic, the use of videoconferencing and video messaging has raised significantly to the point that the network is becoming a scarce resource in many areas, and without better measurement of the combined streams and network usage, it is impossible to improve on the quality of experience. Hence, with this heightened usage comes the need for better video analytics and image analysis to measure traffic and ensure quality [105, 106]. As with many recent developments, these analytics take advantage of combined edge and cloud capabilities, distributed neural networks [103, 107], and camera-equipped devices and embedded CPU in cameras [108] usually used for surveillance [109] but with capabilities that can be ported to a wider set of applications in IoT. Custom streaming protocols that embed the analytics in the video stacks have also been proposed to co-optimize bandwidth and inference.

A typical video analytics application consists of a pipeline of video processing modules [110]. The pipeline configuration is composed of several modules to manage frame resolution and frame sampling rate among others. With the availability of neural networks everywhere in the network [103] and the availability of inexpensive embedded computing resources, local detection and analysis can be performed on Raspberry PI or another general-purpose CPU not just of dedicated hardware. One goal is for the local analytics to dynamically manage, and minimize, bandwidth usage. The combined streaming/analytics Chameleon protocol [111] can reduce the bandwidth by 4–23× while maintaining at least 95% inference accuracy.

10.5.3.2 Extended Reality and Multimedia

Augmented and virtual reality as they related to computing in the network will be combined in this section as XR except when mentioned otherwise, as while they target different applications, they all share a number of stringent delay and bandwidth requirements mostly for immersive experiences [81]. XR is a major example of the multisource multidestination problem that combines video, haptics, and tactile experiences. Generally, XR has been delivered mostly locally via combinations of computers and headsets with some cloud implementations being limited to time invariant imaging in one direction. XR is difficult to deliver with a client-server and cloud-based solution in interactive or networked mode multiparty and social interactions as it requires a combination of stream synchronization, low delays, and delay variations. However, many applications and services are being developed and deployed that could enhance the XR experience with sensor fusion and real-time interaction in virtual space: mobile and tactile immersive applications with time-sensitive data and high bandwidth for high-resolution images [112].

Hence, one main argument for adding computing in the path of XR flows is interactivity [113] and networked experiences in real-time (including multiparty gaming). To achieve networked XR, it is necessary to carefully evaluate the functionality that can be located in network nodes to enhance the experience within latency and data rates requirements, as well as the capability of programmable elements to respond to the dynamics of the services in an efficient, resilient, and secure manner [112, 114]. Other requirements include means to recover from losses and provide optimized caching in the cloud and rendering as close as possible to the user at the network edge [113]. The ICE-AR [14] project already introduced in this chapter (see Section 10.2.2) uses some in-network resources in its architecture for "acceleration as a service."

The localization of the networking resources in order to provide the service becomes an essential component of the overall architecture: computing and storage in the edge network allows to support cloud-based services for XR with much lower latency. Resource discovery and the federation of nodes to provide the required experience when one location needs added functionality, and intelligent coordination and analytics can also be envisaged and borrowed from next-generation IoT [95]. For example, network programmability could enable the use of joint learning algorithms across DC, edge computers, headsets, google, and glasses to allocate functionality and the creation of semipermanent datasets for display support and analytics for usage trending.

10.5.4 Melding AI and Computing for Measuring and Managing the Network

Advanced network management combines the top to bottom networked softwarization of SDN and NFV with ML [115]. It is now also possible to add computing in network devices to these innovations as well as resource discovery at the DC, the aggregation points, and the edge. The match-action feature of the P4 language on PISA switches (Section 10.3.2) enables to measure network performance (telemetry), fast reaction to critical events in data-driven management for systems and networks. It will be seen that this allows existing network services to get better performance and novel services and applications can be developed enabling better network performance.

10.5.4.1 Telemetry

Network telemetry is essential to monitor data traffic in a network-wide manner [116]. Telemetry is used to gather data on the use and performance of applications and application components, including feature usage, execution and function timing, and usage statistics and/or user behavior enabling advanced network analytics for network troubleshooting, congestion control, and path tracing. Telemetry information is essential for both operators and developers to receive data from network nodes while in service to forecast traffic and resource utilization. Users can also use telemetry data ability to provide users with control of what is happening with their data while in transit [117]. However, including telemetry in packets adds significant overhead and increased flow completion times and lower application-level performance [116–119].

More and more telemetry needs to encompass many types of traffic (including video and machine to machine) while keeping a low overhead low even with consistent updating necessary for real-time analytics and traffic management [73]. Hence, telemetry solutions need to trade-off resource usage especially overhead and flow-level measurement accuracy in a dynamic environment and among different network entities, including end-hosts and switches in the data plane as well as in centralized controllers in the control plane. Novel solutions based on programmable data planes such as PINT [118], OmniMon [120], and Unimon [96] as well as some commercial solutions [40] aim to optimize both accuracy and overhead reduction especially in DCs by using in-network computing and message distribution and coordination. Hence, telemetry is moving from "on the network" to "in the network": in-band measurement inside data packet using commodity network devices.

Without going into implementation details, these new telemetry systems all use the programmability and ensuing collaboration of the network nodes to execute telemetry operations and coordinate the results in such a way as meeting resource constraints while keeping high accuracy. Since these operations are fairly simple, P4 has been shown to enable this in band telemetry [121]. Experiments, testbeds, and early commercial deployments on commodity servers and PISA switches have shown the promise of the approach.

10.5.4.2 AI/ML for Network Management

Beyond the decision-making in IoT for production management, the network itself may profit from AI and ML executed into the nodes: intelligent in-network management. Predicting complex system behavior is a very challenging and difficult task because of the requirements of the different contributing elements: this is one challenge for ML [122] especially in large networks. The Internet is such a system and its management demands to respond dynamically to events. AI in support of network operations involves anticipating directions, performing trend analysis, and responding to the impact of mobility, and this is the closest as possible to where the decisions need to happen, hence in the network nodes themselves.

The goals of adding computation and AI to network management and the control plane deal with resource allocation, traffic forecasting, troubleshooting (including attacks) and local diagnostics, and the need for network to automatically identify courses of actions by creating datasets interpreting new data via trained ML algorithms [123]. Many network solutions rely increasingly on forwarding that is not based on traditional routing protocols with least cost routing but on computation of paths that are optimized for certain criteria – for example, to meet certain level objectives, to result in greater resilience, to balance utilization, to optimize energy usage, etc. Many of those solutions can be found in SDN, where a controller or path computation element computes paths that are subsequently provisioned across the network. However, such solutions generally do not scale to millions of paths and cannot be recomputed in sub-second time scales to take into account dynamically changing network conditions.

In this context, AI and ML are helping to create tools to analyze, model, and predict behaviors [124, 125]. While linear regression and other linear prediction algorithm can be used when the network is small, optimization algorithms derived from game theory and recommender systems that have been successfully applied to the stock market according to a set of preferences using strategies derived from gaming theory [100] can be applied to competing flows in the Internet. Research on recommender systems has focused on algorithms for recommending items for individual users, but the approach can also be used to determine a set of preferred features or to highlight a preferred solution with a larger set of inputs. Taking into consideration the dynamics and diversity of the requirements behavior, ML can be used with agent acting on behalf of one set of required outcomes. This can be associated with many one-to-one bilateral management schemes. The ML strategy can first be applied to obtain the maximum utility offer for each category of inputs and generate the most appropriate ranking for each requirement within the category. This is the cloud section of the management system. For action recommendation, it produces the list of ratings among all ranked priorities. As a result, the proposed actions will be based on the available data at that point dynamically and in the managed node ensuring that those actions can be taken in a timely manner and evolve over time as each dataset evolves with the support of the cloud infrastructure. That action can then be executed directly in the network node.

For ML to fully support network management, the match-action functionalities of programmable data planes need to include more complex actionable events. For example, real number operations are not available by default while they are primordial in many AI algorithms. Hence, one challenge facing embedded AI-related computations in the network node is both the capacity of the hardware to support network management operations and the limitations of the hardware programming languages. Virtualization helps but speed requirements in DCs and aggregation nodes limit the scope and the type of algorithms to be executed locally. The allocation of resources to management tasks is another challenge, and they need to be added to the already critical tasks performance in the node (forwarding, monitoring, filtering, etc.). In that case some functional decomposition of AI tasks between network elements, edge servers, or controllers has been also proposed [116]. This should be particularly useful for edge-based and IoT systems and implemented in the common data plane (Section 10.3.2). Optimization and adaption of AI algorithms for constrained environments are also possible [96, 117] with neural networks being one example [126].

10.5.5 Network Coding

Random linear network coding (RLNC) [127] supports what is called "in-network re-encoding," which means that streams and packets can be recombined in the network without having to decode them first.

RLNC operations involve combining the payload of multiple packets and multiplying them by random coefficients at any aggregation node in the network. However, this feature of RLNC is difficult to implement without more network computing advances such as the data plane programming. On one hand, limited programming resources targeted to store and forward limit the capability of adding the Galois field operations that are necessary for the coding and decoding operations. On the other hand, software implementations cannot be operated at line speed and limit the adoption of coding as a performance and security enhancing mechanism. Recent work [128] use programmable networking hardware to implement an RLNC implementation in data plane written in P4 for all packet algebraic functions.

10.6 Conclusion: 6G, the Network, and Computing

The networking community has struggled for quite a long time now with the ideas of a next-generation Internet [99]. Designing a next-generation network is made harder by the current reach of the Internet and the monetary interests of competing stakeholders. Hence, what we have experienced is mostly incremental changes and limited architectural advances. There have been several research projects proposed such as the well-known clean slate ICN [5], Mobility First [129], and the emerging FlexNGIA [22] that are facing the challenges of migrating from testbeds to a large-scale, high-performance production network.

However, the programmability and joint softwarization and computerization of networking functions now offer a timely solution to propel networking ahead [82]. The implementation of innovative architectural approaches in high-speed hardware improves their prospect for Internet-wide deployment. Combined with advances in SDN and virtualization, this clearly demonstrates that programmability in networks is here for many services and applications in a variety of domains.

As the new 6G networks are emerging, it becomes necessary to investigate how to harness and benefit from computing in the network to improve the next generation of networks. While adding computing inside network nodes may represent a disruption to the Internet architecture, it also provides new operational models to improve network and application performance as well as user experience. One way to better grasp the evolution of data plane programmability is to monitor the ongoing shift from controlled environments such as DCs toward edge computing. This can be regarded not as independent elements but as a cloud-edge continuum used to improve service availability and performance.

Thanks to recent research and product development, it is now feasible for networking to move beyond packet interception as the basis of network operations. There is a need for bridging the current divide, still inherently part of the 5G architecture, between the way server computing complexes are implemented and how switches and other smart networking devices such as NICs are programmed. While traditional routers employ rudimentary languages for programming, which could prevent the adding-on of functionality, there are also potential avenues with the additions of white boxes peripherals providing the richer programmability required to support emerging workloads.

The development of the PISA switches and the P4 language has generated both economic (including new companies) and academic (chairs and research programs) activities dedicated to the development of fast packet processing attached to network nodes and enable edge network analytics, ML, and deep learning. In the future such applications may also access to more general-purpose languages and underlying operating system facilities currently being developed. There is also a need to accommodate local and remote caches, dynamic control points, and various forms of data stewardship. These multiple data transformations raise important issues in security, privacy, and data provenance. The simple end-to-end source-destination model now fails to comply to the requirements of next-generation services in the 6G era. How the existing layering of protocols is affected by these considerations remains an open question, one that will need to be addressed: what should (or should not) be exposed across layers and APIs, including parameters that might affect QoS/quality of experience (QoE), orchestration dynamics, and mobility. All of these were briefly addressed in this chapter with some avenues for solutions and future research.

Discussing 6G while 5G is still very much a topic of academic research may seem preposterous. But it is important to note that 5G innovation was driven mainly by a set of services requiring high bandwidth, low delay, and high reliability but with until recently a fairly traditional approach to networking. In particular, the end-to-end principle paradigms of the past 20 years are not challenged in current implementation of 5G. In-network computing devices inserted on the ingress–egress path are currently not included in 5G architectures. As services are becoming more and more inherently distributed, with some elements running at the edge including the end device, some in the DC, 6G will requires an approach to the core-edge continuum that is decoupled from any single infrastructure- based platform and able to support thousands of sites with different capabilities. The ability to manage and orchestrate that many edge sites does require a fundamental rethinking of network management to include function execution and AI. This is still an open discussion topic and research challenges [130]. It is not outside the realm of logic to think that by the time 6G will be deployed not only current industrial deployment of computing in the network and research in academic testbeds will have clearly established the need for and the usefulness of in-network computing.

Hence, the softwarization and computerization of connected devices in networking, industry, and everyday life will continue. Robotics, automation, and even agriculture is increasingly dependent on the acquisition, processing, and interpretation of often large and more and more complex datasets. The continuing impact of large non-traditional operators in networking, such as GAFA, and their global network offerings, on earth and in space, will enable many of these data-driven services. In turn, this will also drive the demand for local and cloud processing to bring critical decision-making close to its needed location. Service providers may become "computation providers" [44] as the economics of connectivity may move from simply providing means for packets to move from point A to point B but also how and when these packets require computing and storage resources in transit.

Knowing the speed with which new computing services are deployed and the intense research activities driven by it, in-network computing will become an inherent part of the "network of the future." This chapter has thus attempted to review how the addition of computing elements in network nodes throughout the core-edge continuum is changing the networking landscape and bringing it closer to becoming a large data computing infrastructure. 6G may be the first deployment of the "network as a computing platform" [130].While innovation has for a long time focused above (e.g. the web) and below the network layer (wireless being a major focus) and at the end points (servers and associated technologies), it may now be time for inside the network programmability to promote networking innovation. In the upcoming 6G era, the network and the computer may become one again.

Acknowledgments

This chapter and the work that lead to it were made possible by the work, discussions, and contributions of the IRTF COIN community of researchers and developers. Thank you to Noël Crespi for the invitation to write this chapter. Finally, special thanks to Dirk Kutscher, Eve Schooler, Jianfe He, Lars Eggert, Noa Zilberman, David Oran, Faten Zhani, Edgar Ramos, Roberto Morabito, Laurent Ciavaglia, Alessandro Bassi, Marc Leclerc, Yves Daoust, and Ethan Mora for support, valuable inputs, and discussions about many of the topics included in this chapter.

References

1 QUIC Internet Working Group. https://quicwg.org/.

2 Soulé, R. (2018). Netcompute Workshop Keynote Address at Sigcomm 2018. Netcompute Worshop. ACM Sigcomm Conference, Budapest, Hungary, 2018. HYPERLINK "http://conferences.sigcomm.org/sigcomm/2018/files/slides/netcompute/2018-%2008-%2020-%20sigcomm.pdf" http://conferences.sigcomm.org/sigcomm/2018/files/slides/netcompute/2018-08- 20- sigcomm.pdf Rexford, J. (2018). Sigcomm 2018 Keynote Address. SACM Sigcomm Conference, Budapest, Hugary, 2018. https://youtu.be/t_5__v6CNYE?t=4652.

3 Fitzek, F., Granelli, F., and Seeling, P. (eds.) (2020). *Computing in Communication Networks: From Theory to Practice*, 1e. Elsevier.

4 5G Network Slicing. https://www.sdxcentral.com/5g/definitions/5g-network-slicing/.

5 Information Centric Networking (ICNRG), IRTF Working Group. https://datatracker.ietf.org/wg/icnrg/about/.

6 Rexford, J. (2018). Sigcomm 2018 Keynote Address. https://youtu.be/t_5__v6CNYE?t=4652.

7 Feamster, N., Rexford, J., and Zegura, E. (2014). The road to SDN: an intellectual history of programmable networks. *ACM SIGCOMM Computer Communication Review* 44 (2): 87–98.

8 Galis, A., Denazis, S., Brou, C., and Klein, C. (2004). *Programmable Networks for IP Service Deployment*. London: Artech House Books.

9 Tennenhouse, D. and Wetherall, D.J. (1996). Towards an active network architecture. *Computer Communication Review* 26: 5–18.

10 Mosko, M., Solis, I., and Wood, C. (2019). Content-centric networking (CCNx) semantics. RFC 8569 (July 2019). https://www.rfc-editor.org/info/rfc8569.

11 Named Data Networking (NDN). https://named-data.net/.

12 Yuan, X., Wang, X., Wang, J. et al. (2016). Enabling secure and efficient video delivery through encrypted in-network caching. *Journal on Selected Areas in Communications (JSAC)* 34 (8).

13 Marxzer, C., Scherb, C, and Tschudin, C. (2017). Introduction to named function networking. *Proceedings of the ACM ICN2017 Conference.*

14 Burke, J. (2018). ICN- enabled secure edge networking with augmented reality: ICE-AR. NDNCOM, Meeting, US NIST, Gaithersburg, Maryland, USA, 2018. https://www.nist.gov/news- events/events/2018/09/named- datanetworking-community- meeting- 2018, http://ice- ar.named- data.net/.

15 Król, M., Mastorakis, S., Oran, D., and Kutscher, D. (2019). Compute first networking: distributed computing meets ICN. *Proceedings of the 2019 ACM ICN Conference.* https://conferences.sigcomm.org/acm-icn/2019/proceedings/icn19-16.pdf.

16 Król, M., Habak, K., Kutscher, D., and Psaras, I. (2018). RICE: remote method invocation in ICN. *Proceedings of the 2018 ACM ICN Conference.* https://conferences.sigcomm.org/acm-icn/2018/proceedings/icn18-final9.pdf.

17 Kubernetes. Kubernetes.io

18 Kutscher, D. (2019). Compute- first networking (CFN): new perspectives on integrating computing and networking. Keynote Address, IEEE Consumer Communications & Networking Conference,11-14 January 2019, Las Vegas, USA https://ccnc2019.ieee- ccnc.org/program/keynotes.

19 Kutscher, D. Protocol design and socioeconomic realities. Personal blog. http://dirk-kutscher.info/blogroll/great- expectations/?fbclid=IwAR1jCHSqMtIcMHj1Ie_6cVPLEzSArFTN7YHc76tTt1Wp- a5fNZk6rxcPwxQ.

20 Kutscher, D., Karkkainen, T., and Ott, J. (2020). Directions for computing in the network. Working Internet Draft. Internet Research Task Force, Computing in the Network Research Group. https://datatracker.ietf.org/doc/draft- kutscher- coinrg- dir/.

21 Open Networking Foundation (ONF). Software-Defined Networking (SDN) Definition. https://www.opennetworking.org/.

22 FlexNGIA. https://www.flexngia.net/.

23 Arashloo, M.T., Koral, Y., Greenberg, M. et al. (2016). SNAP: stateful network-wide abstractions for packet processing. *Proceedings of the 2016 ACM SIGCOMM Conference*, 29–43.

24 NetCompute 2018: In-Network Computing Workshop, Sigcomm 2018. https://conferences.sigcomm.org/sigcomm/2018/workshop-netcompute.html.

25 Tokusahi, Y., Dang, H.T., Pedone, F. et al. (2019). The case for in-network computing on demand. In: *Proceedings of the Fourteenth EuroSys Conference 2019 (EuroSys '19)*, 1–16. New York, NY, USA, Article 21: Association for Computing Machinery. doi: https://doi.org/10.1145/3302424.3303979.

26 Zilberman, N., Moore Andrew, W., and Jon, C. (2016). From photons to big data applications: terminating terabits. *Philosophical Transactions of the Royal Society A* **374**: 20140445. http://doi.org/10.1098/rsta.2014.0445.

27 Automotive Edge Consortium. https://aecc.org/.

28 Agrawal, D. (2018). Data on the edge: leveraging the network edge for internet applications. https://www.cse.ust.hk/pg/seminars/S18/agrawal.html.

29 Schooler, E., Srikanteswara, S., and Foerster, J. (2017). ICN-WEN information centric-networking in wireless edge networks. Presentation at ICNRG@IETF-98 (March 2017). https://www.ietf.org/proceedings/98/slides/slides-98-icnrg-information-centric-networking-in-wireless-edge-networks-eve-schooler-00.pdf.

30 Hadoop Distributed File System. http://hadoop.apache.org/.

31 IPFS: Interplanetary File System. https://ipfs.io/.

32 Memcached. https://memcached.org.

33 Mapreduce. MapReduce: simplified data processing on large clusters. googleusercontent.com.

34 Tensorflow. tensorflow.org.

35 Spark. Spark.apache.org.

36 Pregel. github.com/igrigorik/pregel.

37 DryadLinq. https://www.microsoft.com/en-us/research/project/dryadlinq/.

38 Deterministic Networking (DETNET), IETF Working Group. https://datatracker.ietf.org/wg/detnet/about/.

39 Sigcomm (2020). https://conferences.sigcomm.org/sigcomm/2020/program.html.

40 Noviflow. www.noviflow.com.

41 Mastorakis, S. and Mtibaa, A. (2018). Towards service discovery and invocation in data-centric edge networks. *Proceedings of the IEEE 27th International Conference on Network Protocols (ICNP18)*. https://ieeexplore.ieee.org/abstract/document/8888081.

42 McBride, M., Kutscher, D., Schooler, E., and Bernardos, C. (2020). Edge data discovery for COIN. Working Internet-Draft, draft-mcbride-edge-data-discovery-overview. https://datatracker.ietf.org/doc/draft-mcbride-edge-data-discovery-overview/.

43 Salaht, F., Desprez, F., and Lebre, A. (2020). An overview of service placement problem in fog and edge computing. *ACM Computing Surveys* 53: 1–35.

44 Zilberman, N. (2017). Revolutionising computing infrastructure for citizen empowerment. *Proceedings of the Data for Policy Conference.*

45 Computing in the Network (COINRG), IRTF Research Group. https://datatracker.ietf.org/rg/coinrg/about/.

46 Echeverría, S., Klinedinst, D., Williams, K., and Lewis, G.A.. (2016). Establishing trusted entities in disconnected edge environments. *Proceedings of the IEEE/ACM Symposium on Edge Computing (SEC)*, 51–63.

47 Fink, I. and Wehrle, K. (2020). Enhancing security and privacy with in-network computing. Working Internet Draft, draft-fink-coin-sec-priv-00. https://tools.ietf.org/html/draft-fink-coin-sec-priv-00.

48 Kang, Q., Morrison, A., Tang, Y. et al. (2020). Programmable in-network security for context-aware BYOD policies. *Proceedings of the 29th USENIX Security Symposium (USENIX Security 20)*. https://www.usenix.org/conference/usenixsecurity20/presentation/kang.

49 Boyle, E., Couteau, G., Gilboa, N. et al. (2017). Homomorphic secret sharing: optimizations and applications. *Proceedings of the 2017 ACM SIGSAC Conference on Computer and Communications Security (CCS'17)*, 2105–2122.

50 Dwork, C. Differential privacy: a survey of results. In: *Theory and Applications of Models of Computation. Lecture Notes in Computer Science*, vol. 4978 (eds. N. Agrawal, D. Du, Z. Duan, et al.), 1–19. Berlin Heidelberg: Springer.

51 Decentralized Internet Infrastructure (DINRG), IRTF Research Group. https://datatracker.ietf.org/rg/dinrg/about (accessed 15 June 2021).

52 Peterson, L. (2019). 5G networks a system approach. https://5g.systemsapproach.org/index.html#.

53 Jeyakumar, V., Alizadeh, M., Geng, Y. et al. (2014). Millions of little minions: using packets for low latency network programming and visibility. *Proceedings of SIGCOMM 2014*. http://conferences.sigcomm.org/sigcomm/2014/program.php.

54 Goldschmidt, A. (2020). What's beyond 5G? Invited talk, ACM Sigcomm Conference (online), 2020. https://www.youtube.com/watch?v=iq1TF4NHhwg

55 COAP: constrained application protocol. coap.technology.

56 MQTT: Message Queuing Telemetry Transport. mqtt.org.

57 Storm. https://storm.apache.org/ (accessed 15 June 2021).

58 Madhavapeddy, Mortier, R., Rotsos, C. et al. (2013). Unikernels: library operating systems for the cloud. In: *Proceedings of the eighteenth international conference on Architectural support for programming languages and operating systems (ASPLOS '13)*, 461–472. New York, NY, USA: Association for Computing Machinery. doi: https://doi.org/10.1145/2451116.2451167.

59 Madhavapeddy, A., Leonard, T., Skjegstad, M. et al. (2015). Jitsu: Just-In-Time Summoning of Unikernels. *Proceedings of NSDI15*. https://www.usenix.org/system/files/conference/nsdi15/nsdi15-papermadhavapeddy.pdf.

60 Manzalini, A. and Crespi, N. (2016). An edge operating system enabling anything-as-a-service. *IEEE Communications Magazine* 54 (3): 62–67.

61 RIOT. riot.org.

62 Bosshart, P., Gibb, G., Kim, H.-S. et al. (2013). Forwarding metamorphosis: fast programmable match-action processing in hardware for SDN. *Proceedings of ACM SIGCOMM 2013*.

63 Tofino Switch. https://barefootnetworks.com/products/brief-tofino/.

64 Cavium XPliant. https://www.openswitch.net/cavium/.

65 Trident 4. https://www.broadcom.com/products/ethernet-connectivity/switching/strataxgs/bcm56880-series.

66 Cisco Nexus. https://www.cisco.com/c/en/us/products/collateral/switches/nexus-3000-series-switches/datasheet-c78-740836.html.

67 Bosshart, P., Daly, D., Gibb, G. et al. (2014). P4: programming protocol-independent packet processors. *ACM SIGCOMM Computer Communication Review* 44: 87–95.

68 Sivaraman, A. Cheung, M. Budiu, C. et al. (2016). Packet transactions: high-level programming for line-rate switches. *Proceedings of the 2016 ACM SIGCOMM Conference*, 15–28 (22–26 August 2016).

69 Jin, X., Li, X., Zhang, H. et al. (2017). NetCache: balancing key-value stores with fast in-network caching. *Proceedings of SOSP2017*. https://www.cs.jhu.edu/~xinjin/files/SOSP17_NetCache.pdf.

70 P4 Language Consortium. https://p4.org.

71 P4 Brainstorming (2020). https://github.com/p4lang/p4-spec/wiki/LDWG-Brainstorming.

72 Chen, Y., Yen, L., Wang, W. et al. (2019). P4-enabled bandwidth management. 2019 20th Asia-Pacific Network Operations and Management Symposium (APNOMS), 1–5.

73 Luo, S., Yu, H., and Vanbever, L. (2017). Swing state: consistent updates for stateful and programmable data planes. *Proceedings of ACM SOSR 2017*.

74 Gao, J., Zhai, E., Liu, H.H. et al. (2020). Lyra: a cross-platform language and compiler for data plane programming on heterogeneous ASICs. In: *Proceedings of the Annual conference of the ACM Special Interest Group on Data Communication on the applications, technologies, architectures, and protocols for computer communication (SIGCOMM '20)*, 435–450. New York, NY, USA: Association for Computing Machinery. doi: https://doi.org/10.1145/3387514.3405879.

75 Van Meter, R. (2014). *Quantum Networking*. Wiley.

76 Caleffi, M., Cacciapuoti, A.S., Bianchi, G. (2018). Quantum internet: from communication to distributed computing! *Proceedings of the 5th ACM International Conference on Nanoscale Computing and Communication (NANOCOM'18)*. arXiv:1805.04360.

77 Kozlowski, W., Wehner, S., Van Meter, R. et!al. Architectural principles for a quantum internet. Working Internet Draft, Internet Research Task Force, Computing in the Network Research Group. Quantum Internet Research Group, https://datatracker.ietf.org/doc/draft-irtf-qirg-principles/ (accessed August 15 2021)

78 Montpetit, M.J. (2018). How will the next generation network? TEDX Beacon Street Salon, Cambridge MA, August 2018. https://tedxbeaconstreet.com/videos/how-will-the-future-generation-network/ (accessed August 15 2021).

79 Things to Thing (T2T), IRTF Research Group. https://datatracker.ietf.org/rg/t2trg/about/.

80 Graham, R.L., Bureddy, Devendar, Lui, Pak et al. (2016). Scalable hierarchical aggregation protocol (SHArP): a hardware architecture for efficient data reduction. *Proceedings of the 2016 COM-HPC*. https://ieeexplore.ieee.org/document/7830486.

81 Hinds, A. (2018). The near future of immersive experiences: where we are on the journey, what lies ahead, and what it takes to get there. SIGCOMM 2018 Workshop on AR/VR (August 2018). http://conferences.sigcomm.org/sigcomm/2018/workshop-arvr.html.

82 Sapio, A., Abdelaziz, I., Aldilaijan, A. et al. (2017). In net computing is a dumb idea whose time has come. Hotnets 2017. https://dl.acm.org/citation.cfm?id=3152461.

83 Li, X., Sethi, R., Kaminsky, M. et al. (2016). Be fast, cheap and in control with SwitchKV. NSDI'2016. https://dl.acm.org/citation.cfm?id=2930614.

84 Fa, B., Lim, H., Andersen, D.G. and Kaminsky, M. (2011). Small cache, big effect: provable load balancing for randomly partitioned cluster services. *Proceedings of the 2011 ACM SOCC*. www.istc-cc.cmu.edu/publications/papers/2011/loadbal-socc2011.pdf.

85 Zsolt, I., Sidler, D., Alonso, G., and Vukolić, M. (2016). Consensus in a box. *Proceedings of NSDI 2016*. https://dl.acm.org/citation.cfm?id=2930639.

86 Dang, H., Canini, M., Pedone, F., and Soulé, R. (2016). Paxos made switch-y. ACM Sigcomm Computer Communication Review 2016. https://www.sigcomm.org/sites/default/files/ccr/papers/2016/April/0000000-0000002.pdf.

87 Dang, H.T., Sciascia, D., Canini, M. et al. (2015). NetPaxos: consensus at network speed. SOSR15. https://mcanini.github.io/papers/netpaxos.sosr15.pdf.

88 Li, J., Michal, E., and Ports, D.R.K. (2017). Eris: coordination-free consistent transactions using in-network concurrency control. SOSP 2017. https://syslab.cs.washington.edu/papers/eris-sosp17.pdf.

89 Ports, D.R.K., Li, J., Liu, V. et al. (2015). Designing distributed systems using approximate synchrony in data center networks. *Proceedings of NSDI 2015*. https://www.usenix.org/node/188949.

90 Ramos, R.M. and Kainulainen, J.P. (2019). Distributing intelligence to the edge and beyond [research frontier]. *IEEE Computational Intelligence Magazine* 14: 65–92.

91 Li, H., Ota, K., and Dong, M. (2018). Learning IoT in edge: deep learning for the internet of things with edge computing. *IEEE Network* 32 (6): 96–101. https://doi.org/10.1109/MNET.2018.1700202.

92 Ogrodowczyk, Ł., Belter, B., and Leclerc, M. (2016). IoT ecosystem over programmable SDN infrastructure for Smart City applications. *Proceedings of EWSDN16*. https://noviflow.com/wp-content/uploads/EWSDN16_demo_paper_final-copy.pdf.

93 Ren, J., Guo, H., Xu, C., and Zhang, Y. (2017). Serving at the edge: a scalable IoT architecture based on transparent computing. *IEEE Network* 31 (5): 96–105.
94 Zhao, J., Jianxin, T., Tiplea, R. et al. (2018). Data analytics service composition and deployment on IoT devices. *Proceedings of MobiSys'18: 16th Annual International Conference on Mobile Systems, Applications and Services.*
95 Ramos, E., Schneider, T., Montpetit, M.J., and De Meester, B. (2020). Semantic descriptors for intelligence services. Under review.
96 Liu, Z., Manousis, A., Vorsanger, G. et al. (2016). One sketch to rule them all: rethinking network flow monitoring with UnivMon. *Proceedings 2016 ACM SIGCOMM Conference.*
97 Rüth, J., Glebke, R., Wehrle, K. et al. (2018). Towards in-network industrial feedback control. Netcompute Workshop, Sigcomm 2018.
98 Vestin, J., Kassler, A., and Åkerberg, J. (2018). FastReact: in-network control and caching for industrial control networks using programmable data planes. https://arxiv.org/abs/1808.06799.
99 Peterson, L., Anderson, T., Culler, D., and Roscoe, T. (2003). A blueprint for introducing disruptive technology into the internet. *ACM SIGCOMM Computer Communication Review* 33 (1).
100 Ben-Porat, O. and Tennenholtz, M. (2018). A game-theoretic approach to recommendation systems with strategic content providers. https://papers.nips.cc/paper/7388-a-game-theoretic-approach-to-recommendation-systems-with-strategic-content-providers.pdf.
101 Cohen, J.P., Boucher, G., Glastonbury, C.A. et al. (2017). Count-ception: counting by fully convolutional redundant counting. *Proceedings of the ICCV2017 Workshop on BioImage Computing.* arXiv:1703.08710.
102 Saha, S. (2018). A comprehensive guide to convolutional neural networks. https://towardsdatascience.com/a-comprehensive-guide-to-convolutional-neural-networks-the-eli5-way-3bd2b1164a53.
103 Teerapittayanon, S., McDanel, B., and Kung, H. (2017). Distributed deep neural networks over the cloud, the edge and end devices. *Proceedings of the IEEE 37th International Conference on Distributed Computing Systems (ICDCS17)*, 328–339.
104 Open Neural Network Exchange. https://onnx.ai.
105 Ananthanarayanan, G., Bahl, V., Bodík, P. et al. (2017). Real-time video analytics: the killer app for edge computing. *IEEE Computer* 50 (10): 58–67. https://www.microsoft.com/en-us/research/publication/real-time-video-analytics-killer-app-edge-computing.
106 Howard, G., Zhu, M., Chen, B. et al. (2017). Mobilenets: efficient convolutional neural networks for mobile vision applications. arXiv:1704.04861.
107 Chang, M.A., Panda, A., Bottini, D. et al. (2018). Network evolution for DNN. SysML. https://www.sysml.cc/doc/182.pdf.
108 Naderiparizi, S., Zhang, P., Philipose, M. et al. (2016). Glimpse: a programmable early-discard camera architecture for continuous mobile vision. *Proceedings of the 15th Annual International Conference on Mobile Systems, Applications, and Services (Mobisys)*, 292–305.
109 Zhang, T., Chowdhery, A., Bahl, V. et al. (2015). The design and implementation of a wireless video surveillance system. *Proceedings of Mobicom'15: 21st Annual International Conference on Mobile Computing and Networking*, 426–438.
110 Jiang, J., Ananthanarayanan, G., Bodik, P. et al. (2018). Chameleon: scalable adaptation of video analytics. *Proceedings of Sigcomm 2018.*

111 Pakha, C., Chowdhery, A., and Jiang, J. (2018). Reinventing video streaming for distributed vision analytics. *Proceedings of HotClouds 2018.*

112 Montpetit, M.J. (2020). In network computing enablers for extended reality. Working Internet Draft. https://datatracker.ietf.org/doc/draft-montpetit-coin-xr/.

113 Bastug, E., Bennis, M., Médard, M., and Debbah, M. (2017). Towards interconnected virtual reality: opportunities, challenges and enablers. https://arxiv.org/pdf/1611.05356.pdf.

114 Soh, L., Burke, J., and Zhang, L. (2018). Supporting augmented reality (AR): looking beyond performance. *Proceedings of the ACM SIGCOMM 2018 Workshop on Virtual Reality and Augmented Reality (VR/AR Network 2018).*

115 Clemm, M., Zhani, F., and Boutaba, R. (2020). Network management 2030: operations and control of network 2030 services. Journal of Network and Systems , Vol. 28, pp. 721–750. https://doi.org/10.1007/s10922-020-09517-0

116 Gupta, A., Harrison, R., Canini, M. et al. (2018). Sonata: query-driven streaming network telemetry. *Proceedings 2018 ACM SIGCOMM Conference.*

117 Yang, T., Jiang, J., Liu, P. et al. (2018). Elastic sketch: adaptive and fast network-wide measurements. *Proceedings 2018 ACM SIGCOMM Conference.*

118 Basat, R.B., Ramanathan, S., Li, Y. et al. (2020). PINT: probabilistic in-band network telemetry. *Proceedings of SIGCOMM2 2020.*

119 Satyanarayanan, M., Simoens, P., Xiao, Y. et al. (2015). Edge analytics in the internet of things. *IEEE Pervasive Computing* 14 (3): 24–31.

120 Huang, Q., Sun, H., Lee, P.P.C. et al. (2020). OmniMon: re-architecting network telemetry with resource efficiency and full accuracy. *Proceedings of SIGCOMM 2020.*

121 P4 Applications Working Group: In-band Network Telemetry (INT) Dataplane Specification. https://github.com/p4lang/p4-applications/blob/master/docs/telemetry_report.pdf.

122 Sapio (2019). Scaling distributed machine learning with in-network aggregation. https://arxiv.org/abs/1903.06701.

123 Rusek, K., Suárez-Varela, J., Mestres, A. et al. (2019). Unveiling the potential of Graph Neural Networks for network modeling and optimization in SDN. *Proceedings of the 2019 ACM Symposium on SDN Research*, 140–151.

124 Cohen, J.P., Lo, H.Z., and Ding, W. (2016). RandomOut: using a convolutional gradient norm to rescue convolutional filters. International Conference on Learning Representations (ICLR 2016). arXiv:1602.0593.

125 Samek, W., Wiegand, T., and Müller, K.R. (2017). Explainable artificial intelligence: understanding, visualizing and interpreting deep learning models. arXiv:1708.08296.

126 Liang, S., Yin, S., Liu, L. et al. (2018). FP-BNN: binarized neural network on FPGA. *Neurocomputing* 275: 1072–1086.

127 Lun, D.S., Médard, M., Koetter, R., and Effros, M. (2008). On coding for reliable communication over packet networks. *Physical Communication* 1 (1): 3–20.

128 Gonsalves, D., Signorello, S., Ramos, F.M.V., and Médard, M. (2019). Random linear network coding on programmable switches. *Proceedings of IEEE-ANCS 19.* https://arxiv.org/pdf/1909.02369.

129 Mobility First Future Internet Architecture. http://mobilityfirst.winlab.rutgers.edu/.

130 McKeown, N. (2020). The network as a programmable platform: fertile new ground for networking research. Invited Talk, Sigcomm 2020.

11

An Approach to Automated Multi-domain Service Production for Future 6G Networks

Mohamed Boucadair, Christian Jacquenet, and Emmanuel Bertin

Orange Innovation, France

11.1 Introduction

11.1.1 Background

The advent of telco cloud infrastructures, where network, storage, and computational resources may be disseminated all over the Internet and hosted in various cloud infrastructures is a business catalyst that should facilitate the dynamic production of services at large scale. In addition to traditional connectivity services that may span multiple domains, the scope of multifunction services extends beyond the borders of a single domain that is operated by a single administrative entity.

As a reminder, the roaming services provided by mobile operators since the very beginning of mobile communication networks is one example of inter-domain connectivity services, where the user can place a call from any country as long as the local operator has contracted a roaming agreement with the home operator that provides the mobile connectivity service that is subscribed by the aforementioned user.

Likewise, inter-domain Virtual Private Network (VPN) services that rely upon a basic Border Gateway Protocol/Multi-Protocol Label Switching (BGP/MPLS) design *a la* RFC 4364 [1] have been proposed by many incumbent operators for quite some time. Such VPNs assume the allocation of Provider edge (PE) resources that are operated by partnering (VPN) service providers, and which connect the customer premises to the inter-domain VPN, wherever these premises may be. The typical downside of such services is the time to deliver the service. Since inter-domain services involve different service providers by definition, the negotiation that takes place to reserve, allocate, and operate resources operated by different parties takes time. More than 20 years of inter-domain VPN experience shows that such time is usually in the magnitude of weeks, if not months before the VPN service is deployed, up and running. Such delivery time is also aggravated by the need to set up physical attachment points to serve local sites.

Shaping Future 6G Networks: Needs, Impacts, and Technologies, First Edition.
Edited by Emmanuel Bertin, Noel Crespi, and Thomas Magedanz.

The recent introduction of network automation with the deployment of Software Defined Networking (SDN) [2] techniques in the service delivery procedures along with Network Functions Virtualization (NFV) aims at improving the time it takes to deploy a service. Service delivery times can be optimized if automation techniques can be globally applied, from the service parameter exposure and negotiation to service operation. Focusing on the service provisioning phase is essential but may not be sufficient to optimize the overall service activation and delivery if it is not coupled with optimization means to expose service capabilities to customers, capture their requirements, and map them with a set of network capabilities.

With the help of artificial intelligence that may contribute to the smooth transition between one service operation step to another (Figure 11.1), the automated production of services at the scale of a domain is becoming a reality.

Many network operators worldwide have disclosed their SDN/NFV deployment strategy, and some (embryonic) forms of automated service deployment have bloomed all over the world.

But automating the production of inter-domain services for 6G networks is a completely different story.

11.1.2 The Need for Multi-domain 6G Networks

6G networks should heavily rely on automation techniques for the production of services, like those introduced in the previous chapters. However, these future 6G services will be mostly operated in a multi-domain environment. While legacy mobile networks used to only consider the multiple domains operated by carriers for the delivery of roaming services (and of course for interconnection purposes), 5G networks introduce the split between public and private domains, where private 5G networks might be operated by

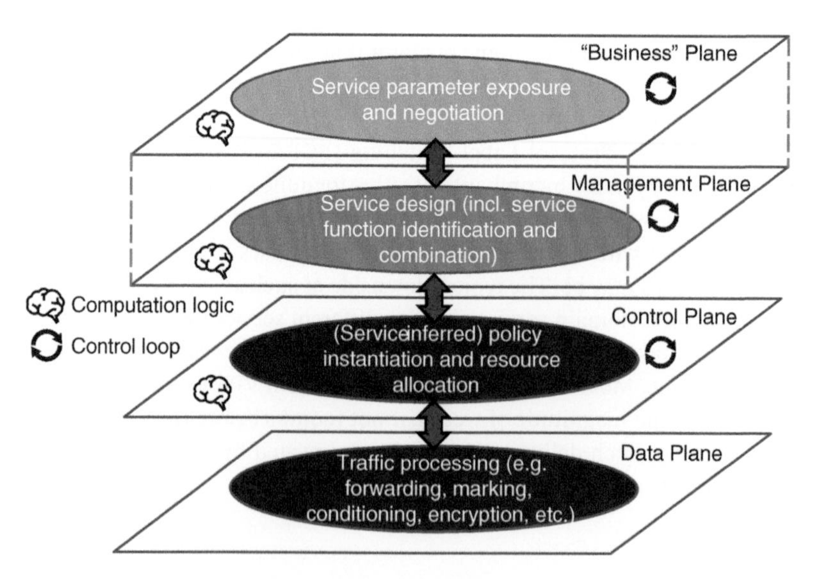

Figure 11.1 Automated service delivery procedure.

carriers but also industrial actors to address their own needs. 6G networks should amplify this disaggregation trend, with a multitude of players providing connectivity and resources, wherever they may be: access networks (e.g. integration of third-party access providers for local connectivity or relying on specific technologies like satellite), edge resources, and core networks.

Thus, future 6G services are unlikely to be restricted to a single domain but should rather span multiple domains, whatever the actors operating these domains. Let us discuss some of the use cases introduced in the first chapter of this book:

- Digital twin: a machine located in a factory relying on a private industrial 6G network is remotely operated or inspected by a teleworker through a public 6G network.
- Health services: For example, a telemedicine service involves a private hospital 6G network and accesses specific computing resources hosted in edge-based secured infrastructures, a public carrier network, and a local provider for last mile connectivity purposes.
- Intelligent transport services: Data generated through a city-owned 6G network of urban sensors are delivered with an ultralow latency to various vehicles (e.g. unmanned cars or aerial vehicles) or devices (e.g. for vulnerable road users). The aforementioned data are thus forwarded over different public 6G networks.

These examples draw the perspective of multi-domain 6G networks, where the value of a carrier will rely not only on its radio coverage but also on its capacity to deliver in nearly real-time a secure service across multiple domains. Depending on the nature of the service, these various domains should be dynamically selected, based on technical, business, and operational considerations (cf. Section 11.2.4). An important issue raised by the design of 6G services is thus the ability to dynamically identify the domains (hence the partners) that will be involved in the production and the operation of the resources that will compose these 6G services. This raises several challenges.

11.1.3 Challenges of Multi-domain Service Production and Operation

Whenever multiple parties are involved in the provisioning of a service, various issues arise. To name but a few:

- How to identify/discover the most relevant partners, i.e. the service providers that can provide the resources that are required to best accommodate the service's requirements and constraints within a given delivery scope (e.g. geographical footprint)?
- How can a trust model be enforced, where all involved parties can reliably collaborate to deliver the service (e.g. by preventing data leaks and enforcing consistent security policies)?
- What kind of information needs to be exchanged between partners to operate the service (e.g. assurance notifications, event notifications, and planned maintenance)?
- What are the boundaries in terms of respective responsibility whenever an event (link failure, traffic congestion, resource overload, etc.) affects the operation of a service? This is exacerbated when a third party is involved (e.g. Voice over Internet Protocol [VoIP] interconnect between two partners, each relying on different underlying connectivity providers).

- What kind of guarantee can be provided as far as the consistency of the decisions (in terms of resource allocation or policy enforcement, for example) made by the involved computation logics is concerned?
- What procedure(s) should be enforced for testing and troubleshooting purposes?

This chapter aims at exploring some of these issues. It also discusses the technical options that may address some of them.

The remainder of the chapter is as follows:

- Section 11.2 introduces frameworks, roles, and assumptions related to the dynamic delivery of multiparty, multi-domain services.
- Section 11.3 discusses the various steps that are required for the delivery of multiparty, multi-domain services. It also sheds some light on the techniques that may be used to proceed with the execution of these steps.
- Section 11.4 presents an application of multi-domain service design, delivery, and operation concepts to the specific case of a hierarchical, multi-domain service function chaining (SFC) [3].
- Section 11.5 identifies a set of pending issues and details the corresponding research challenges.
- Section 11.6 concludes the chapter with a few comments on the evolution perspectives.

11.2 Framework and Assumptions

11.2.1 Terminology

This chapter uses the following terms.

- **Business owner**: the administrative entity (e.g. a service provider, a network operator, or a cloud provider) that is the privileged contact for a customer to subscribe to a (multi-party, multi-domain) service (e.g. an inter-domain VPN service or an inter-domain network slice). The customer is likely to be unaware that the service he/she requests involves resources located in distinct administrative domains.
- **Business partner**: an administrative entity (a network operator, a cloud provider, a service provider, etc.) that partners with a business owner for the delivery of a multi-domain service. A business partner typically establishes an agreement with the business owner of a particular multi-domain service.
- **Customer**: an entity that subscribes to a service. Within the context of this chapter, a customer only interfaces with the business owner.
- **Digital sovereignty**: from a technical standpoint, digital sovereignty implies the exclusive competence of a network operator, a service provider, etc. for the design, deployment, and operation of resources that compose its infrastructure according to its footprint.
- **Domain**: a set of resources (e.g. routers, virtual machines, virtual network functions, and servers), which are operated by a single administrative entity such as a network operator, a cloud provider, or a service provider.

- **Multi-domain service**: a service that relies upon a set of resources spread over multiple domains operated by different, possibly competitive, entities. The delivery and the operation of such service thus involves multiple parties (the business owner and its business partners for that particular service).
- **Resource**: refers to the network, CPU, and/or storage resources.
- **Single-domain service**: a service that relies upon a set of resources that reside within a single domain operated by the business owner of the service.
- **Tenant**: a legal entity that can be a customer, a business owner, or a business partner. A tenant is responsible for the resources that it owns. A tenant may also be granted access (either read-only or read-write) to resources operated by another tenant (e.g. a business owner), depending on various criteria that include (but are not limited to) the nature of the service, the tenant's policy and digital sovereignty, the terms of the contract that may have been established between two tenants, etc.; of course, the rights and duties of a given tenant may very well raise specific issues as far as the delivery and the operation (including the reconfiguration) of a multi-domain service is concerned. For example, granting read-write access rights to an inter-domain VPN service customer may affect the consistency of decisions made by business partners. If VPN traffic redirection is in effect because the customer decided so, this decision may affect the traffic forwarding policies enforced within a given domain.

11.2.2 Assumptions

11.2.2.1 SDN-enabled Domains

Each domain is SDN-enabled. This means that dynamic resource management (whatever the resource and whether it is a physical or a virtual resource) is handled by an SDN computation logic. Such entity, often designated by SDN controller, is responsible for:

- Maintaining a global, network-wise, accurate, and up-to-date vision of the available resources and capabilities.
- Mapping service requirements to available resources.
- Dynamically allocating resources and instantiating policies that pertain to the delivery of a service, either single domain or multi-domain.
- Dynamically forwarding policy-provisioning information (including configuration instructions) to the resources that are involved in the delivery of a single- or multi-domain service.
- Dynamically readjusting decisions (e.g. selection of new or additional Service Function (SF) instances) and triggering resource allocation cycles according to several criteria that include (but are not limited to) the enforcement of a multi-year network planning policy, network-originated notifications (e.g. detection of a link failure or a traffic congestion), and the number of service subscription requests that need to be processed over a given period of time.

There may be several SDN computation logics per domain. Each of them may be responsible for the dynamic management of a specific set of resources, according to local deployment considerations (e.g. organic or organizational considerations). Typically, a network operator could use an SDN controller for the dynamic computation and establishment of optical paths while another SDN controller is used for the dynamic computation and establishment of Internet Protocol (IP) routes, assuming a federative Generalized Multi-Protocol Label Switching (GMPLS) control plane that facilitates the consistency of the decisions made by both controllers.

Within the context of an inter-domain service, the interaction between an SDN controller and the resources it manages is local to each domain. From that standpoint, it is not required that the communication interfaces used between an SDN controller and the resources it manages should also be used between SDN controllers of different domains.

11.2.2.2 On-service Orchestrators

Likewise, the SDN computation logic of a domain may be assisted by an orchestrator that typically resides in the management plane and which is responsible for processing the customer's service orders, by structuring the single- or multi-domain service. Structuring a service means the identification of the elementary SFs (forwarding, routing, Quality of Service (QoS), security, etc.) that will compose the service. The orchestrator may also be responsible for dynamically structuring the service-inferred SFC, i.e. identify the order in which the aforementioned elementary SFs need to be invoked so that traffic forwarding within the domain should be achieved to meet the service objectives.

In addition, the orchestrator may use the outcomes of a successful yet dynamic negotiation [4] between the customer and the business owner to feed its computation logic so that dynamic service structuring can be facilitated and eventually make sure the designed service exactly matches the terms of the completed negotiation (e.g. in terms of latency, maximum packet loss rate, and inter-delay packet variation).

11.2.2.3 Any Kind of Multi-domain Service, Whatever the Vertical

No specific assumption is made in this chapter about the nature of the multi-domain service. That being said, the typical market segment is the corporate market whatever the vertical – retail, finance, industry, etc.

Of course, one could consider the access to the Internet as a very basic form of a multi-domain (connectivity) service, but the fact is that such service does not assume any specific commitment from business partners. Internet traffic is forwarded *a la* best effort and any specific guarantee for such service remains the sole responsibility of the business owner (e.g. guarantees about the access rate may be contractually provided to the customer).

Yet, there are two typical multi-domain services that will likely be supported by future 6G networks:

- Inter-domain VPN services: These services have been deployed for quite some time. Their design typically assumes BGP/MPLS capabilities. Forwarding and routing functions are at the core of the service, but they are very often complemented by a set of capabilities that include security capabilities (traffic encryption, traffic filtering, [outsourced] firewalling) and QoS capabilities (traffic marking and classification, traffic scheduling, and conditioning). Such additional capabilities reflect customer's traffic prioritization and security policies. Inter-domain VPN customers also solicit cloud resources (either private, public, or both) for workflow management purposes. Several features thus compose a VPN service package (e.g. customized access to cloud realms for application-specific needs and storage)

- Network Slice-as-a-Service (NSaaS) services [5]: The concept of slice has been reintroduced with the 5G specification effort. It is generally defined as an overlay that refers to a customized set of resources over a shared physical infrastructure. An NSaaS service may include those that have the most stringent constraints like immersive services that rely upon augmented reality (AR) and virtual reality (VR) techniques and that are known to be bandwidth- and CPU-greedy.

11.2.3 Roles

As introduced in Section 11.2.1, the delivery and the operation of multi-domain services involves several tenants. Namely:

- The customer that subscribes to the multi-domain service.
- The business owner that directly interfaces with the customer and that is the legal entity that is responsible before the customer for the delivery and the operation of the service. Traditionally, a contractual agreement is established between the customer and the business owner: this is the service-level agreement (SLA).
- The business partners that interface with the business owner and that are involved in the allocation of the resources that are required by the multi-domain service. The relationship between a business owner and a business partner can also be defined as a customer–business owner relationship (to some extent). Such relationship indeed involves a commitment of each business partner toward the business owner, so that the allocated resources accommodate the requirements and constraints of the multi-domain service, as expressed by the customer. This model can be seen as a brokering model where the actual data to be handled by a business partner in the context of a given SLA can be received from (or forwarded to) domains that are not managed by the business owner. The SLA should include the information that identifies traffic that will be granted access to the network facilities that support the service subscribed by the customer. Other models can be considered (stitching model, where business partners operate adjacent domains), but such models are not elaborated in this chapter.

The design and the delivery of a multi-domain service thus leads to a multi-stage approach, as depicted in Figure 11.2.

Note that each party involved in the delivery of the service is responsible for the management (configuration, invocation, operation) of the resources it owns according to its digital sovereignty and which will contribute to the delivery of the service. In other words, one basic assumption is that the business owner does not have a direct read-write access to the resources allocated by a business partner: during the multi-domain service operation, any event that may affect some of the resources allocated by a business partner should dynamically trigger a resource allocation cycle that is placed under the sole responsibility of the aforementioned business partner.

But the processing of such event should be reported to the business owner along with any other relevant information (e.g. nature, location, status of the resources that have been

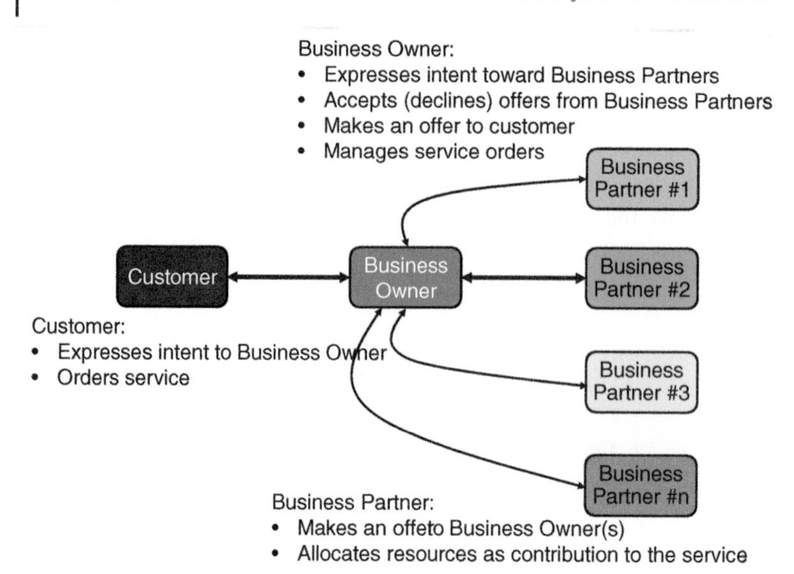
Business Owner:
• Expresses intent toward Business Partners
• Accepts (declines) offers from Business Partners
• Makes an offer to customer
• Manages service orders
Business Partner #1
Customer
Business Owner
Business Partner #2
Customer:
• Expresses intent to Business Owner
• Orders service
Business Partner #3
Business Partner #n
Business Partner:
• Makes an offeto Business Owner(s)
• Allocates resources as contribution to the service

Figure 11.2 Roles of and interactions between tenants of a multi-domain service.

affected, as well as the nature of the decisions made by the SDN/orchestration computation logics operated by the business partner).

11.2.4 Possible Multi-domain Service Delivery Frameworks

The design, delivery, and operation of a multi-domain service assume a business owner and one or several business partners. The collaboration between the business owner and the business partners for the delivery and the operation of a given multi-domain service leads to several possible frameworks.

11.2.4.1 A Set of Bilateral Agreements

For the delivery of the multi-domain service, the business owner establishes bilateral agreements with the business partners involved. This is the current practice for inter-domain VPN services, where the business owner usually contacts the incumbent operator of the countries where PE resources need to be allocated to connect the local customer premises. The set of bilateral agreements eventually leads to the allocation of PE resources wherever they are needed. These agreements may also comprise additional key performance indicators (KPI), like QoS parameters.

11.2.4.2 A Set of Bilateral Agreements by Means of a Marketplace

A marketplace can be defined as a community of tenants assuming a customer/business partner (including business owner) heuristic, meaning that a business owner for a given multi-domain service may become a business partner for the delivery and the operation of another service. The business owner may also become the customer of another business owner: this would be the case of subscribing to a wholesale capacity service that would span over multiple domains, for example.

The difference between a marketplace and the framework to establish a set of bilateral agreements between the business owner and several business partners resides in the fact that the marketplace assumes a bidding procedure, whereas the framework described in Section 11.2.4.1 assumes that the business owner specifically contacts target business partners with whom bilateral agreements will be established.

Participation to the marketplace often assumes "entrance" fees, a bit like Internet Exchange Points (IXPs), where peering network operators pay the IXP provider for hosting their routers (along with the required connection resources) so that they can establish peering relationships with the routers operated by the other IXP customers to get the full Internet routing table.

11.2.4.3 A Set of Bilateral Agreements by Means of a Broker

Unlike the marketplace, the delivery and the operation of a multi-domain service that relies upon a brokering facility assume that the broker solicits business partners on behalf of the business owner, possibly through a bidding system. In that case, the broker may solicit business partners as a function of the nature of the service to be delivered, its scope, or the requirements and constraints that may have been expressed by the business owner (e.g. according to the reputation of the possible business partners that may be involved or the business owner's business policy).

As such, the broker establishes a series of bilateral agreements with the various business partners that have been selected, e.g. according to the aforementioned criteria.

11.3 Automating the Delivery of Multi-domain Services

11.3.1 General Considerations

The delivery of a multi-domain service relies upon three basic stages:

- Stage 1: Once a customer expresses service (subscription) intent, the business owner needs to identify the business partners. In addition, the receipt of the customer's request triggers a negotiation cycle that is further detailed in Section 11.3.3.
- Stage 2: Once business partners are identified, the aforementioned negotiation cycle involves them until the negotiation with the business partners gets successful, eventually. Such successful negotiation conditions the completion of the negotiation with the customer. Whether a customer's request triggers a negotiation cycle with other business partners is hidden to the customer. The service guarantees are endorsed by the business owner.
- Stage 3: Assuming the successful completion of the negotiation, the business owner and its partners then proceed with dynamic resource allocation, policy enforcement, and service fulfillment until the multi-domain service delivery is completed. Upon such completion, the multi-domain service is up and running, which triggers the operation phase. The procedure for sharing service assurance information is agreed upon during the negotiation phase.

Stage 1 is about partner identification. Such identification can be straightforward if the business owner decides to solicit the partners it is used to work with, based upon several years of legacy service delivery practice. But partner identification can be dynamic. In that case, dynamic identification schemes may consist of dynamically acquiring reachability information (e.g. IP addresses along with a set of administrative information).

In a brokering environment (Section 11.2.4.3), the dynamic acquisition of such information can rely upon a VPN facility: business partners interested in providing multi-domain services may connect to the VPN operated by the broker, which therefore acquires the required identification information. It is then up to the broker to engage negotiation with the relevant business partners, i.e. the partners that may/will be involved in the delivery of the multi-domain service as per (i) the customer's requirements and (ii) the interpretation of such requirements by the business owner.

This interpretation (namely, the identification of the multi-domain service that needs to be delivered, possibly augmented by the identification of the resources that will compose the service) will then be passed by the business owner to the broker, which triggers the aforementioned negotiation cycle with the business partners.

In the case of a marketplace environment (Section 11.2.4.2), the business owner can acquire both the reachability scope and any other administrative information that pertains to each business partner by means of networking features, as further elaborated in Section 11.3.2.

Dynamic multi-domain service parameter negotiation occurs at the service order management (SOM) plane, as depicted in Figure 11.1. This means that SDN controllers may not be involved at Stage 2, and negotiation can be handled by orchestrators operated by the various business partners, the business owner, and the broker.

Stage 3 is when SDN controllers are involved. Each SDN controller proceeds with the dynamic allocation of the resources it is responsible for and which pertain to the multi-domain service to be delivered. But they also need to coordinate one with each other for the sake of global consistency. Such coordination obviously assumes communication between SDN controllers. Given the likely sensitivity of the information to be exchanged (e.g. report about the execution of a configuration task by the SDN controller operated by a business partner to the SDN controller operated by the business owner), such communication must be safe and secure. This communication may involve the dynamic establishment of tunnels (especially in the case where the various domains are not adjacent from a BGP routing perspective, Section 11.3.2), and also encourage explicit authentication means, so that hard guarantees about the identity and the credentials of peering SDN controllers can be provided.

11.3.2 Discovering Partnering Domains and Communicating with Partnering SDN Controllers

SDN computation logics collaborate for the delivery of multi-domain services. First and foremost, this collaboration assumes that the SDN computation logics that will potentially collaborate need to know each other. In other words, SDN controllers need to know how to reach each other. This information can be included in the SLAs that have been established between the business partners and the business owner in the case where Stage 1 (Section 11.3.1) is completed in a "traditional" fashion where the business owner knows the business partners a priori, but more dynamic approaches could be useful to facilitate the completion of Stage 1.

Among such options is the ability to announce the reachability information of the SDN controllers by means of the BGP protocol [6]. This option has a couple of implications:

- Business partners may operate domains that are not necessarily adjacent from a BGP routing perspective. The processing of the BGP UPDATE messages that carry the SDN controller reachability information by BGP routers of domains that are not involved in the delivery of the multi-domain service must imply that information is not altered, filtered, let alone discarded. The use of a specific transitive attribute as the Specific Extended Community introduced next (Figure 11.3) can address such constraint. A third-party authorization model may be required to decide whether a discovered SDN controller from a peer network is allowed to maintain a secure association with local SDN controllers. The information to establish an encrypted communication channel with other SDN controllers must also be provided.
- Traffic that is specific to Stage 3 should explicitly refer to the multi-domain service that is being delivered, for the sake of proper management operation execution (e.g. the dynamic instantiation of the service-inferred policies that will be enforced within each domain). This means that service identification information should be carried some way, possibly in every packet. The RESTCONF protocol can be used to address security concerns and the underlying HTTP (Hyper-Text Transfer Protocol) can be used to carry the service identification information.

The Extended Community BGP attribute [7, 8] is a means to label information that can be carried by BGP. This attribute could thus be used to carry the IP reachability information of the SDN controller of a given domain as well as any other relevant information, such as the name of the business partner that operates the SDN controller whose reachability information is carried in the Extended Community attribute. A domain name that points at a business partner (different from the entity that operates the SDN controller) could also be provided.

More specifically, either the IPv4 Address-Specific Extended Community attribute or the IPv6 Address-Specific Extended Community attribute can be used to provide the IP reachability information of an SDN controller operated by a business partner. Figure 11.3 provides an example of an IPv6 Address-Specific Attribute.

The 16-octet Global Administrator field encodes the IPv6 address of the SDN controller, whereas the 2-octet Local Administrator field encodes an information (defined by the

<table>
<tr><td>0x00 or 0x40</td><td>Sub-Type</td><td colspan="2">Global Administrator</td></tr>
<tr><td colspan="4">Global Administrator (cont.)</td></tr>
<tr><td colspan="4">Global Administrator (cont.)</td></tr>
<tr><td colspan="4">Global Administrator (cont.)</td></tr>
<tr><td colspan="2">Global Administrator (cont.)</td><td colspan="2">Local Administrator</td></tr>
</table>

Figure 11.3 Using the IPv6 Address-Specific Extended Community attribute to carry IPv6 reachability information of an SDN controller operated by a business partner.

Sub-Type of the attribute) that can, for example, consist in naming the business partner. The attribute should be transitive so that information can be propagated across BGP domains.

Note that the BGP sessions established between the various domains should be secured, meaning that they should be digitally signed and that some guarantees about the integrity of the route announcements should be provided. The latter thus suggests the signing of BGP attributes by the originating domain as well as any upstream autonomous system (AS), i.e. any AS that is further along the AS path [9, 10].

The combined usage of digitally signed TCP (Transmission Control Protocol) connections (that support the BGP session) and attributes may thus be considered as a possible selection criterion for a business owner to negotiate a deal with a business partner.

11.3.3 Multi-domain Service Subscription Framework

The subscription to a multi-domain service involves three parties: the customer of the service, the business owner that is directly solicited by the customer to provide such service, and the business partners that will be involved in the delivery of the service by means of appropriate resource allocation. Figure 11.4 depicts the multi-domain subscription framework according to a few assumptions:

- Contacts between the customer and the business owner and between the business owner and the business partners assume an IP communication, possibly based upon the IP reachability information of the (SDN/orchestration) computation logics that will be involved in the design and the delivery of the multi-domain service. For example, the customer may contact the business owner through a dedicated portal that allows the customers to express their requirements, constraints, and any other relevant information that may be used to design the multi-domain service.
- The nature of the information to be exchanged between the different parties encourages the use of a reliable and secure transport mode (e.g. TCP, possibly combined with the use of the IPsec protocol suite).

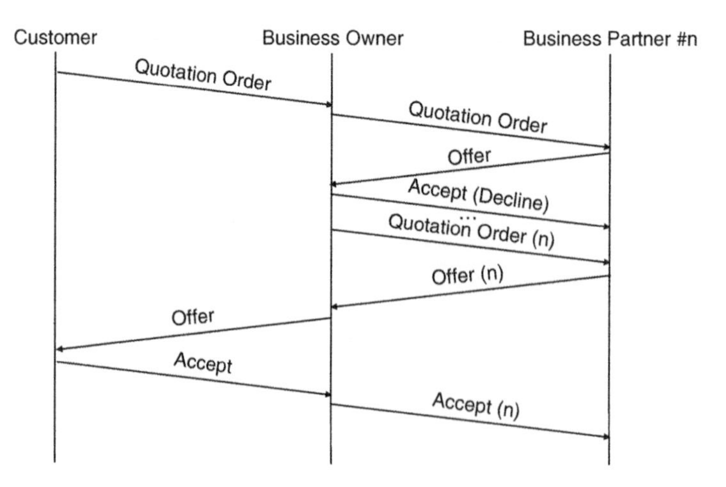

Figure 11.4 Multi-domain service subscription framework with "n" business partners and a CPNP-based successful negotiation.

- Exchanges between the customer and the business owner and between the business owner and the business partners may include negotiation cycles to best accommodate customer's and business owner's requirements. From this perspective, Figure 11.4 illustrates a multi-domain SOM procedure that allows room for dynamic negotiation based upon the Connectivity Profile Negotiation Protocol (CPNP) [11] as an example. Other SOM frameworks may be considered [12, 13], but they should leave room for dynamic negotiation.

On the right-hand side of Figure 11.4, only the exchanges between the business owner and business partner "n" are represented. These exchanges assume a negotiation that is eventually successful. Since the business owner accepts the proposals from the "n" business partners, it gets back to the customer with an offer that is accepted with no further negotiation.

Note that the model depicted in Figure 11.4 is centralized and not recursive. A more recursive negotiation model can be considered where each peer business partner directly contacts other business partners to satisfy the parent service request from the business owner. In such a model, the business owner does not have to necessarily maintain a full visibility on the involved "child" partners; it establishes a single agreement with one business partner.

The completion of this successful multi-domain service subscription (i.e. both the business owner accepts the offers made by the "n" business partners and the customer accepts the offer made by the business owner and therefore decides to subscribe to the multi-domain service) triggers the multi-domain service delivery procedure.

11.3.4 Multi-domain Service Delivery Procedure

The processing of the multi-domain service subscription order includes the triggering of the multi-domain service delivery procedures. Now that a deal has been concluded between the customer and the business owner based upon the deals that have been concluded between the business owner and the business partners, the time has come for both the business owner and the business partners to proceed with resource allocation and policy enforcement.

The computation logics in the resource allocation and policy enforcement processes support interfaces that allow communication with (i) the participating resources and (ii) the partnering computation logics, as illustrated in Figure 11.5. Of course, the business owner's computation logic interfaces with the customer to handle service subscription requests and companion negotiation cycles.

Figure 11.5 suggests the following comments:

- To deliver a multi-domain service, the various computation logics need to communicate between themselves. This may read like a trivial statement, but it implies a few, possibly challenging conditions like:
  - The need to speak the same language (communication protocol) over a secure channel. From that standpoint, the use of RESTCONF [14] appears to be one of the most straightforward options.
  - The need to understand each other. This is not only a matter of speaking the same language but also a matter of understanding the data carried in RESTCONF messages. From that standpoint, the use of standard data models (e.g. the YANG data model for L3VPN [Layer 3 Virtual Private Network] service delivery [15]) is critical.

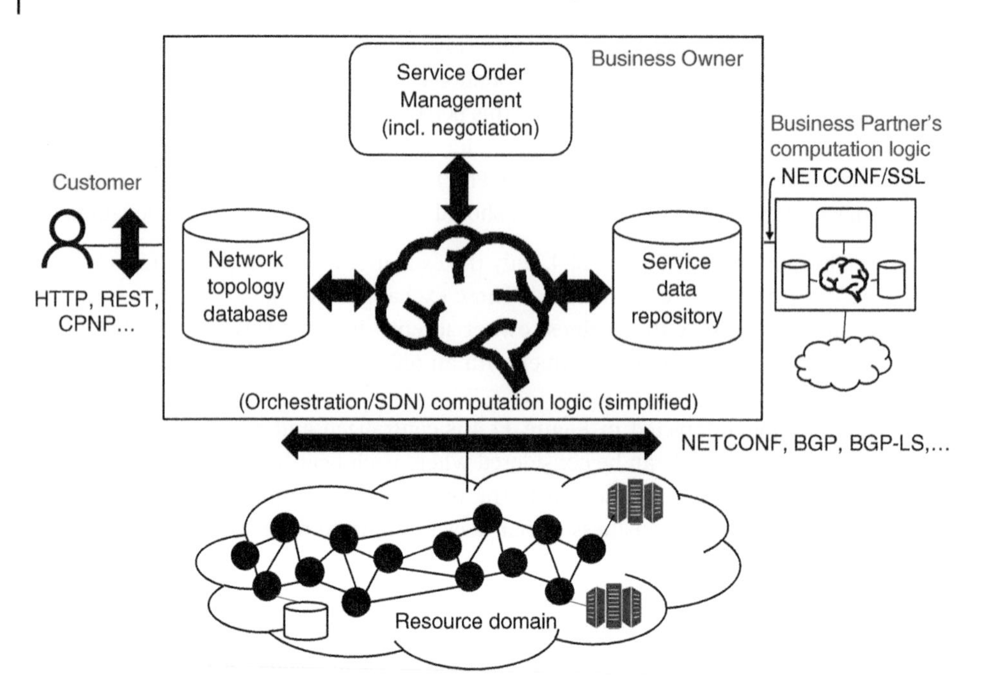

Figure 11.5 Interfaces of a business owner's computation logic.

- Because of the digital sovereignty principle, each computation logic will proceed with multi-domain service-inferred dynamic resource allocation and policy enforcement within the strict scope of their responsibility. They may use various protocols for that purpose (NETCONF, BGP-LS, BGP, etc.), but what really matters is that they must report about the execution of the corresponding actions to the business owner's computation logic (and possibly to the other business partners' computation logics), to make sure that (i) the business owner's computation logic can keep a global, service-wise, consistent view of the multi-domain service at any given time and (ii) the business owner's computation logic is informed (in real-time, ideally) about any decision made by any other business partner's computation logic that may affect the multi-domain service operation. Such decision may impact not only the configuration and the operation of the resources of a given business partner's domain but also the whole multi-domain service operation. NETCONF provides an asynchronous notification delivery service [16] that can be used for that very purpose.
- As already mentioned, the use of standard data models is of the utmost importance. But multi-domain services imply the maintenance of data stored and maintained by the various computation logics involved. This raises a critical data synchronization issue that is further aggravated by the risk that participating business partners may not be ready to disclose data that reflect their own *savoir faire* (because of competition considerations). "Data lake" approaches *a la* Gaia-X [17] may address such issue.
- Yet, using standard data models is not enough. One iconic example is the dynamic enforcement of cross-domain QoS policies, based upon a differentiated services (DiffServ) architecture [18]. Traffic marking can indeed rely upon the use of DiffServ Code Points

(DSCP) and corresponding per-hop behaviors (PHBs) that locally instruct routers about what they should do with incoming packets. All business partners and the business owner may thus very well claim they support, say, the Expedited Forwarding (EF) PHB [19]. But the implementation of such PHB may very well vary from one domain to another. For example, the processing of out-of-profile traffic by a token bucket algorithm may be different from one domain to another: one business partner may decide to discard packets in case there's no token left in the bucket (thereby disrupting the connectivity service), whereas another may decide that EF-marked traffic should be stored in another queue until traffic congestion disappears. Different implementations (besides resource allocation) may therefore seriously challenge the proper operation of the multi-domain service.

At the time of writing this document, the automation of multi-domain service delivery procedures remains technically challenging. It also has some business implications as the collaboration of multiple business partners may challenge the digital sovereignty principle. Some of these challenges are not specific to the introduction of network automation techniques. These issues are further discussed in Section 11.5.

11.4 An Example: Dynamic Enforcement of Differentiated, Multi-domain Service Traffic Forwarding Policies by Means of Service Function Chaining

The ever-growing number and complexity of connectivity services that rely upon the invocation of a large variety of service functions may challenge the design and the operation of multi-domain SFC. The variety of multi-domain services and their respective scopes therefore assume several SFC domains that will be involved in the forwarding multi-domain, service-inferred traffic.

11.4.1 SFC Control Plane

SFC has been designed to facilitate the enforcement of differentiated traffic forwarding policies within a network, based upon the ordered invocation of elementary SFs whose nature and complexity can be extremely variable (e.g. network address translators [NAT], deep packet inspection [DPI], TCP acceleration engine, and video cache), and whether these functions are virtualized or not.

The SFC computation logic can be fed with other input data that include (but are not necessarily limited to) the location and the status of the various instances of SFs that belong to the SFC domain managed by the aforementioned computation logic, a set of business guidelines derived from the multi-domain service that has been delivered to the customer, locators assigned to traffic classifiers and SFC proxies, etc. The SFC computation logic is also responsible for defining traffic classification rules that will be applied at the border of an SFC domain to map incoming traffic with the SFC chain it belongs to.

The SFC computation logic is thus responsible for building and monitoring the service-aware topology. It maintains a data repository that stores the various SFC chains that are in operation, the corresponding traffic classification rules, and the information that maps chains with service function paths (SFPs), i.e. paths used to forward packets within the SFC domain according to the chain these packets are associated with. Such paths can be dynamically established in a "hop-by-hop" fashion according to the SFC instructions that would be typically carried in the Network Service Header (NSH) [20] that is bound to each packet.

These paths can also be traffic engineered and preestablished (the path being fully or partially specified) by the SFC computation logic, e.g. by means of a path computation element (PCE) for the computation of inter-domain traffic-engineered label switched paths [21]. The SFC computation logic is also responsible for providing the nodes that embed one or more several SFs as well as traffic classifiers with the SFC forwarding policy tables. Such SFC forwarding table reflects the SFC-specific traffic forwarding policy enforced by SF nodes for every relevant incoming packet that is associated with one of the existing SFCs.

11.4.2 Consistency of Operation

Decisions made by the SFC computation logic operated by the business owner need to be consistent with the decisions made by the SFC computation logics operated by the business partners, meaning that state synchronization between the various SFC computation logics is required.

Consistency between SFC domains involved in the forwarding of multi-domain service traffic also assumes that the SFC computation logics instruct all SFC data plane elements about the behavior they should adopt regarding the selection of the encapsulation scheme (whenever the next SF to invoke is located in another SFC domain), the NSH type to enable, etc.

Such design can also be provided by means of hierarchical SFC [22]. The top-layer SFC controller does not have the information about the decisions made by SDN controllers that reside in "lower" SFC domains.

11.4.3 Design Considerations

The context of multi-domain services again mandates the collaboration between the SFC domains operated by the business owner and each relevant business partner. However, packets that enter a domain may already carry SFC instructions. Domains interface with each other by means of boundary nodes (BNs) (Figure 11.6). BNs are seen as a specific SF from an SFC computation logic's standpoint.

BNs are responsible for stripping SFC encapsulation that pertains to the domain the traffic comes from and then apply traffic classification rules that pertain to the local domain, according to the decisions made by the SFC computation logic operated by the corresponding business partner. These decisions need to be consistent with those made by the SFC computation logic of the business owner.

BNs can be flow-aware and forward packets along the appropriate SFP, based upon the transport layer coordinates of the packets. Such coordinates are typically a 5-tuple (source and destination IP addresses, source and destination ports, transport protocol). BNs will create a state for the corresponding flows. States are indexed by these coordinates. Because

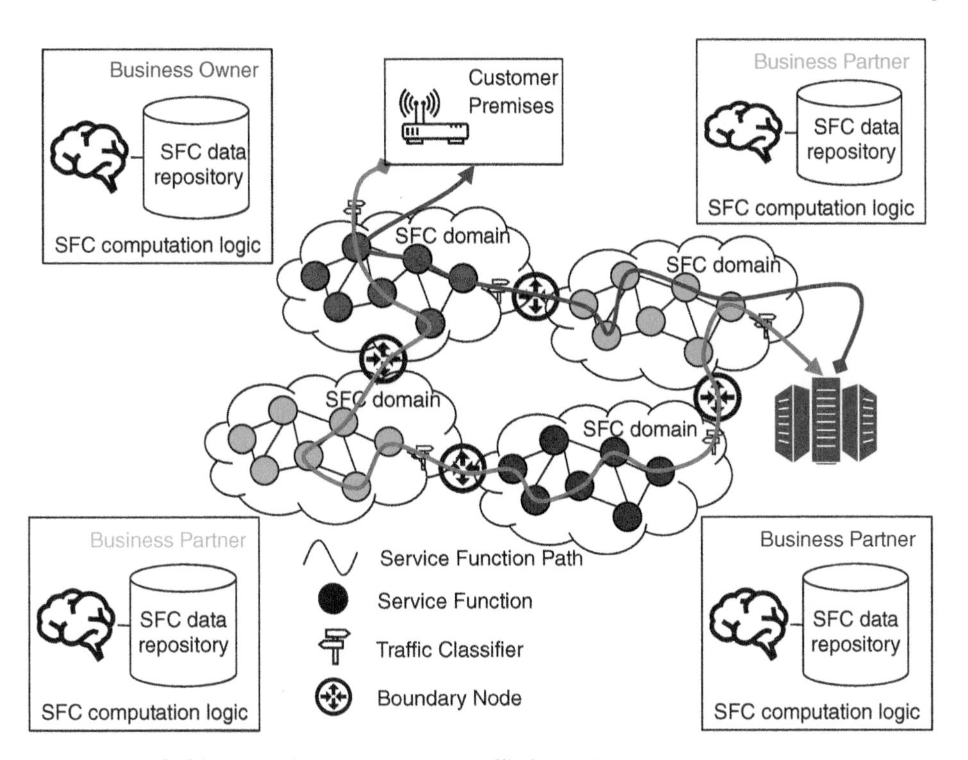

Figure 11.6 SFC-based multi-domain service traffic forwarding.

of this stateful approach, there is no need for a given SFC computation logic to modify an SFP whenever a path thereof that crosses another domain changes because of a decision made by the SFC computation logic that is responsible for establishing the aforementioned part of the cross-domain SFP.

The "glue" between SFC domains resides in the SFC metadata carried in packets forwarded within a domain: when a BN receives such packets, these metadata are used by the BN to make its forwarding decision accordingly. SFC computation logics of each domain do not necessarily need to know about the SFC policies enforced within the other domains.

11.5 Research Challenges

The automation of multi-domain service delivery procedures raises many challenges. Besides the possible business implications of exchanging domain-specific data that may disclose the domain's business partner or owner's strategy or *savoir faire*, the automated production and operation of multi-domain services are one of the most complex challenges that future networks will have to tackle, considering the progressive deployment of hybrid and telco cloud infrastructures that involve multiple stakeholders that may have very different interests.

This section discusses a few research challenges that can be classified into several categories, as introduced in the following subsections.

11.5.1 Security of Operations

From the dynamic exposure and possible negotiation of multi-domain service parameters to service fulfillment and assurance, the collaboration between several computation logics to (i) design the service (identify its elementary resources, their location, their interaction, etc.), (ii) proceed with the subsequent resource allocation and policy enforcement, (iii) and report about the application of the decisions made by the computation logics raises many security issues [23]. In particular:

- Communication between collaborative computation logics needs to be reliable and secure, to prevent any risk of man-in-the-middle, spoofing, or denial-of-service attacks.
- Collaborative computation logics must mutually authenticate so that hard guarantees about their identity and credentials can be provided before exchanging any data.
- Access to provisioning information must be controlled. For example, collaborative computation logics may need to exchange data that relate to the instantiation of a DiffServ QoS policy [24] to make sure that the respective implementations of PHBs are consistent one with another.

11.5.2 Consistency of Decisions

This is probably one of the trickiest issues. One decision made within a domain should be consistent with the decisions made by other domains. Otherwise, the operation of the service may be compromised, hence its availability. For example, if a link failure occurs within a domain, it will trigger a new resource allocation cycle that may affect the operation of a multi-domain service.

The allocation of extra resources or the decision to redirect traffic along other paths should be consistent with the BGP routing policies enforced by each domain, for example. Likewise, event-triggered decisions should not question the SLA that has been contractually established between the customer and the business owner, let alone between the business owner and the business partners.

At the very least, this issue suggests robust feedback mechanisms, so that collaborative computation logics can be fully synced at any time.

11.5.3 Consistency of Data

Unsurprisingly, multi-domain services rely upon the instantiation of a set of data models that are stored and maintained by the collaborative computation logics in specific data repositories. Of course, since each of the involved domains is solely responsible for the resources it manages, the delivery of a multi-domain service is likely to lead to fragmented images: each domain will a priori maintain a partial image of the whole multi-domain service, namely, the image of the resources that have been allocated within the domain.

This again raises data consistency and synchronization issues [25]. Two options may be considered to address this issue:

- Each domain maintains an image of the complete multi-domain service resources and policies. But this may raise issues considering that the data reflect the *savoir faire* and the strategy of each business partner (including the business owner). And these data may not be the kind of data that should be exposed to other parties.

- Multi-domain service-specific data are stored in a data lake. They can be accessed by each business partner under strict conditions that are yet to be defined.

11.5.4 Performance and Scalability

Collaboration between computation logics assumes the exchange of a potentially vast amount of information, let alone its implications on the amount of signaling traffic to be exchanged between the computation logics of each domain and the resources they manage.

This probably means that the application of network automation is restricted to a subset of services, like inter-domain VPNs or network slices: namely, services that remain relatively static overtime and that do not necessarily demand frequent resource allocation or additional negotiation cycles.

11.6 Conclusion

The provisioning of multi-domain services with guarantees has always been a challenge. Even though such services do exist for quite some time, their production takes a long time (read weeks or months), depending on their complexity. However, 6G networks can no more be conceived as isolated domains, where interactions between domains are restricted to predefined, well-known, use cases (e.g. roaming). The design of 6G networks should rather include these multi-domain issues from the very beginning. While the introduction of network automation techniques may very well serve the production and the operation of such services, it raises many technical and business challenges that are hardly addressed (let alone investigated) for the time being.

This chapter introduced an approach to automated multi-domain service production. It in particular highlighted some of the issues and technical options that may address them, mostly from a security standpoint. Admittedly, this is still a long way before all the known issues (and those that are yet to be discovered) can be safely tackled.

The glory days of networking research are not over!

References

1 Rosen, E. and Rekhter, Y. (2006). BGP/MPLS IP virtual private networks (VPNs). RFC 4364. https://tools.ietf.org/html/rfc4364.

2 Boucadair, M. and Jacquenet, C. (2014). Software-defined networking: a perspective from within a service provider environment. RFC 7149. https://tools.ietf.org/html/rfc7149.

3 Halpern, J. and Pignataro, C. (ed.) (2015). Service function chaining (SFC) architecture. RFC 7665. https://tools.ietf.org/html/rfc7665.

4 Boucadair, M., Jacquenet, C., and Wang, N. (2014). IP connectivity provisioning profile (CPP). RFC 7297. https://tools.ietf.org/html/rfc7297.

5 GSM Association (2019). Generic network slice template. Official Document NG.116, Version 1.0. NG.116-v1.0-1.pdf (gsma.com).

6 Rekhter, Y., Li, T., and Hares, S. (eds.) (2006). A broder gateway protocol 4 (BGP-4). RFC 4271. https://tools.ietf.org/html/rfc4271.

7 Sangli, S. and Rekhter, Y. (2006). BGP extended communities attribute. RFC 4360. https://tools.ietf.org/html/rfc4360.

8 Rekhter, Y. (2009). IPv6 address specific BGP extended community attribute. RFC 5701. https://tools.ietf.org/html/rfc5701.

9 Lepinski, M. and Sriram, K. (ed.) (2017). BGPsec protocol specification. RFC 8205. https://tools.ietf.org/html/rfc8205.

10 Sriram, K. (ed.) (2018). BGPsec design choices and summary of supporting discussions. RFC 8374. https://tools.ietf.org/html/rfc8374.

11 Boucadair, M., Jacquenet, C., Zhang, D., and Georgatsos, P. (2020). Dynamic service negotiation. RFC 8921. https://tools.ietf.org/html/rfc8921.

12 ATIS & MEF (2017). Ethernet ordering technical specification. J-SPEC-001/MEF 57, Joint Standard. https://www.mef.net/Assets/Technical_Specifications/PDF/MEF_57.pdf.

13 TM Forum (2020). Service ordering API user guide. TMF641, v4.0.1. TMF641 Service Ordering API User Guide v4.0.1 | TM Forum | TM Forum.

14 Bierman, A. Bjorklund, M., and Watsen, K. (2017). RESTCONF protocol. RFC 8040. https://tools.ietf.org/html/rfc8040.

15 Wu, Q., Litkowski, S., Tomotaki, L., and Ogaki, K. (eds.) (2018). YANG data model for L3VPN service delivery. RFC 8299. https://tools.ietf.org/html/rfc8299.

16 Chisholm, S. and Trevino, H. (2008). NETCONF event notifications. RFC 5277. https://tools.ietf.org/html/rfc5277.

17 GAIA-X – Home (data-infrastructure.eu).

18 Blake, S., Black, D., Carlson, M., Davies, E., Wang, Z., and Weiss, W. (1998). An architecture for differentiated services. RFC 2475. https://tools.ietf.org/html/rfc2475.

19 Davie, B., Charny, A., Bennett, J.C.R., Benson, K., Le Bouidec, J.Y., Courtney, W., Davari, S., Firoiu, V., and Stiliadis, D. (2002). An expedited forwarding PHB (Per-Hop behavior). RFC 3246. https://tools.ietf.org/html/rfc3246.

20 Quinn, P., Elzur, U., and Pignataro, C. (eds.) (2018). Network service header (NSH). RFC 8300. https://tools.ietf.org/html/rfc8300.

21 Vasseur, J.-P. (eds.) et al. (2009). A backward recursive PCE-based computation (BRPC) procedure to compute shortest constrained inter-domain traffic engineering label switched paths. RFC 5441. https://tools.ietf.org/html/rfc5441.

22 Dolson, D. Homma, S., Lopez, D., and Boucadair, M. (2018). Hierarchical service function chaining (hSFC), RFC 8459. https://tools.ietf.org/html/rfc8459.

23 Varadharajan, V., Karmakar, K., and Tupakula, U. (2019). A policy-based security architecture for software-defined networks. *IEEE Transactions on Information Forensics and Security* 14 (4) https://arxiv.org/pdf/1806.02053.pdf.

24 Phemius, K., Bouet, M., and Leguay, J. (2014). DISCO: distributed multi-domain SDN controllers. IEEE Network Operations and Management Symposium (NOMS). https://ieeexplore.ieee.org/stamp/stamp.jsp?tp=&arnumber=6838330.

25 Sakic, E., Sardis, F., Guck, J., and Kellerer, W. (2017). Towards adaptive state consistency in distributed SDN control plane. IEEE International Conference on Communication (ICC). https://ieeexplore.ieee.org/stamp/stamp.jsp?tp=&arnumber=7997164.

12

6G Access and Edge Computing – ICDT Deep Convergence

Chih-Lin I[1], Jinri Huang[1], and Noel Crespi[2]

[1] *China Mobile Research Institute, Beijing, China*
[2] *IMT, Telecom SudParis, Institut Polytechnique de Paris, Paris, France*

12.1 Introduction

4G has already changed our life, while 5G is on the verge of transforming our society. As more and more operators begin to rollout 5G deployment across the globe, pioneering researchers from academia to industry have begun exploring 6G vision profiling, along with a broad investigation of potential key technologies. It is clear that Radio Access Networks (RANs) will continue being a most exciting focal point of the necessary revolutionary evolution in the march toward 6G (Figure 12.1). In this chapter we will first review the intriguing evolutionary path of RANs post 4G long-term evolution (LTE), trying to gain some enlightenment via rethinking the fundamentals, and further establishing that the 6G RAN will be based on the information, communication, and data technology (ICDT) deep convergence. One such ICDT-based RAN architecture will be illustrated, the remaining challenges and opportunities presented, and some of the associated state-of-the-art ecosystem efforts and progress will be highlighted in this chapter.

12.2 True ICT Convergence: RAN Evolution to 5G

The classical concept of cellular systems was first proposed in 1947 by Douglas H. Ring and W. Rae Young, two researchers from Bell Labs [1]. Since the first generation of cellular standards, such cell-centric design has been maintained through every new generation of standards, including 4G. The nature of a homogeneous cell-centric design is that cell planning and optimization, mobility handling, resource management, signaling and control, and coverage and signal processing are all carried out for and by each base station (BS) separately and uniformly.

Shaping Future 6G Networks: Needs, Impacts, and Technologies, First Edition.
Edited by Emmanuel Bertin, Noel Crespi, and Thomas Magedanz.

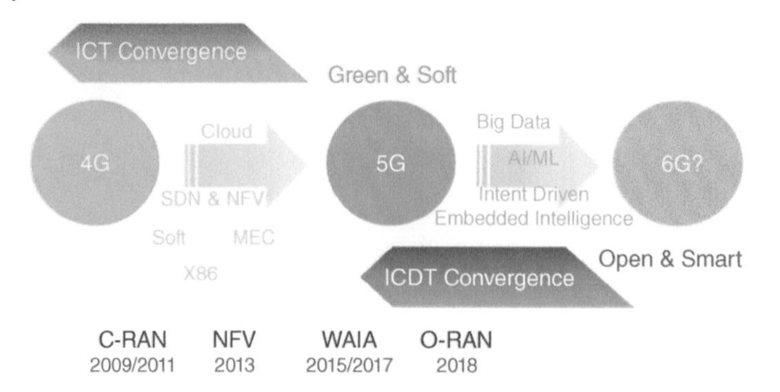

Figure 12.1 Revolutionary evolution of radio access networks.

The start of 5G studies can be traced back to 2011–2012, when LTE development was at its peak. We have identified the key themes of the next generation: Efficiency and Agility, by taking a path toward Green and Soft technologies [2]. In 2014, the Institute of Electrical and Electronics Engineers (IEEE) Communication Society published its first special issue dedicated to 5G topics. It was at this time that people began to grasp the idea, by rethinking the fundamentals, of what the future 5G could look like and what key technical solutions for it could be [3–6]. As early pioneers of the 5G endeavor, the authors in [3, 7] proposed that 5G networks should feature two key themes: "Green" and "Soft," which signifies a true convergence of information and communication technology (ICT). While "Green" emphasizes the importance of the efficient use of all the resources exploited, which is key to energy savings and environmental conservation [8], "Soft" means that all 5G elements, from core to access, must be fast reconfigurable and, learning from the OTT's successful experience of being *software centric*, should move away from the traditional hardware-centric approach. Such approaches facilitate a system's agility, an aspect made necessary by the ever faster changing trends in user and market needs. Soft technology should bring forth much a shorter lead time between generations or even fractional generation upgrades. Many communications technology (CT) professionals initially rejected the notion that efficiency and agility could be achieved jointly, since the conventional CT approach to maximizing efficiency had been very much based on building special purpose hardware with extremely tightly coupled software. Today, this new philosophy, being fulfilled by the persistent movement toward true ICT convergence, has been widely recognized and accepted in the industry [9]. Following that vision, many new technologies have emerged to serve both the standards and the actual implementation of 5G.

A Soft Network is envisioned to bring end-to-end *agility* into the architecture, together with the implementation of each network element from the core network (CN) to the access network, as well as with the building blocks of the air interface. The journey of soft networks in the edge and RAN began with C-RAN, network function virtualization (NFV), and software-defined networking with control and data decoupling.

Network function and resource virtualization should be the core of a soft network. This will decouple software and hardware, control and data, and uplink and downlink to facilitate a converged network synergistic with ICT convergence, multi-radio access technology (RAT) convergence, RAN and CN convergence, cross-layer content convergence, and spectrum convergence. This approach enables a super flat architecture that achieves cost-efficient network deployments and operation and management (OAM). In a soft network, the computing, storage, and radio resources are virtualized, and centralized or hierarchically distributed as necessary, to achieve dynamic user-centric resource management matching service features. Soft networks are expected to be built on a telecom-grade cloud platform to enable network-as-a-service, featuring open network capabilities sharing in scale. They will achieve *network flexibility and scalability* and provide users with a substantial variety of services with required quality of experience (QoE).

The Soft Network concept should be extended to the air interface as well. Instead of a globally optimized air interface, either as a trade-off among many contradicting factors or as an overdesigned solution with unnecessary extreme performance in most scenarios, a software-defined air interface (SDAI) should be considered [10]. Parameters, including spectrum, bandwidth, waveform, multi-input multi-output (MIMO) mode, and so on, are all tunable such that the air interface can be optimized to each individual application scenario. This enables the broad adaptability of future networks to potentially new application scenarios with extremely diverse requirements.

A Green Network is envisioned to maximize the *efficiency* in the utilization of any resources supporting wireless communications, from the network side to the user terminal side.

Green networks target a thousand-fold capacity increase with a minimal burden on spectrum resources [11]. Advanced signal processing to effectively explore spatial resources, centralized coordination to reverse harmful interference to the useful signal, and joint baseband and RF processing to enhance flexible duplexing are some of the key technologies to improve radio resource efficiency. Early analysis has indicated opportunities to improve both the spectrum efficiency (SE) and energy efficiency (EE) from the air interface design in global system for mobile communications (GSM) and LTE [3]. However, achieving a 100-fold EE improvement to reduce power consumption and thus operational expenditure (OPEX) for sustainable operations is too tall an order to expect from merely air interface design improvements. It will require end-to-end energy management and optimization so that the total energy consumption will be minimized while meeting service requirements. Green networks enable network capacity migration, dynamically matching service variations without unnecessary consumption of network resources. Practical cell breathing, "plug and play," effective on–off operation at fine resource granularities of time slots, subcarriers, transceiver chains, and nodes are all essential elements of a green network. An advanced self-organizing network (SON) should build on these technologies from a hardware or software module to a subsystem, a BS, a (single RAT) network, and ultimately to being able to treat a heterogeneous multi-RAT network as one system in terms of dynamic network and topology configuration, as well as near-real-time (near-RT) network optimization.

It is worth noting that a soft and virtualized network at scale will inherently beget opportunities for greener functioning, since it can take full advantage of the temporal and spatial demand variation for resource sharing. Finally, Green Networks should maximize the use of renewable energy, such as wind and/or solar energy, as an alternative power supply for networks and bioelectric, kinetic, and/or thermal energy for terminals.

The past decade has witnessed great efforts and progress in truly converging information technology (IT) and CT, with both moving toward green and soft networks, particularly made visible by the communications industry's embrace of technologies such as softwarization, cloud computing, and virtualization. A vast amount of new experience is being gained by the telecom architects designing future networks. The following subsections introduce four of the most important technical aspects, namely, C-RAN, next-generation fronthaul interface (NGFI), mobile edge computing (MEC), and cloud/NFV, in this journey of ICT convergence.

12.2.1 C-RAN: Centralized, Cooperative, Cloud, and Clean

The concept of C-RAN, shown in Figure 12.2, was first proposed by China Mobile in 2009 [7, 12–14]. It is not one technology, but a technology path. C-RAN breaks away from the classical Ring and Young cell-centric structure and takes on the user-centric vision with the *No More "Cell"* adventure proposed in [3]. The letter "C" here conveys multiple stages and characteristics of the path. It stands for centralized, collaborative, cloud, and clean. C-RAN realization is a stage-by-stage process with different features realized at different stages. The basic idea of C-RAN starts from centralization, which is the aggregation of different baseband units (BBUs), which in a traditional deployment are geographically separated, into the same location. Once *centralized*, it is possible to allow different BBUs to

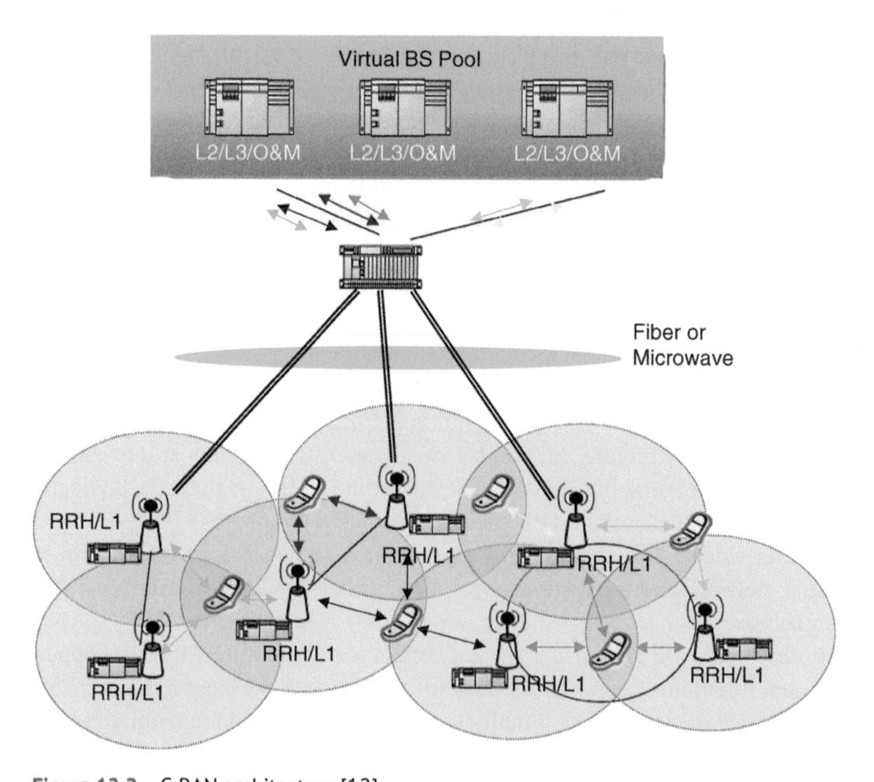

Figure 12.2 C-RAN architecture [12].

communicate with each other more easily and quickly by connecting them with high-speed switch networks, thereby allowing implementation of the *cooperative* algorithms to improve system performance. For example, one of the user-centric features C-RAN can offer is significantly reduced QoE variation between the cell-center and the cell-edge. To alleviate the statistically more severe interference suffered by cell-edge users, coordination among multiple cells is necessary. C-RAN provides an ideal structure to facilitate the implementation of coordinating technologies with the full or partial channel state information of all users available centrally. C-RAN is also an ideal match to the deployment of ultradense networks (UDN), which involves the joint consideration of many issues, including control and user plane decoupling, inter-site carrier aggregation and coordination, and interference mitigation in a heterogeneous network. In this case, C-RAN plays an important role in the internal high-speed low-latency switching mechanism and the central processing to implement those key technologies.

C-RAN presents opportunities to virtualize the processing resources into a pool on common *cloud* platforms so that the resources can be managed and dynamically allocated on demand. Virtualization of RAN embraces evolving standard IT technologies to consolidate traditional customized network equipment onto high-volume commercial off-the-shelf (COTS) servers, switches, and storage devices. It delivers RAN functions in software running on a range of general-purpose platforms (GPPs), that can be moved to, or instantiated in, various locations in the network as required, without the need for the installation of new equipment. There is a unique advantage to supporting multi-RAT with the adoption of GPP and virtualization technology. In C-RAN, different RATs can be virtualized in the form of virtual machines (VM) so that they can operate separately and independently on the same platform. C-RAN can further help with multi-RAT coordination. An even more intriguing potential in GPP-based C-RAN is that RAN functions (communication) and application layer processing (computations) can be cohosted on the same platform in their respective VMs.

C-RAN systems are expected to enjoy greater energy savings than traditional RAN architectures and thus be deemed *clean*. This special benefit is attributed to several factors. Centralization helps to save energy due to facilities savings, especially from sharing air-conditioning loads. The adoption of cloud technologies also contributes to the energy savings. The advantages of C-RAN have been verified in various scenarios. The early trial deployment of a centralized stage C-RAN in a 3G network [12] has demonstrated its advantages in terms of the total cost of ownership (TCO) savings, network deployment speed, and energy savings. The TCO savings have been reported to be as high as 70% compared to the traditional deployment method, where the OPEX can also be greatly reduced [13].

12.2.1.1 NGFI: From Backhaul to xHaul

In the 4G era, C-RAN combined with the centralization feature proved successful, with demonstrated advantages such as network deployment speedup, power savings, OPEX savings, and so on. When it comes to 5G evolution, C-RAN itself needs to evolve, and one of the critical technologies lies in the next-generation *fronthaul* interface (NGFI, aka xhaul). From "Rethinking Ring and Young" in 2011 to proposing NGFI in 2014 [13, 15, 16], the RAN revolutionary path to meet ambitious 5G demands has been charted out.

Traditionally, BSs are *backhauled* to the CN. A BS consists of a BBU and a remote radio unit (RRU), where the baseband-related functions are processed by the baseband unit (BBU) while the RRU processes radio-frequency-related functions. The transport between the BBU and the RRU is the fronthaul (FH), and the traditional FH interface is Common Public Radio Interface (CPRI), a TDM-based point-to-point interface. CPRI fell short both in required bandwidth and architecture flexibility. It is the simple partitioning between the BBU and RRU that leads to the shortcomings of CPRI, as mentioned earlier. Therefore, the NGFI design should start with a paradigm shift by rethinking and redesigning the function split between BBU and RRU. Moreover, the function split between BBU and RRU may be different according to the bandwidth and latency of FH, which is adaptive to different scenarios. Figure 12.3 illustrates multiple functional split options [17, 18] that can be considered, where Option 8 corresponds to the traditional CPRI FH interface.

Based on the preceding considerations, NGFI proposed by China Mobile targeting a packet-based, traffic-dependent, and antenna scale-independent interface played a central role in the 5G RAN revolution, including the establishment of the IEEE 1914 NGFI [19] standard work, the development of the enhanced CPRI (eCPRI) interface (corresponding to a variation of Option 7), and the eventual 5G RAN configuration [20] of centralized units (CUs) and distributed units (DUs), where F1, the interface between CU and DU, corresponds to Option 2.

The new FH as the NGFI between the BBU and RRU is defined with the following features:

- Its data rate should be traffic-dependent and therefore support statistical multiplexing;
- the mapping between BBU and RRU should be one-to-many and flexible;
- it should be independent of the number of antennas; and
- it should be packet-based, i.e. the FH data can be packetized and transported via packet-switched networks.

The standard working group IEEE1914 for the NGFIs was kicked off in 2015 with exactly these features in mind [19].

The work in [17, 18] presented a two-level NGFI architecture and the function split options with their associated requirements in, e.g. latency, bandwidth, and synchronization. The challenges as well as the potential solutions for NGFI realization were also discussed. The function split between the baseband divides a traditional BBU into two logical entities, CU and DU, defined by 3rd Generation Partnership Project (3GPP) [20]. This proposed two-level NGFI is significant in the sense that the exploration of the NGFI solution partially leads to the creation of the new BS structure, i.e. CU-DU-based 5G BS, which is also called next-generation NodeB (gNB), as shown in Figure 12.4. Note that according to 3GPP, a gNB consists of one CU and multiple DUs, and the CU can coordinate among different DUs, while each DU belongs to only one CU. Most of the controlling functionalities are centralized on the CU, while the fast scheduling on the air interface is realized on the DUs. In some sense, a gNB is like a mini-C-RAN.

Traditional 4G eNodeB (evolved NodeB)(eNB) has only one FH interface of CPRI. In comparison, for a 5G gNB, two FH interfaces (i.e. NGFIs) exist, indicated as NGFI I and NGFI II in Figure 12.4. Correspondingly, from the transport network perspective, there are two kinds of NGFI transport networks in 5G RAN. They are indicated as FH and midhaul

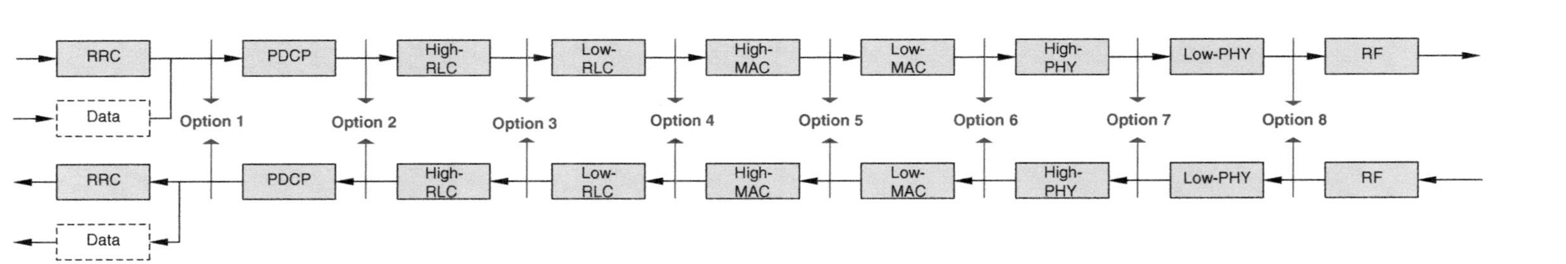

Figure 12.3 xHaul split options [17].

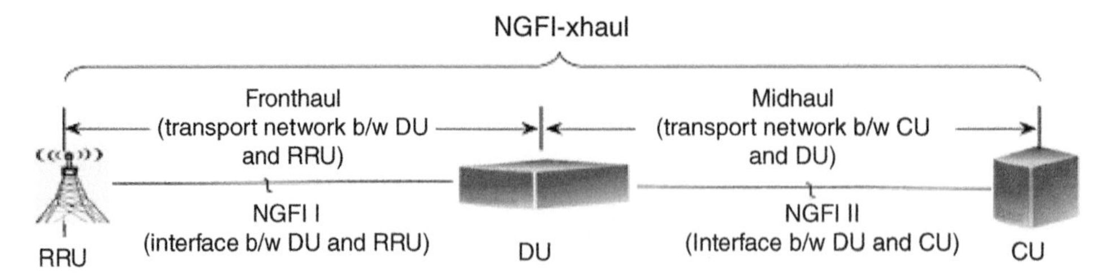

Figure 12.4 FH and MH for a 5G gNB [18].

(MH), where FH connects DU and RRU, while MH is between CU and DU. They are collectively referred to as "NGFI-xhaul" since either MH or FH originates from the idea of NGFI. It should be noted that 3GPP has chosen Option 2, i.e. a split between Packet Data Convergence Protocol (PDCP) and radio link control (RLC), as the MH split between a CU and a DU, whereas the FH split between DU and RRU is left open from the 3GPP point of view.

Diverse 5G applications and technologies have imposed new requirements on NGFI transport. For example, different user plane (UP) latency from different services as well as the Hybrid Automatic Repeat Request (HARQ) will limit possible NGFI transport delay budgets. In addition, compared to 4G, NGFI-xhaul network for 5G would have much higher synchronization accuracy, mainly to support key technologies such as carrier aggregation (CA) and joint transmission (JT) and to shorten radio frame as well as positioning service. Furthermore, the design of NGFI should take into account support of network slicing. Proposed in 2014, the concept of NGFI, featuring packet-based, traffic-dependent, and antenna scale-independent is now widely recognized as the key design principles for the 5G FH solution [21, 22]. Combined with the 5G gNB structure, the CU-DU-based C-RAN architecture with NGFI is viewed as a key feature of the 5G RAN revolution.

12.2.1.2 From Cloud to Fog

Proponents of fog computing and networking propose that "Fog rises as cloud descends to be closer to the end users" [23, 24]. Fog computing is an end-to-end horizontal architecture that distributes computing, storage, control, and networking functions closer to users along the cloud-to-thing continuum. Fog-RAN shown in Figure 12.5 extends the network into the end devices. Cloud-based wireless networking systems employ centralized resource pooling at various hierarchical levels to improve operation efficiency and latency. Fog-based wireless networking systems further reduce latency by incorporating end devices to form ad hoc network extensions. The work in [24] demonstrated a vision to realize a seamless continuum of computing services from the cloud to things and showed how context-aware collaboration among network servers and end-user devices can help achieve low-latency, high-bandwidth, and agile mobile services for 5G. However, engaging the end devices of multiple users in one ad hoc network extension remains an elusive vision, while engaging multiple end devices from the same user as the user's own ad hoc network extension appears to be worth pursuing.

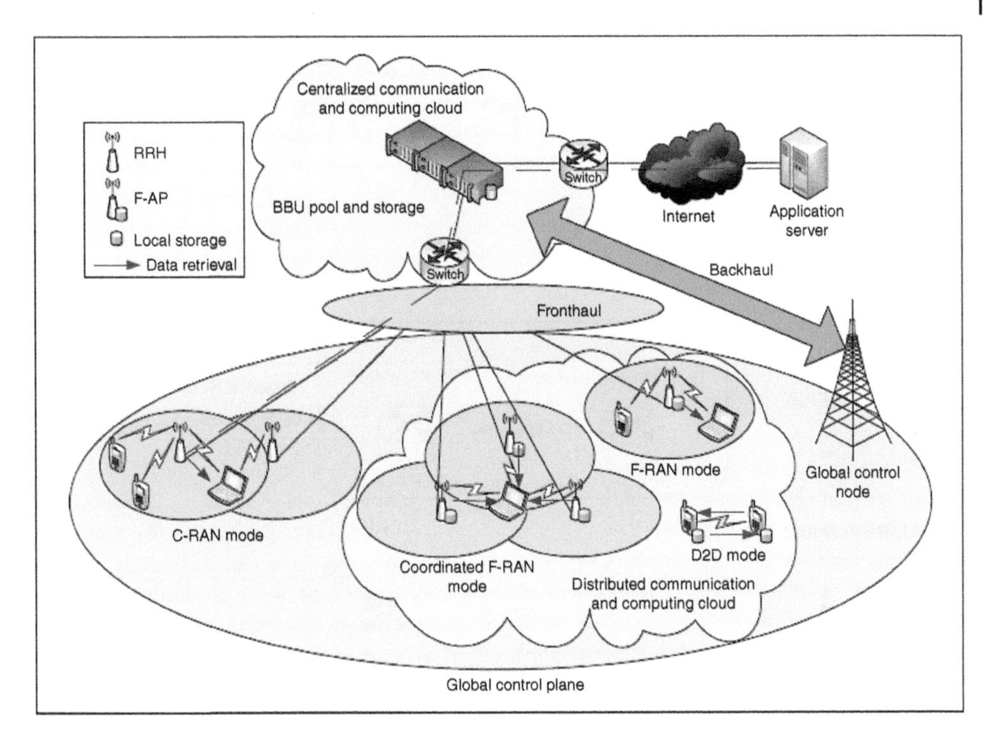

Figure 12.5 Fog arch [24].

12.2.2 A Turbocharged Edge: MEC

It was proposed in [5] that the network transformation from 4G to 5G would feature (i) core network and radio access network (CN-RAN) repartitioning, (ii) RAN restructuring, (iii) turbocharge edge, and (iv) Network Slice-as-a-Service, as shown in Figure 12.3. That work envisioned how, with the expected restructuring of the RAN architecture and the repartitioning of both CN and RAN functions, some of the BS functions will move upward, whereas some of the CN functions will move downward, thus creating a clearly identified edge domain in between. This new domain, namely, the turbocharged edge in the figure, will be critically important, as it will cohost important communication and networking functions as well as potentially valuable services and applications. The turbocharged edge proposed has become a key new feature of the 5G network, often referred to as MEC [5]. The figure clearly illustrates the value of MEC in early-stage trials, with the following perspectives:

- A radio data center close to the edge can be deployed to handle local breakout traffic. Meanwhile, the RAN edge controller should select optimal routing for specific Internet Protocol (IP) flows provisioned by third-party applications, including both local and remote gateways. Moreover, the edge controller also handles user mobility among multiple gateways, including local gateways.
- The RAN capabilities and network contexts can be exposed to third-party service providers so that service experiences can be optimized by matching network capabilities and

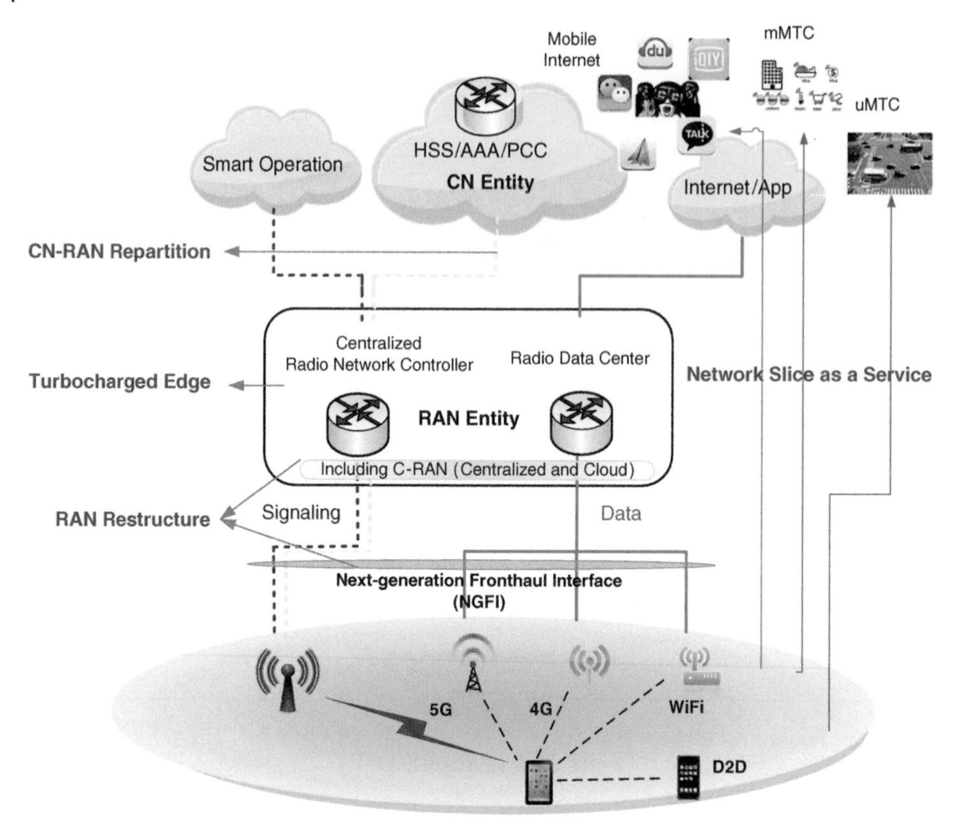

Figure 12.6 New framework of radio access network [5].

conditions. As an example, the medium access control (MAC) or RLC layer throughput of a user is exposed for the optimization of Transmission Control Protocol (TCP) window control on the server side.

- In addition to communications capabilities, computing resources can be open and available at the network edge. Such edge computing helps improve the quality of service (QoS) with stringent requirements on the delay and bandwidth. For instance, the local network resources would be very useful for user experience improvement in virtual reality/augmented reality (VR/AR)-based interactive online applications. Of course, additional novel network functions could also be generated therein as needed (Figure 12.6).

From another perspective, the essence of MEC [25, 26] is indeed very simple: to deploy services at the edge so that they are closer to the users, as opposed to traditional networks where services are deployed behind the CN. The benefits are abundant. First, getting closer to the users reduces the delay of the delivered applications, which in turn improves the user experience. Second, the backhaul traffic between the BS and the CN can be reduced, since the BS can directly interact with the services on the same edge. Last but not least, the concept of MEC brings with it the opportunity of open edges, where RAN information can be exposed through suitable open interfaces to third-party service providers, which can help with service optimization and innovation.

In the 5G standalone network, the user plane function (UPF) from a service-based architecture (SBA) CN is indeed distributed to the edge. The CU/DU architecture of 5G RAN sees the CU moving up to the edge and enables a great many opportunities and flexibilities when its integration with MEC cohosting the UPF is considered.

12.2.3 Virtualization and Cloud Computing

Proposed by China Mobile and other operators in a joint white paper under the European Telecommunications Standards Institute (ETSI), NFV, launched in 2013 [27], has been widely accepted in the industry as one of the most important trendsetting movements to realize 5G and beyond. As illustrated in Figure 12.7, the basic idea of NFV is to "consolidate many network equipment types onto industry standard high-volume servers, switches and storage, which could be located in Data centers, network nodes and in the end user premises" [27]. We extended the concept of NFV from the onset to include RANs as well as CNs, so that the benefits from NFV, including flexibility, scalability, EE, and speed of innovation, can be fully exploited throughout the whole network. This extension is key to providing the important synergy between NFV and C-RAN. The RAN delay and jitter requirements often demand hypervisor upgrades [7]. However, some RAN functions still require accelerators to complement a pure GPP platform. Typical accelerators are field-programmable gate array (FPGA) based, with digital signal processor (DSP), graphics processing unit (GPU), and application-specific integrated circuit (ASIC) vying for the role.

The concept of cloud computing originates from the Internet industry. Cloud computing has the advantages of cost reduction, rapid expansion, and flexible deployment, making it the focus of investment for major Internet companies. The telecom industry has been

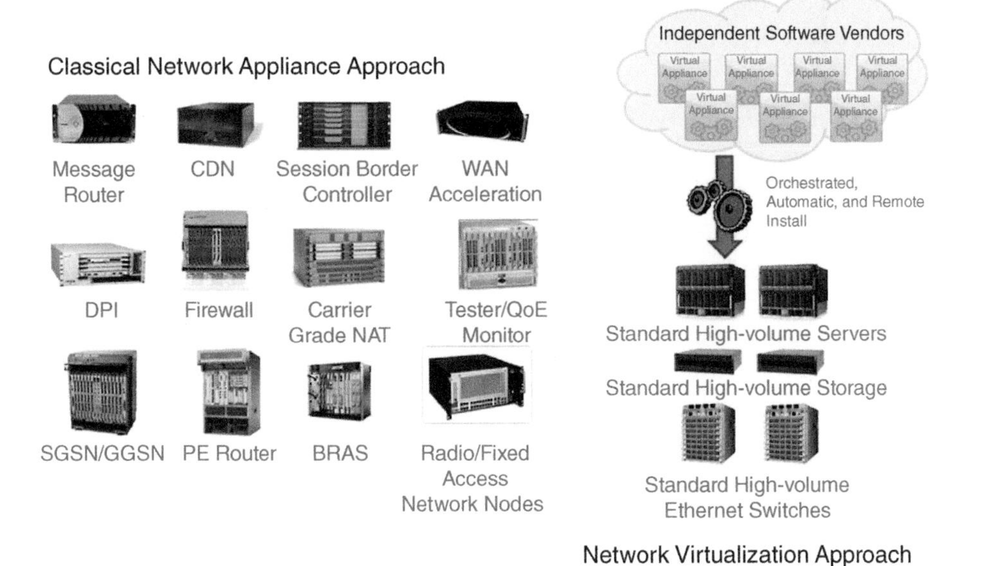

Figure 12.7 Network function virtualization (NFV) [27].

exploring end-to-end virtualization/cloud deployment of the whole end-to-end (E2E) network for multiple perceived advantages, including the following:

1) The decoupling of hardware and software, RAN included, supports the independent evolution and procurement of software and hardware and enables quick and flexible adjustment of the network.
2) The standardization and simplification of the hardware components simplify equipment inventory operation and spare parts management and improve facilities efficiency.
3) The infrastructure and service automation enabled by cloudification can minimize the expenditure and maintenance costs for the network deployment and operation, eliminate problems caused by human errors, support faster response to problems, and gradually realize automatic network repairs.

5G-compatible design, including CU/DU split, NFV, virtualized RAN (vRAN), software-defined networking (SDN), IPv6 transmission, the packet CN of control user plane separation (CUPS), edge computing, and network slicing, at scale and with dynamic capacity providing 4G and 5G services through flexible software upgrades, has been the holy pursuit of the mobile industry. Expanded mobile network ecosystems also help to encourage new business models.

In February 2019, Rakuten, a Japanese operator, successfully tested the world's first cloud-native 4G network and announced its commercial service in October 2019. Rakuten took only one year from initial planning to completion of construction and at significant lower cost than earlier systems. The Internet companies' disruptive innovation renders the traditional telecom industry barriers no longer indestructible. Today, major operators around the world are gradually deploying cloud architecture networks. Facing the huge cost investment pressure of 5G construction and to address the vastly unmet vertical industry's needs, exploring how to reap the benefit of cloud network architectures while maintaining the delicate balance of cost and performance has become a major focus of operators' research and development (R&D) efforts.

12.3 Deep ICDT Convergence Toward 6G

12.3.1 Open and Smart: Two Major Trends Since 5G

With the advent of 5G, wireless networks are becoming more and more complex than in previous generations. There are several reasons behind that, including the much higher network densification, the larger network scale, the more demanding applications with extremely diverse requirements, and so on. However, traditional networks are developed in a closed, instead of open, manner with tight coupling among different elements. Such tight coupling brings difficulties in achieving quick system upgrades, speedy service rollouts, multi-vendor compatibility, and fast network and service innovation. It's foreseeable that big challenges would lie ahead for large-scale deployment of 5G and beyond should we proceed with the traditional approach.

As elaborated in the previous section of this chapter, a broad scope of true ICT convergence has evolved and greatly influenced the design of 5G networks, including the 5G RAN. Going forward, the pursuit of higher flexibility, scalability, agility, and manageability continues. We expect ever increasing levels of ICT convergence, and in the meantime, we have seen growing success of big data analytics, machine learning (ML), and artificial intelligence (AI) in many applications since 2014. It's obvious that the deep convergence of IT, CT, and DT has emerged as an essential enabler of future networks [28]. We believe it will be native in the 6G network! In addition to being green and soft, ICDT deep convergence embedded 6G networks will feature two additional themes: openness and intelligence.

Intelligence encompasses both the network as a whole as well as individual network components. Next-generation networks will take full advantage of ML and AI technologies to deliver efficient, dynamic resource and service optimization, and ultimately enable multitier, including RT, closed-loop automation and even creation of new features (functions, capabilities, and services).

Openness is broadly defined, as opposed to a tightly coupled system, as an attribute in which the network is disaggregated into self-contained, interoperable hardware or software modules that may be sourced from different solution providers in a greatly expanded, inclusive, robust, and innovative ecosystem.

An ICDT integrated RAN that's open and intelligent will disaggregate the future RAN in multiple dimensions, including radio disaggregation, compute disaggregation, management plane disaggregation, and control plane (CP) disaggregation.

12.3.1.1 RAN Intelligence – Enabled with Wireless Big Data

Embedded RAN intelligence through the introduction of ML/AI techniques, which could not only help optimize and simplify the network management and orchestration but also help improve system performance optimization [29–33]. Key to exploit potential RAN intelligence lies in wireless big data (WBD) [34, 35]. Mobile networks generate massive amount of vast varieties of WBD, including traffic distribution, resource utilization, user information, user application usage statistics, network configuration, FCAPS (fault, configuration, accounting, performance, and security) data, as well as private service data. WBD can be found in business, operation, and management (BOM) domains and on value-adding service platforms.

5G CN adopts SBA, separates the CP and the UP, facilitates NFV, and enables flexible deployment of the CN functions. The network data analytic (NWDA) was introduced into the 5G CN standard [36]. NWDA is used to collect massive data on each function of the CN, including user mobility data, service flow data, billing information, and network entity status data. The NWDA can analyze the collected data and improve the QoS of user services through the policy control function (PCF), in order to enhance the user experience. NWDA can also collect and store a large number of user service data, based on which the operators can derive the user profile, precision service targeting, and cross-domain high value services.

The RAN architecture includes a CU and a DU of the next-generation NodeB (gNB) [20]. The RAN exhibits the characteristics of hierarchical architecture and distributed deployment. The gNB-CU and gNB-DU can collect massive wireless network data, e.g. wireless channel data, wireless measurement reports, wireless access layer data, wireless traffic data, and wireless network operation and maintenance data. Based on the collected big data of the RAN, they can perform channel model optimization, wireless resource scheduling optimization, agile physical layer (PHY) transmission optimization, and generation of a big data-enabled full spectrum map. Among the aforementioned, both the physical layer channel optimization modeling and prediction-based radio resource management (RRM) optimization can be characterized by short validity duration of the data source, which requires RT data collection and strong RT characteristics of the processing data at the network data resource. Moreover, prediction can be conducted according to the prominent RT characteristics of the processed data using ML technology, and parameter optimization can be performed in the corresponding network entity based on the predicted results, in order to achieve automatic RT and closed-loop network optimization.

When centralized big data analytics network architecture is directly introduced in the RAN, it will face the following challenges:

1) Data privacy challenges: Data collected from the RAN may carry user privacy information and user privacy may be exposed.
2) High data transmission cost: round trip time (RTT), transmission time interval (TTI), and other data generated by the RAN, especially the DU site, has extremely high data rate (>1 Gbps). If all the unprocessed data is uploaded to the centralized big data analysis node, the data transmission from DU to the centralized big data analysis node will be very expensive.
3) Difficult online training/prediction in RT: The validity duration of data in RAN is very short, and data backhauling delay cannot realize RT (<10s) training/prediction of data in centralized data analysis nodes.
4) The data processing is too computationally expensive: Each site of the RAN generates a large amount of data and aggregates the data of all sites to a centralized data analysis node, which consumes large amounts of computing resources and becomes a bottleneck of the data analysis. In summary, a radio access network big data analysis network architecture (RANDA) needs to be introduced to the wireless RAN, which consists of various analysis modules including data acquisition, data storage, data processing, training models, model execution, and parameter tuning optimization [30, 31]. The RANDA module includes the centralized unit data analysis (CUDA) function for the wireless network center and the distributed unit data analysis (DUDA) function for the wireless network.

In order to meet the RT requirements of the applications, the data analysis module must be deployed at the data source, to reduce the duration between the data processing, prediction/decision-making, and network parameter optimization. ML technology requires the use of existing big data training prediction/decision models to improve the accuracy and precision of prediction. However, in a wireless network, the independent gNB-DU devices typically suffer limited computational resources (even some gNBs may also be the same), such that the data analysis modules deployed on these devices can hardly support such high computational complexity tasks. Therefore, it is necessary to adopt the models to separate the training from the prediction method. For such separation, the parts with concentrated computational resources can undertake the tasks with large computational complexity including the model training, while independent gNB-DU and resource-constrained gNB can achieve the prediction tasks with smaller computational complexity.

For the application of network optimization and planning, the data analysis module of the data source can store the RAN data, which is uploaded to the network operation and maintenance entities for further processing. Such application requires the data service interface to Business and Operation Support Systems (BOSS) and other network entities' data analysis function in order to accept the data subscription for BOSS and other network entities, to provide the appropriate data services (Figure 12.8).

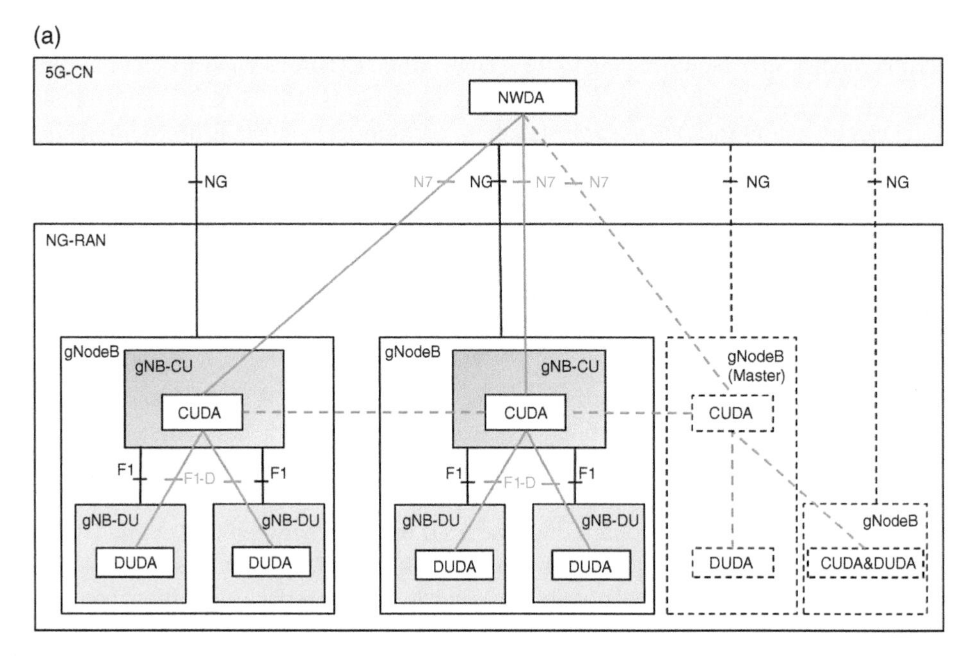

Figure 12.8 (a) WAIA architecture of wireless big data analytics [29].

12.3.1.1.1 Big Data Module Function Definition NWDA is a carrier-controlled network data analysis logic entity that can provide network data analysis results to PCF, which further performs the QoS policy decision based on the network data analysis results. NWDA also needs to provide the data subscription service for the BOSS and upload the preprocessed subscribed data to the BOSS, for further big data operations and services.

RANDA is primarily adopted to support data-driven intelligent RRM and physical layer or higher layer optimization in the RAN [29, 38]. The RANDA should interact both with the CP and UP of the RAN to realize the data collection and policy configuration. It also needs to provide data subscription services for NWDA and BOSS/OSS and upload preprocessed subscription data to NWDA and BOSS/OSS for further big data operations and services. RANDA can also subscribe to the NWDA data analysis results for the RAN-side service optimization. RANDA includes CUDA and DUDA.

CUDA is used in quasi-RT optimization for radio resource control (RRC), Service Data Adaptation Protocol (SDAP), PDCP, and other protocol layers (such as multi-connection, interference management, and mobility management). More specifically, it includes the data analysis, the quasi-RT predicting, decision-making model training, the online model prediction, and the strategy generation and configuration based on the predicted results, in order to provide the DUDA data feature and model subscription distribution. CUDA can support both master and slave modes. The slave CUDA can request the master CUDA to perform some computational tasks, such as model training; while the master CUDA can conduct some computationally intensive model training for the slave CUDA and offer some network-level collaborative optimization recommendations.

DUDA is adopted for the RT RAN data collection and preprocessing, prediction, parameter optimization, and training tasks with low computational complexity in DU (e.g. PHY\MAC\RLC). DUDA needs to offer data features needed for training prediction/decision models after preprocessing to the CUDA, while CUDA can assist DUDA to conduct some computationally intensive model training tasks. Assisted by CUDA, the trained model can be sent to the DUDA for installation, perform RT prediction/decision-making based on the RT collected data, and generate the corresponding strategy based on the prediction results, in order to perform RT closed-loop control for the DU's process (such as scheduling, link adaptation, etc.).

NWDA and RANDA (CUDA and DUDA) form a hierarchical and distributed deployment of data-driven intelligent network architecture, in order to improve the big data service in the wireless network.

12.3.1.1.2 Interface The N23 interface connects NWDA and PCF, which transmits PCF and NWDA subscription information. PCF receive the network data analysis results through the N23 interface from the NWDA, such as the congestion information of a specific slice.

For RANDA, it is necessary to introduce the F1-D interface between CUDA and DUDA for data subscriptions/distributions between CUDA and DUDA, as well as the subscriptions and distributions of predictive/policy models. The interface definition between CUDA and CU-C and CU-U can provide more flexibility in supporting multi-scene deployments, and thus facilitates rapid iterative updates of network capabilities. The N7 interface needs to be introduced between RANDA and NWDA to subscribe to and distribute processed datasets and data analysis results.

The architecture above proposed by Wireless AI Alliance [29, 38] in early 2017 failed unfortunately to make it into the 5G standard by 3GPP. Else we would have seen the 5G RAN born with embedded intelligence already.

12.3.1.2 OpenRAN

Openness here brings forth multiple implications, including open architecture, open interfaces, open application programming interfaces (APIs), open-source software, and open hardware reference design, also known as white box. Open interfaces are essential to enable smaller vendors and operators to quickly introduce their own services or enable operators to customize the network to suit their own unique needs. It also enables multi-vendor deployments, enabling a more competitive and vibrant supplier ecosystem. While open-source software enables faster product development and encourages innovation, opening the reference design for hardware could further reduce the R&D cost and the entry barrier for small vendors, which further spurs the innovation throughout the industry and brings prosperity to the ecosystem.

Open interfaces are essential for multi-vendor interoperability. Traditional RAN interfaces are not fully open and rely on specific vendor implementations, thus leading to single vendor lock in and difficulty introducing new features available from other suppliers; e.g. the interface between RAN BBU and RRU had long been a "standard interface" that was not fully specified, thus not "open." Truly open interfaces mean fully specified and standardized radio network key interfaces that are not open originally. It will energize the innovation and reduce the cost, while bringing about a vibrant industrial ecosystem and cutting down the cost of various module units.

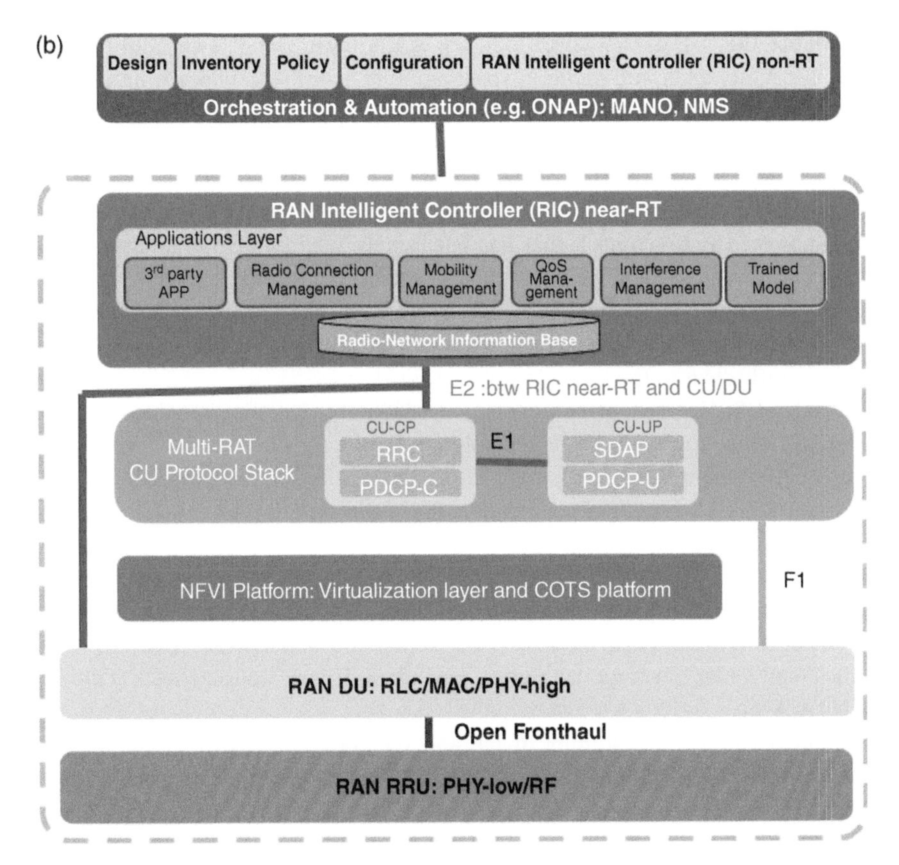

Figure 12.8 (b) O-RAN reference architecture [37].

Open-source software brings in agility and diversity. Traditional radio equipment is developed by vertically integrated manufacturers independently and separately in a considerably long period and at a relatively high cost. The purpose of open-source software is to leverage a much broader community including vast IT expertise to build up a common radio network platform using open-source modules, such as the operating system, protocol stack software, and so forth. It can lower the cost and accelerate the progress of R&D with many more ecosystem players involved in contributing and sharing the open-source codes. Based on the open-source software, every industrial partner can easily focus its R&D on own strength, be it core algorithms that reflect its special ability and value or unique new functionalities that provide differentiated benefits. It begets a win-win outcome through lowering the cost of the overall industrial R&D as well as reflecting more efficiently the differentiated value of each technology provider.

Open reference design (white box) hardware lowers entry barriers for emerging ecosystem players. Significant technical and resource challenges in the R&D for traditional radio equipment prohibit the participation of emerging small and medium-sized innovative companies. The idea of white box hardware is to use the open reference designs to form the scale effect, reduce the cost of the whole industry chain, lower the cost and difficulty of R&D, and attract more small and medium-sized enterprises into the telecom industry to promote competition and innovation.

The O-RAN Alliance (Section 12.3.1) architecture is designed to enable next-generation RAN infrastructures. Empowered by principles of intelligence and openness, the O-RAN architecture is the foundation for building the virtualized RAN on open hardware, with embedded AI-powered radio control, that has been envisioned by operators around the globe. The architecture is based on well-defined, standardized interfaces to enable an open, interoperable supply chain ecosystem in full support of and complementary to standards promoted by 3GPP and other industry standards organizations. The openness theme is addressed abundantly clear in the following aspects.

Decoupled User/CP: The core principle of the O-RAN architecture is to decouple the CP from the UP. This extends the CP/UP split of CU being developed within 3GPP through the E1 interface to further separate selective RRM functions from CU into the near-RT and non-RT RAN intelligent controller (RIC), through the A1 and E2/B1 interfaces (Section 12.2.1.1). The first benefit this offers is to allow the UP to get more standardized, since most of the variability is in the CP, and allows for easy-scaling and cost-effective solutions for the UP. The second benefit of such a decoupling is to allow for advanced control functionality for increased efficiency and better RRM. These control functionalities can then leverage analytics and data-driven approaches including advanced ML/AI tools.

Software-defined CP: The decoupling of CP and UP through open interfaces in the O-RAN Alliance reference RAN architecture allows the migration of existing RAN infrastructure, currently wound in tightly controlled vendor-specific boxes, into decoupled software implementations running on generic hardware (HW) platforms. Leveraging the software-defined CP to take advantage of AI in the RAN network, including several aspects from the RAN architecture and orchestration to the entities of RAN algorithm optimization. Specification of the function block and standardization of the related interface to build an open platform to enhance the operator's management and control of the RAN network by using AI and big data processing technologies will lead to highly efficient and optimized device management and RRM in complex network environments.

RAN Virtualization, Modularity, and Open-source Components: The O-RAN architecture enables cloudification and promotes open-source HW/software (SW) components. This includes defining the Network Functions Virtualization Infrastructure (NFVI)/Virtual Infrastructure Management (VIM) requirements to allow for virtualization of the various splits, e.g. high layer split PDCP/RLC, low layer split PHY, CP–UP split, and virtualized CU.

Open Interfaces: The O-RAN reference architecture specifies a decoupled RAN architecture by standardizing various interfaces between the decoupled RAN components. This includes enhancing the 3GPP-defined interfaces (E1, F1, V1, X2) to make them truly interoperable between vendors and defining additional new interfaces like NGFI-I (FH interface between the DU and RRU), E2 interface (between the CU/DU and the RIC near-RT), and the A1 interface (between the RIC near-RT and the network management system (NMS)/RIC non-RT to RIC near-RT).

White Box Hardware: Based on the decoupling of hardware and software and the development of common components and platforms, specify and release complete reference designs of high performance, spectral and energy-efficient white box BSs, including the reference hardware and software architectures, detailed design schematics for both BBU and RRU. It not only lowers the entry barrier to new ecosystem players but also takes advantage of scaling benefit and reduces the industry cost.

Open-source Software: Drive the open-source strategy of complete O-RAN software including the radio intelligent control software framework, protocol stack, baseband processing, and virtualization platform that should be driven through existing open-source communities.

12.3.1.3 Scope of RAN Intelligence Use Cases

Data analytics and ML are expected to empower the network with the following capabilities [39]:

- Reliable prediction: The vast multidimensional data collected from the network enables the capability to predict the network traffic, network anomaly, service pattern/type, user trajectory/position, service QoE, radio fingerprint, interference, etc. These predictions will undoubtedly empower the proactive network management and control, leading to significantly improved network resource and EE and customized user experience assurance.
- Advanced network optimization and decision: Driven by the collected data from the real network, data analytics and ML can help to efficiently solve massive problems in the 5G network that are usually hard to model or suffering great computation complexity due to the extremely high dimensions or non-deterministic polynomial-time hardness (NP-hardness).

Focusing on the RAN, the use cases can be roughly categorized into four types: intelligent network management and orchestration, intelligent MEC, intelligent RRM, and intelligent radio transmission technology. It may also further extend to optimize the radio frequency area, e.g. AI-assisted digital pre-distortion. Figure 12.9 shows a natural

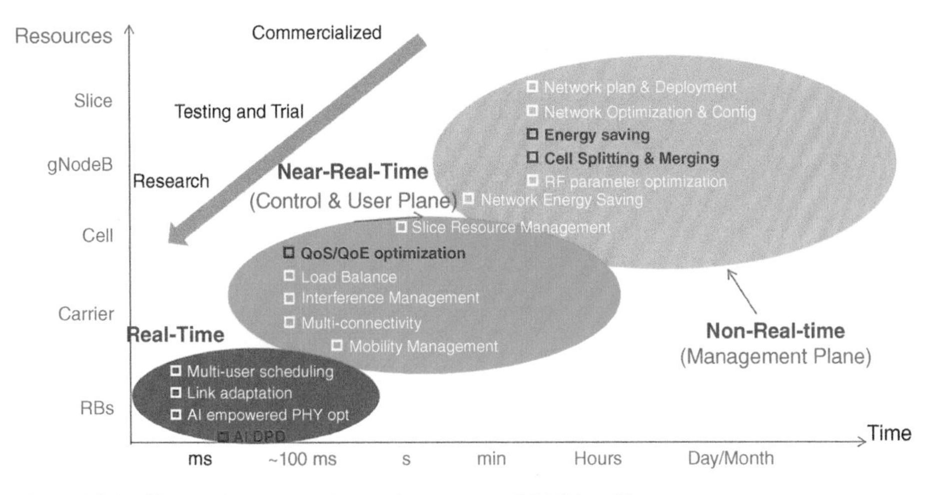

Figure 12.9 The application realms and use cases of RAN intelligence.

clustering of typical RAN intelligence uses cases according to the time and resource resolution, i.e. granularity. There is a clear positive correlation in these two dimensions. The larger in scale the set of resources to be optimized is, the longer the optimization time required will be; whereas the faster the time response needed is, the smaller the set of resource optimized can be. Generally speaking, the realms of management plane optimization, CP optimization, and scheduling and transmission domain optimization are associated with non-RT, near-RT, and RT processing, respectively.

In the following, four typical use cases will be shared in detail from the operator's perspective based on the commercialized and test trial experience. For other use cases, readers could refer to [40].

12.3.1.3.1 Energy Saving To cope with the energy challenges caused by mobile network expansion, the multi-RAT cooperation energy-saving system (MCES) was developed by China Mobile to improve the EE of mobile networks. MCES interacts with the RAN in RT and can support 2G/3G/4G RAN equipment from multiple vendors.

Specifically, the MCES has three major technical features.

1) Network-level energy saving
2) Energy-saving cell discovery function based on big data
3) Cell turn-off/on at a lower time granularity

MCES has been deployed in 19 provinces, including one million cells. In 2020, the annual energy saving was over 230 million KWh. Now the MCES is also evolving to incorporate 5G system to enable the coordinated energy saving of both 4G and 5G networks [28, 41].

12.3.1.3.2 Automatic Anomaly Analysis Anomaly detection (AD) and analysis has always been an important part of the OAM system. By introducing a dynamic ML-based AD algorithm, the number of manual rules can be reduced and more accurate root cause

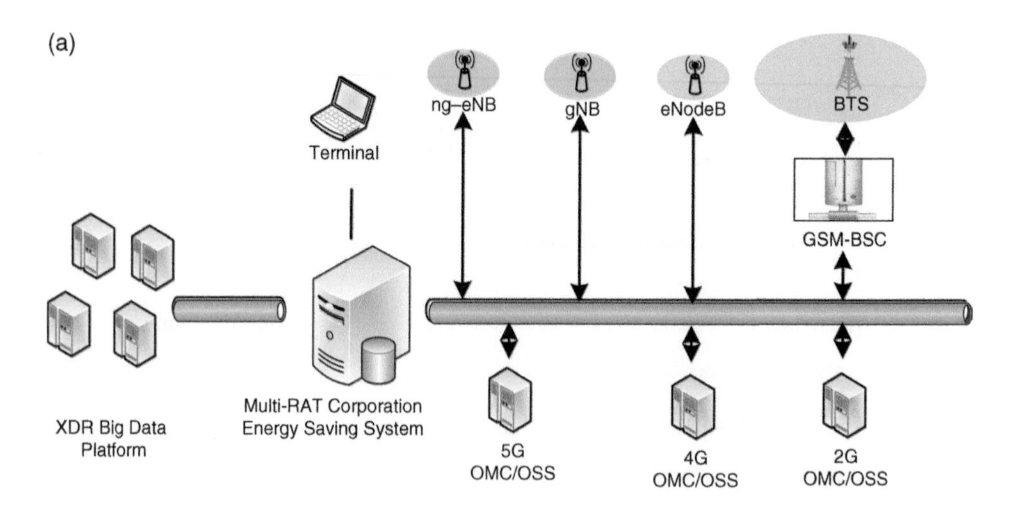

Figure 12.9 (a) Topology of multi-RAT cooperation energy-saving system (MCES).

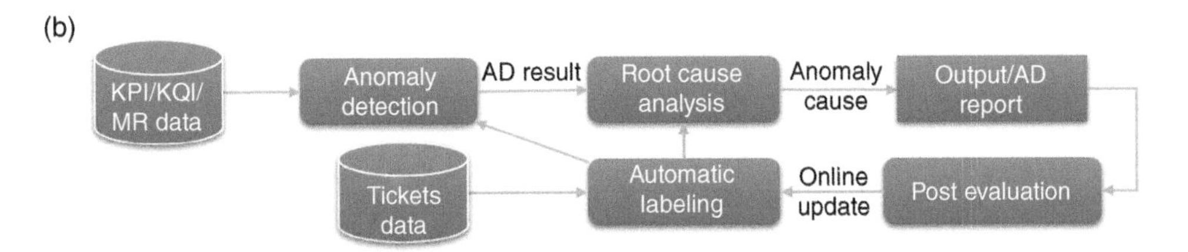

Figure 12.9 (b) Workflow of the automatic anomaly analysis.

information can be revealed. Meanwhile, the AI-based root cause analysis (RCA) algorithm directly replaces part of human effort in the anomaly analysis process.

12.3.1.3.3 Near-real-time QoE Optimization The 5G business model is changing from "volume" to "value." User QoE is playing a key role in the commercialization of 5G, and thus network optimization targets are shifting from key performance indicator (KPI)- to key quality indicator (KQI)-related QoE. RIC is positioned as a data-driven platform to provide customized RAN capabilities and exposure especially for vertical industries and OTTs. High-definition video streaming, cloud VR, and cloud gaming are expected to be the first majorly popular services in the 5G era. Multidimensional data can be acquired and processed via ML algorithms to support traffic recognition, QoE prediction, and finally guiding closed-loop QoS enforcement decisions. ML models can be trained offline, while model inference will be executed RT. The interface helps to deliver the policy/intents/AI models and RAN control to enforce the QoS for the QoE optimization.

In 2019 China Mobile Communications Corporation (CMCC) conducted trials in the Shanghai 5G network and the following features have been verified: (i) AI/ML-based QoE prediction and guarantee for cloud VR and (ii) radio bandwidth estimation for cloud VR adaptive coding selection.

12.3.1.3.4 Radio Fingerprint-based Traffic Steering Traffic steering, also termed mobile load balancing, is a widely used network solution to distribute the traffic load among cells or to transfer traffic in order to achieve network performance. This solution is to enhance the performance of traffic steering by constructing a radio fingerprint that divides the cell into grids by the serving cells and neighboring cells radio signal levels, in order to locate the user equipment (UE's) grid and perceive UE's coverage information, which can largely reduce the numbers of UE inter-frequency measurement and speed up traffic steering.

China Mobile with partners also carried out trial tests of radio fingerprint-based traffic steering in the commercial network. Based on the optimization of radio fingerprint-based traffic steering, the test results show, that compared with traditional load balancing, the duration of high loading is reduced by 13% and the measurement reconfiguration from BS and the measurement report signaling overhead from UE are reduced by 54 and 83%, respectively. Moreover, with the radio fingerprint-based load balancing, the average IP delay of the tested cells is reduced by 20%.

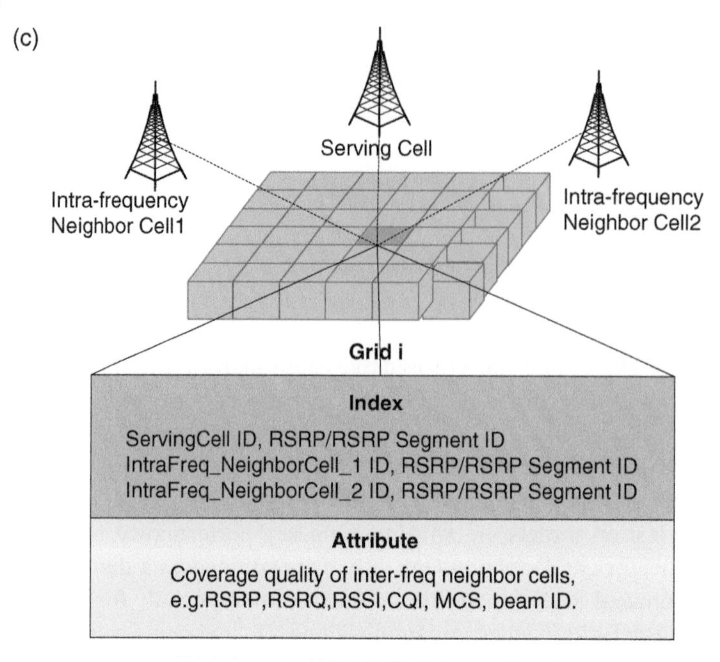

Figure 12.9 (c) 1 Virtual grid of radio fingerprint.

12.3.2 An OpenRAN Architecture with Native AI: RAN Intelligent Controller (RIC)

In this section, we will elaborate in detail an AI-embedded RAN architecture with hierarchical and distributed intelligence. It introduces the RIC in line with the O-RAN architecture defined in [37, 42].

Figure 12.10 shows this two-layered intelligent RAN architecture, which consists of two key entities: non-RT RAN intelligent controller (NRT-RIC) and near-RT RAN intelligent controller (nRT-RIC). The NRT-RIC is usually on the same level as traditional network management systems to provide enhanced intelligence for network management. The nRT-RIC is closer to the BSs to provide more fine-grained intelligent control and scheduling. Compared with the RT processing at TTI level in wireless networks, the control loop of nRT-RIC is in the order of 10–1000 TTI, while that of NRT-RIC is not less than 1000 TTI.

The NRT-RIC is introduced to enable closed-loop automation and optimization of RAN elements and resources, making it more intelligent (ML/AI), more granular (per-UE or group of UEs), and more flexible (intents/policies). It can provide policy-based guidance, ML model management, and enrichment information to the nRT-RIC function so that the RAN can optimize, for example, RRM under certain conditions. It can also perform intelligent RRM functions in non-RT interval (i.e. greater than 1 second). NRT-RIC can use data analytics and AI/ML training/inference to determine the RAN optimization actions. It may also utilize enrichment information of users, services, and environment, either acquired from external systems or generated from network management data, to assist the nRT-RIC.

Featuring inter-cell coordination and user-level optimization, nRT-RIC is introduced into the wireless cloud network to enhance RRM and resource allocation. nRT-RIC decouples RRM from air interface protocol processing and may acquire near-RT wireless resource data and control capability on demand. With the help of ML and service/user enrichment

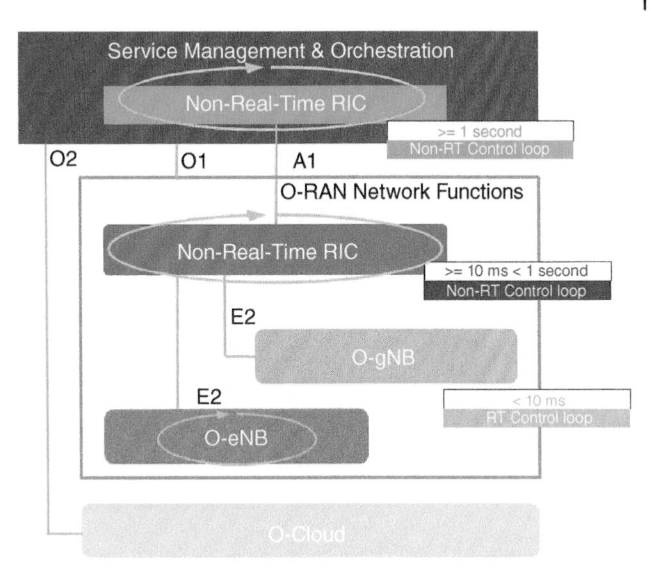

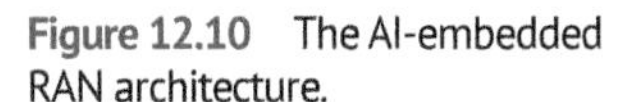
Figure 12.10 The AI-embedded RAN architecture.

information, the efficiency of intra- and inter- cell wireless resources can be enhanced, and differentiated services to different users are enabled, leading to a comprehensive optimization of network and services [43].

12.3.2.1 NRT-RIC Functions

NRT-RIC is a logical function that resides within the Service Management and Orchestration (SMO) framework and that has a direct association with the A1 interface toward the nRT-RIC. The primary goal of NRT-RIC is to support intelligent RAN optimization by providing policy-based guidance, ML model management, and enrichment information to the nRT-RIC function. It can also perform the intelligent RRM function in non-RT interval (i.e. greater than 1 second). NRT-RIC can use data analytics and AI/ML training/inference to determine the RAN optimization actions for which it can leverage management services such as data collection and provisioning services of the O-RAN nodes.

NRT-RIC functional architecture, shown in Figure 12.11, includes three key components: radio applications (rApps), NRT-RIC framework, and open APIs for rApps.

- rApps are modular applications that leverage the functionality exposed by the NRT-RIC to provide added value services relative to intelligent RAN optimization and operation. Examples of such added value services include: providing policy-based guidance and enrichment information across A1 interface; performing data analytics, AI/ML training, and inference for RAN optimization or for the use of other rApps; and recommending configuration management actions, such as network energy saving.
- NRT-RIC framework is a collection of NRT-RIC framework functions. It includes the set of inherent NRT-RIC framework functions to support A1 interface and rApps, other NRT-RIC framework functions if any, and "implementation variable" functions that are deployed in NRT-RIC. NRT-RIC framework functions provide services to rApps via the open APIs.
- Open APIs for rApps (R1 interface) are NRT-RIC internal interface between rApps and the NRT-RIC framework. It exposes the NRT-RIC framework and SMO services to rApps. It's desirable to guarantee the rApps portability in the API design.

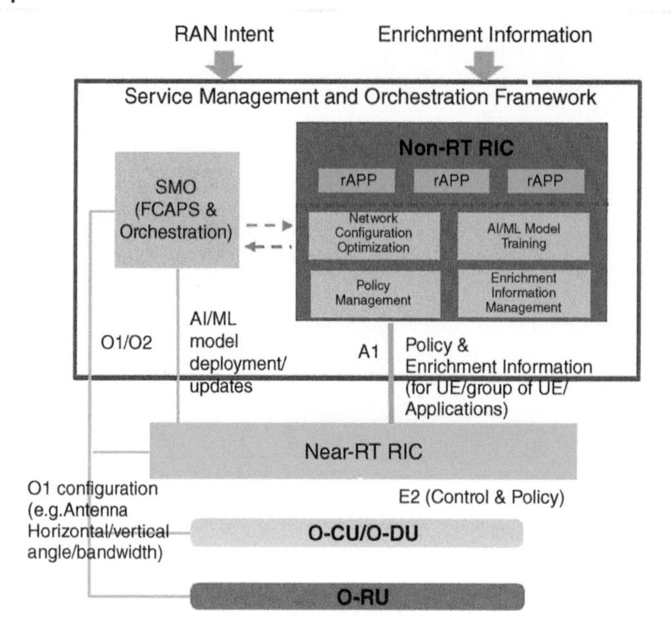

Figure 12.11 The non-RT RAN intelligent controller.

NRT-RIC framework functions include:

- rApps supporting functions, to provide services for the rApps, e.g. rApps service exposure functions, rApps conflict mitigation, etc.
- A1 functions, to enable the A interface related management, e.g. A1 logical termination, A1-policy coordination and catalog, A1-EI coordination and catalog, etc.
- AI/ML workflow functions, for introducing AI/ML functionalities, e.g. AI/ML model Life Cycle Management (LCM) and catalog and AI/ML continuous operation functions.
- Other logical terminations, e.g. external EI termination, external AI/ML termination, human–machine termination, etc.
- "Implementation variable" functions, e.g. data sharing, AI/ML model management functions, AI/ML data preparation functions, AI/ML modeling/training functions, which is left to implementation. If an "implementation variable" function is deployed in the NRT-RIC in a particular implementation, then this function is regarded as a part of NRT-RIC framework, and its service is exposed to rApps via "NRT-RIC framework service exposure function."

NRT-RIC enables the wireless data analytics and AI/ML capability in the management layer and also opens the opportunities to bring in external service demands/intent and features to the RAN. It may include the AI/ML model management functions, AI/ML data preparation functions, AI/ML modeling/training functions, and it can also leverage the external AI/ML platform capabilities. There are three terminations of external interfaces for an NRT-RIC with external connections. External EI termination is connected to external EI sources to import enrichment information for NRT-RIC applications. External AI/ML termination is connected to the external AI/ML server for ML model importation. Human–machine termination is used to inject RAN intent manually.

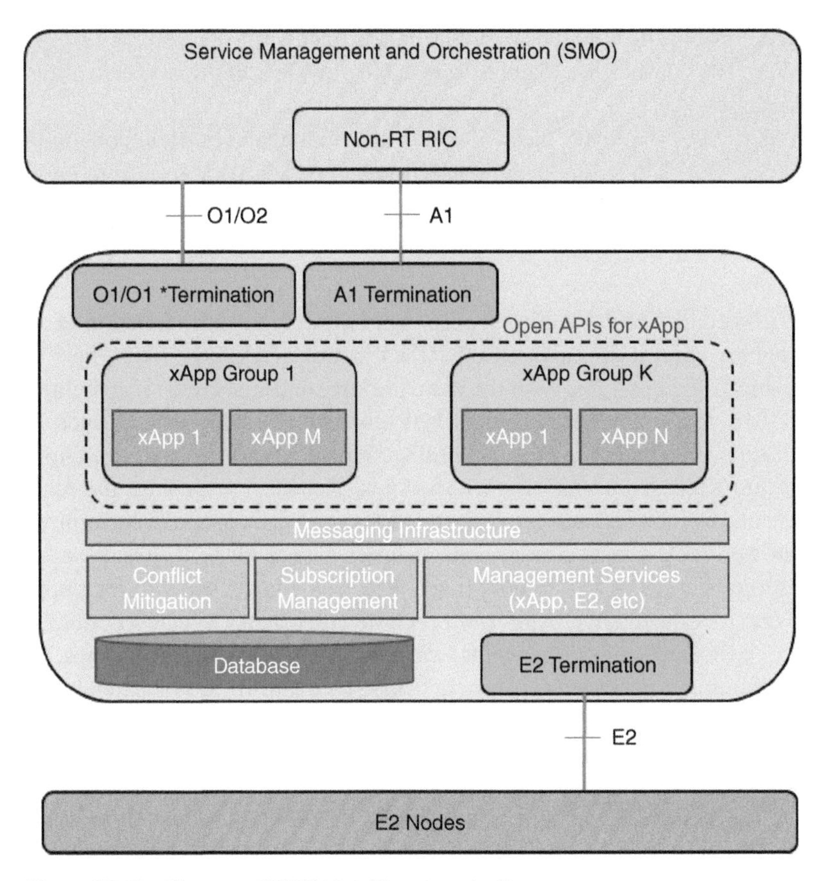

Figure 12.12 The near-RT RAN intelligent controller.

12.3.2.2 nRT-RIC Functions

Connected with BSs via the E2 interface (see Figure 12.12), the nRT-RIC could obtain the radio resource data from and issue control signaling to BSs. In the meantime it could interact with the NRT-RIC, receive the policies or models, and send feedback information via the A1 interface. It also supports the FCAPS via the O1 interface.

The nRT-RIC consists of two parts: nRT-RIC platform and x-applications (xApps). The two parts interact via the open API provided by the nRT-RIC platform. The nRT-RIC platform provides different xApps with the public platform functions. The basic platform functions include conflict mitigation, subscription management, database, messaging infrastructure, management services, security, and interface termination (E2/A1/O1).

- Conflict mitigation: to resolve the conflict between multiple RRC signaling messages generated by different xApps.
- Subscription management: to manage radio resource data subscriptions of xApps, avoiding duplicated subscriptions to the same radio resource data on E2 interface.
- Database: to store radio resource data and other types of data needed by nRT-RIC, which can be grouped into semi-static parameters and dynamic data. The semi-static parameters are mainly the configurations from network management systems and NRT-RIC and

the status parameters of cell/users. The dynamic data is mainly the measurements from cell/users. A shared data layer (SDL) can be designed to provide uniform access to different types of databases.
- Messaging infrastructure: to provide messaging transport among internal components of nRT-RIC.
- Management services: for FCAPS management of nRT-RIC.
- Security: to ensure security in nRT-RIC, e.g. preventing xApp from abusing radio resource data.

Based on diverse scenarios and different service/user requirements, the function of an xApp is customized to a specific use case, such that the function of nRT-RIC can be flexibly tailored by deploying different xApps. With the radio resource data and control capabilities provided by nRT-RIC, xApp produces characteristic customized radio services, such as QoS/QoE optimization, and load balancing. According to the input/output features, xApps can be classified into RRM xApps and non-RRM xApps. The RRM xApps use the radio resource data provided by nRT-RIC and generate the RRC signaling to meet the customized requirements. The non-RRM xApps use the radio resource data of nRT-RIC to expose the radio capabilities to other parties but do not provide RRC signaling. Open API for xApps allows the deployment of third-party xApps from vendors other than those of BSs and the nRT-RIC platform, laying a good foundation for a rich application ecosystem of xApps.

With the enrichment and opening of xApp, the nRT-RIC platform can be extended to provide an enhanced RRM algorithm warehouse, an AI model warehouse, and an AI model inference. The purpose of the enhanced RRM algorithm warehouse is to provide the common RRM algorithms required by different xApps and complete the translation from RRM requirements or decisions in xApps to RRC messages. The AI model warehouse is to provide common AI models required by different xApps. Compared with the way that AI models are packaged in xApp software package, the AI model warehouse may provide runtime AI models to xApps that match the actual hardware resources of the cloud platform, so as to improve the matching degree between running AI models and cloud platform. AI model inference aims to provide common AI model inference required by different xApps. It provides optimized AI model inference service for xApps according to platform resources and shortens the time of AI model inference.

12.3.3 Key Challenges and Potential Solutions

12.3.3.1 Customized Data Collection and Control

The data collection of traditional RAN mainly refers to the alarm, configuration, and performance data provided for network management. One of the main characteristics is the unified data model of the whole network. In current networks, the resource requirements and data transmission bandwidth of the BS equipment are considered. Generally, the time granularity of the network data collection is one minute or more, and the cell level performance data are the main ones. nRT-RIC adopts use cases based data subscription to subscribe data from the BS by triggering conditions and data items. At the same time, the data subscription items supported by nRT-RIC include cell level information and user level information. This data subscription can not only meet the needs of data with smaller time granularity in use cases but also greatly reduce the processing resource overhead and data transmission bandwidth requirements of BS equipment.

Early verification shows that for 71 cell-level and user-level data items of 1 s–1 ms, taking 14 000 cells and 100 users in each cell as an example, the volume of data collected in one day by traditional networks is 14 PB (Petabytes). Using the use case based data subscription of nRT-RIC, via the customization of trigger conditions, data items, cells, and users, the data collection can be accurately specified as a 1 ms UE-level data item of a user in a cell. With data collection during the 9:00–10:00 time interval, the corresponding data volume will be reduced to 0.413 GB (Gigabytes). Therefore, the use case based, customizable data collection of nRT-RIC, which can acquire radio resource data in a more flexible and low-cost manner, is the basis of opening up wireless capabilities on demand, supporting differentiated services for users, and realizing collaborative optimization of network and services.

The RRM of nRT-RIC can be divided into two types: "condition + action" policy or pure control instruction. When the nRT-RIC sends a "condition + action" policy to the BS, the BS judges whether the condition is met according to the internal information. If the condition is met, the action will be executed. This type of RRM can provide near-RT guidance information for RT radio resource scheduling of BSs, so as to achieve near-RT optimization of performance indicators such as wireless bandwidth and delay. As mentioned before, because of the introduction of enrichment information and ML, nRT-RIC has more comprehensive information and better optimized algorithms. When the nRT-RIC directly sends control instructions to the BS, the nRT-RIC has determined the timing of RRM according to the received data, so the BS is only responsible for the execution of the instructions.

12.3.3.2 Radio Resource Management and Air Interface Protocol Processing Decoupling

The functions of RAN can be divided into air interface protocol processing and RRM. The air interface protocol processing function refers to the functions, procedures, and information elements clearly defined by 3GPP to realize the air interface interoperability between RAN and UE. The RRM function is defined as the functions, processes, and information units to generate the radio resource allocation and adjustment information needed in the air interface protocol processing. The goal of traditional RRM is to improve the SE. With the evolution of wireless networks and rich application scenarios, the demand for EE optimization and refined wireless resource customization is increasing. Therefore, the RRM function aims to provide optimized radio resource allocation schemes for SE, EE, and customized radio resource requirements. The RRM function highlights customization and inter-cell features. It needs to obtain the measurement from the connected cell and to send control information.

The air interface protocol processing function has demanding RT requirements, focusing on single cell protocol processing. From the perspective of network deployment and application, the protocol processing function is relatively stable, and each software and hardware function upgrade typically is valid for a long period. However, the processing of the RRM function in nRT-RIC is less RT than that of the air interface, and it does more on the management of radio resource across cells and on customized requirements. Compared to the traditional network deployment approach, the software function iterates fast, and the functions supporting new business get online quickly. The decoupling of RRM and air interface protocol processing functions is conducive to the introduction of enrichment information of services and environment at user level and the introduction of ML technology, so as to achieve "centralized" inter-cell collaboration and flexible customized needs.

12.3.3.3 Open API for xApp

Open API is the interface between nRT-RIC platform and xApps. Open API showcases the characteristics and technical trends of ICDT convergence in nRT-RIC and treats the functions of nRT-RIC platform and xApps as reusable services. The open API interface between services uses lightweight interface communication, which aims to achieve efficiency, softwarization, and openness of such a system. Based on the open API interface, xApps can be customized to use cases and deployed using the services of nRT-RIC platform. At the same time, based on the open API, xApp can be provided by a third party other than the vendors of BS equipment and the nRT-RIC platform, which is conducive to a prosperous xApp ecosystem, as well as the collaborative optimization of network and services.

According to the nRT-RIC architecture, open API will support E2-related subscription and control functions, xApp management functions, and A1-related policies and enrichment information. With E2-related subscription and control functions, xApps may subscribe to dynamic radio resource data and issue control requests on demand, and the nRT-RIC platform may distribute dynamic radio resource data and control request feedbacks to xApps. The management functions of xApp support service discovery, registration, and activation. A1-related policies and enrichment information function support nRT-RIC platform to distribute A1 policy and enrichment information to xApps and receive feedback.

12.4 Ecosystem Progress from 5G to 6G

A particular entity stands out, announced in Barcelona during Mobile World Congress (MWC)2018. True to its launch statement: O-RAN represents a comprehensive embodiment of Communication 4.0 and is empowered by the true ICDT convergence. It will drive the RAN from Green and Soft to open and smart! The section will enlist O-RAN Alliance as well as related development in the global community.

12.4.1 O-RAN Alliance

The O-RAN Alliance was launched in 2018, and currently it has more than 275 members globally and over 3000 active technical contributors committed to promote ICDT-integrated RAN development together. With its focus toward openness and intelligence, the Alliance has so far delivered more than 100 technical spec and reports based on the reference architecture in Figure 12.13 below [42]. The scope covered include open xhaul (both FH and MH), open interfaces (A1, O1, O2, E2), open cloud with RAN appropriate accelerator abstraction, open reference designs hardware for indoor small cells and outdoor macrocell, SMO framework (OAM, Management and Network Orchestration (MANO), and NRT-RIC), nRT-RIC, three sets of open-source software [44],integration and testing spec, etc. It's worth noting that the O-RAN Open Testing and Integration Center (OTIC) initiative is triggering an array of globally distributed open facility being set up for any hardware or software technology and solution provider to enjoy open and robust environment for O-RAN conformance testing and interoperability testing.

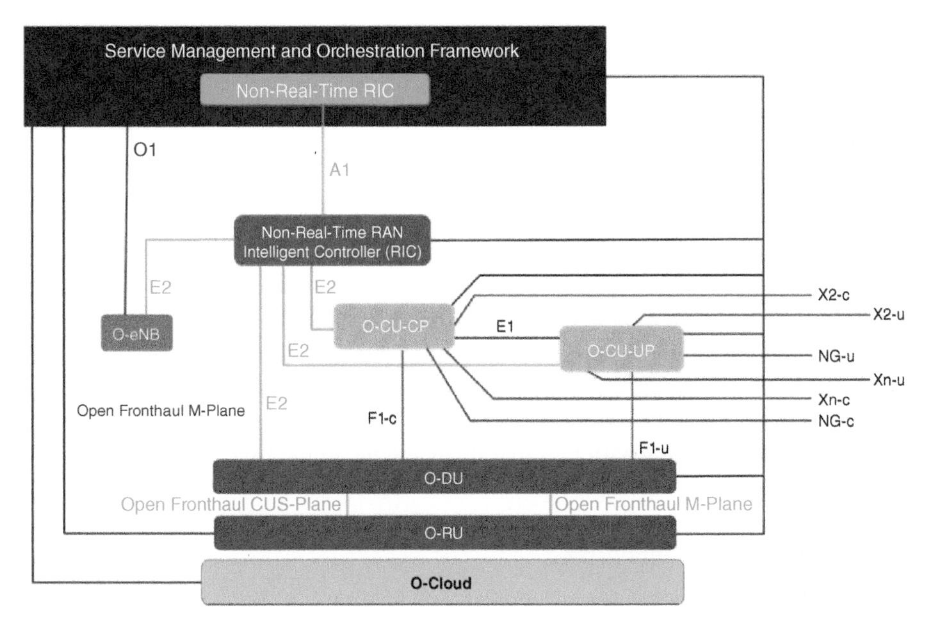

Figure 12.13 O-RAN overall logical architecture.

Collaboration and partnership with various global standards developing organizations (SDOs) and industry fora have been a mainstay of O-RAN [45]. It has been collaborating with Open Networking Foundation (ONF), Small Cell Forum (SCF), Telecom Infrastructure Project (TIP), Global System for Mobile Association (GSMA), Next Generation Mobile Networks Alliance (NGMN), ETSI, etc. and interacting with 3GPP, International Telecommunication Union - Telecommunication Standardization Sector (ITU-T), IEEE, Internet Engineering Task Force (IETF), Internet Assigned Numbers Authority (IANA), Open Compute Project (OCP), Open Air Interface (OAI) [46], etc. Through these partnerships, a coherent revolutionary evolution is taking place smoothly in the global RAN ecosystem.

12.4.2 Telecom Infrastructure Project

The OpenRAN project under TIP is another endeavor on the network openness [47]. OpenRAN focuses on realization and deployment of RAN solutions on a common processing platform. It will help to foster an open ecosystem of complete RAN solutions and solution components that can take advantage of the latest capabilities of the common processing platform, both at the software level and through programmable mechanisms. OpenRAN focuses on two fundamental aspects of RAN development: software-defined radio-based (SDR-based) software and GPP-based hardware. The radical change in the building blocks will make OpenRAN cheaper to build and operate in a more flexible and powerful way. Interfaces and operating software separate RAN components and build a modular BS software stack that can run on GPP hardware. This architecture means that BBUs, radio units, and remote radio heads can be assembled from any vendor to construct a network. OpenRAN project will be delivering prototype and early-stage field trials based on O-RAN Alliance technical specifications.

12.4.3 GSMA Open Networking Initiative

GSMA has also recognized the trend toward openness. In 2019, GSMA set up an initiative called Open Networking Initiative, exploiting the key solutions, strategy, and potential roadblocks and solutions to fulfill open networking [48]. In their initial study, it is concluded that the virtualized infrastructure, together with the open interface would be the potential solutions to operators' challenges in the 5G era and beyond. The cloud infrastructure could bring flexibility and increase the operation and utilization efficiency of mobile networks. The openness based on the open interface could reinvigorate the ecosystem by introducing more innovative suppliers. However, on the path toward true openness, several challenges lie ahead. For example, security issues will become more noticeable in the context of multi-vendors compared with traditional single-vendor environments. In the meantime, the intellectual property regime needs to be revisited and rearranged for the support of a promising new ecosystem with multiple small and medium-sized suppliers.

12.4.4 Open-source Communities

Unlike traditional SDOs, developing technical specification is not the only focusing area for the O-RAN Alliance. From the first day when the O-RAN Alliance was formed, it has already adopted open source as a key enabler. In order to realize the O-RAN commercialization as early as possible, the O-RAN Alliance has deemed that software development is as important as specification development and therefore turned to open-source community, a powerful asset successfully demonstrated long as a core feature of the IT industry.

O-RAN Software Community (OSC) was officially established jointly with the Linux Foundation (LF) on 2 April 2019. The mission of the OSC is to develop open-source software based on the specification developed in the O-RAN Alliance with the aim to eventually build a modular, open, intelligent, efficient, and agile disaggregated RANs.

Figure 12.14 shows the OSC project structure. Two repositories were set up for OSC, with Apache 2.0 and O-RAN Software License (fair, reasonable, and non-discriminatory [FRAND] based) [49] intellectual property rights (IPR) licenses, respectively, to accommodate both IT and CT communities' preference. As of the end of 2020, the OSC contains 12 projects plus one Requirements and Software Architecture Committee.

The OSC utilizes upstream projects such as Open NFV (OPNFV) [50], Akraino, Open Network Automation Platform (ONAP), etc. to accelerate the software development. For example, the Akraino is a set of open infrastructures and application blueprints for the edge [51], which could be utilized for RIC platform. As of the end of 2020, the OSC has released three open-source releases, namely, Amber, Bronze, and Cherry to the industry. With the contribution from dozens of companies, the community has delivered over 2M lines of codes. The key O-RAN components, including the NRT-RIC, nRT-RIC, RIC platform, etc., have all been developed to demonstrate a key use case, traffic steering, which has been identified among the top-priority use cases facilitating mobility management, interference management, load balancing, etc. Open Testing Framework (OTF) in Figure 12.15, as part of project INT, is the infrastructure in OSC integration and testing lab. OTF provides a set of virtual test heads (type of micro services), which can be used

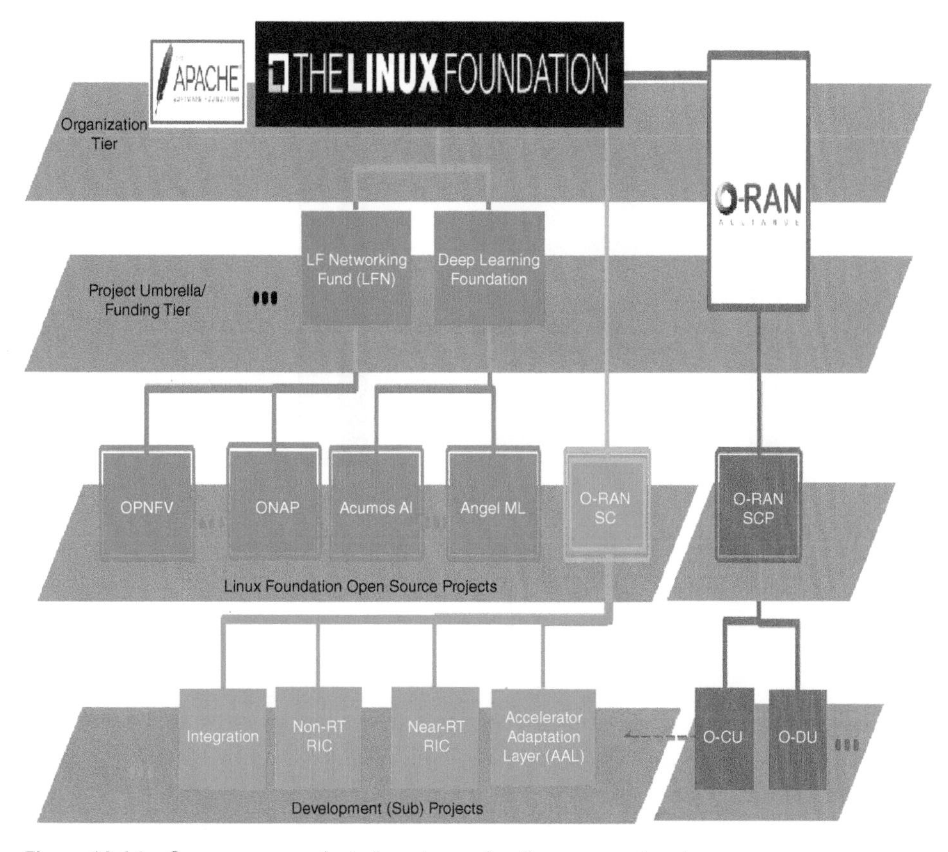

Figure 12.14 Open-source projects for edge and radio access network.

to build various OTF test strategies for OSC component or service testing. It's an open platform used actively for testing Cherry release use cases currently.

12.5 Conclusion

In the past 10 years, 4G has transformed our lives, but it is still struggling with ICT convergence as an afterthought. 5G is destined to transform our society by enabling the digital transformation of verticals. It does enjoy the benefits of a moderate degree of ICT convergence, but it is still only playing catch up on true ICDT deep convergence. In the meantime, nontraditional SDOs and alliances, together with the open-source community, are taking great strides forward in the continuous revolutionary evolution of network access and edge. The 6G era will be upon us in 8–12 years. We can be certain that 6G will be built with full ICDT deep convergence, and thus, be truly green, soft, open, and smart. Therefore, it seems that O-RAN may very well be the essential DNA of 6G access and edge networks. Furthermore, it should be obvious that open source culture and AI/ML integration are bound to transform traditional SDO operations, as well as the resultant standards in the 6G era.

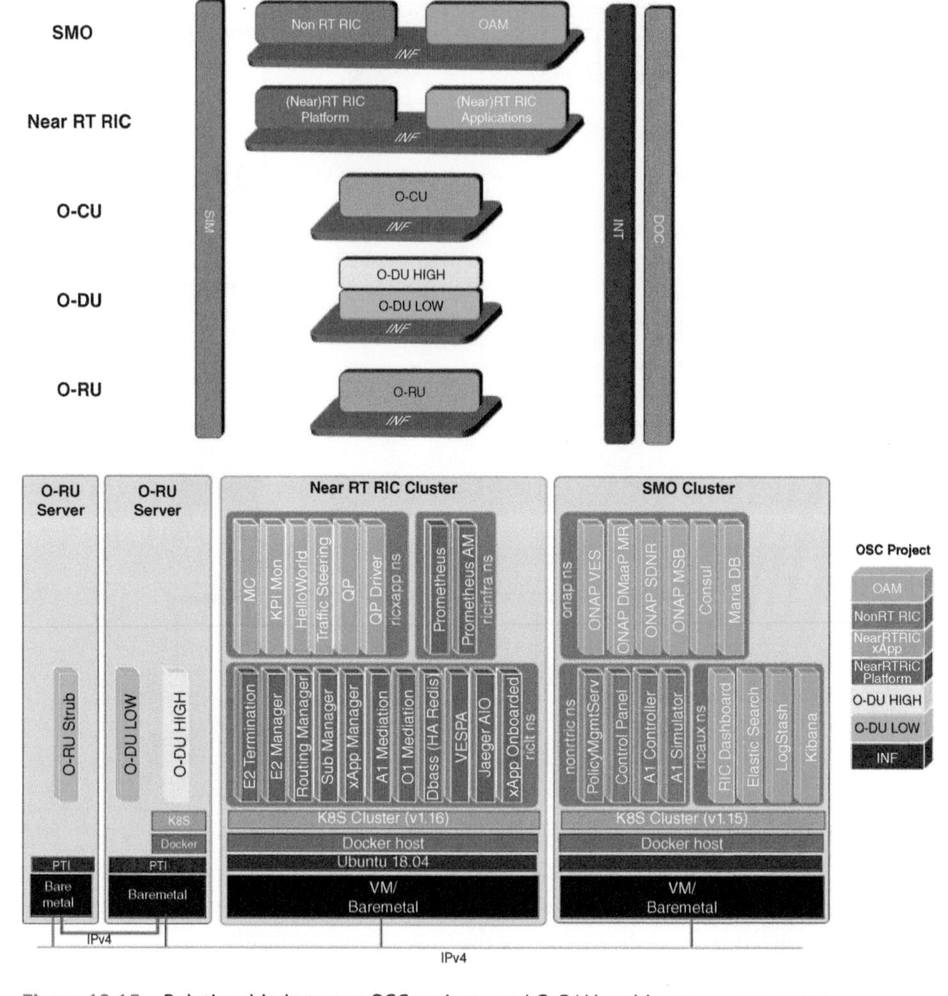

Figure 12.15 Relationship between OSC projects and O-RAN architecture components.

Acknowledgments

The authors would like to express their gratitude to Yannan Yuan, Weichen Ni, Xiaofei Xu, Qi Sun, and Ran Duan for their contribution to this work over the years in the Green Communications Research Center of China Mobile Research Institute.

References

1 Ring, D.H. and Young, W.R. (1947). Mobile telephony – wide area coverage – case 20564. https://web.archive.org/web/20120207062016/http://www.privateline.com/archive/Ringcellreport1947.pdf.

2 Chih-Lin, I. (2013). Towards green and soft. Plenary, IEEE WCNC.
3 Chih-Lin, I., Rowell, C., Han, S. et al. (2014). Towards green & soft: a 5G perspective. *IEEE Commun. Mag.* 52 (2): 66–73.
4 FuTURE FORUM (2014). 5G: rethink mobile communications for 2020+.
5 FuTURE FORUM (2015). 5G: rethink mobile communications for 2020+ Version 2.0.
6 GSMA Intelligence (2014). Understanding 5G: perspectives on future technological advancements in mobile. white paper.
7 Chih-Lin, I., Yu, G., Han, S., and Li, G.Y. (2019). *Green and Software-defined Wireless Networks: From Theory to Practice*. Cambridge.
8 Chih-Lin, I., Han, S., and Bian, S. (2020). Energy-efficient 5G for a greener future. *Nat. Electron* 3: 182–184.
9 Chih-Lin, I. (2017). SDX: how soft is 5G? Plenary, IEEE WCNC.
10 Chih-Lin, I., Han, S., Xu, Z. et al. (2016). New paradigm of 5G wireless internet. *IEEE JSAC* 34 (3): 474–482.
11 GreenTouch Consortium. http://www.bell-labs.com/greentouch/index-page=about-us.html.
12 CMRI WP (2014). C-RAN: the road towards green RAN. http://labs.chinamobile.com/cran.
13 Chih-Lin, I., Huang, J., Duan, R. et al. (2014). Recent progress on C-RAN centralization and cloudification. *IEEE Access* 2: 1030–1039.
14 Quek, T.Q.S., Peng, M., Simeone, O., and Yu, W. (2017). *Cloud Radio Access Networks*. Cambridge.
15 CMRI WP (2015). Next generation fronthaul interface. http://labs.chinamobile.com/cran.
16 Chih-Lin, I., Yuan, Y., Huang, J. et al. (2015). Rethink fronthaul for Soft RAN. *IEEE Commun. Mag.* 53 (9): 82–88.
17 Chih-Lin, I. and Huang, J. (2017). RAN revolution with NGFI (xHaul) for 5G. OFC (19–23 March 2017).
18 Chih-Lin, I., Li, H., Korhonen, J. et al. (2018). RAN revolution with NGFI (xhaul) for 5G. *IEEE J. Lightwave Technol* 36 (2): 541–550.
19 IEEE NGFI. https://sagroups.ieee.org/1914/.
20 3GPP TR38.801 (2017). Study on new radio access technology: radio access architecture and interfaces. V1.2.0 (2017-02).
21 Wong, V.W.S., Schober, R., Ng, D.W.K., and Wang, L.-C. (2017). *Emerging Technologies for 5G Wireless Systems*. Cambridge.
22 Tornatore, M., Chang, G.-K., and Ellinas, G. (2017). *Fiber-wireless Convergence in Next-generation Communication Networks*. Springer. [1] LF OPNFV. https://www.opnfv.org/.
23 Chiang, M., Ha, S., Chih-Lin, I. et al. (2017). Fog computing and networking. *IEEE Commun. Mag.* 55 (4 and 8): 16–20.
24 Ku, Y.-J., Lin, D.-Y., Lee, C.-F. et al. (2017). 5G radio access network design with the fog paradigm: confluence of communications and computing. *IEEE Commun. Mag.* 55 (4): 46–52.
25 Taleb, T., Samdanis, K., Mada, B. et al. (2017). On multi-access edge computing: a survey of the emerging 5G network edge cloud architecture and orchestration. *IEEE Commun. Surv. Tutor.* 19 (3): 1657–1681.
26 Mao, Y., You, C., Zhang, J. et al. (2017). A survey on mobile edge computing: the communication perspective. *IEEE Commun. Surv. Tutor.* 19 (4): 2322–2358.

27 ETSI NFV ISG (2012). Network functions virtualisation. [Online]. http://portal.etsi.org/portal/server.pt/community/NFV/367.

28 I, C.-L., Liu, Y., Han, S. et al. (2015). On big data analytics for greener and softer RAN. *IEEE Access* 3: 3068–3075.

29 FuTURE Forum 5G SIG (2017). Wireless big data for smart 5G. http://www.future-forum.org/dl/171114/whitepaper2017.Rar.

30 Han, S., Chih-Lin, I., Wang, S., and Sun, Q. (2017). Big data enabled Mobile Network Design for 5G and beyond. *IEEE Commun. Mag.* 55 (9): 150–157.

31 Chih-Lin, I., Sun, Q., Liu, Z. et al. (2017). The big-data-driven intelligent wireless network: architecture, use cases, solutions, and future trends. *IEEE Veh. Technol. Mag.* 12 (4): 20–29.

32 Kibria, M.G., Nguyen, K., Villardi, G.P. et al. (2018). Big data analytics, machine learning, and artificial intelligence in next-generation wireless networks. *IEEE Access* 6: 32328–32338.

33 Alexiou, A. (2017). 5G wireless technologies. IET.

34 Cheng, X., Fang, L., Yang, L., and Cui, S. (2017). Mobile big data: the fuel for data-driven wireless. *IEEE Internet Things J.* 4 (5).

35 Cao, X., Liu, L., Cheng, Y., and Shen, X.S. (2018). Towards energy-efficient wireless networking in the big data era: a survey. *IEEE Commun. Surv. Tutor.* 20 (1): Q1.

36 3GPP (2017). Study of enablers for network automation for 5G. 3GPP SA2, Hangzhou, China, S2-173827.

37 O-RAN Alliance (2018). O-RAN: towards open and smart RAN. www.o-ran.org.

38 WAIA (2018). Mobile AI for Smart 5G – Empowered by WBD. WAIA/FuTURE/WWRF. http://www.future-forum.org/dl/190202/5G2018MWC.pdf.

39 FuTURE FORUM (2019). Open ICDT: convergence of IT, CT and DT for 5G RAN and beyond.

40 O-RAN Alliance (2020). O-RAN use cases detailed specification 2.0. https://www.o-ran.org/specifications.

41 Chih-Lin, I. (2021). AI as an essential element of a Green 6G. *IEEE TGCN* 5 (1): 1–3.

42 O-RAN Alliance (2020). O-RAN architecture description v1.0. https://www.o-ran.org/specifications.

43 Chih-Lin, I., Kuklinski, S.A., Chen, T., and Ladid, L. (2020). Perspective of O-RAN integration with MEC, SON, and network slicing in the 5G era. *IEEE Network. Mag.* 53 (9): 3–5.

44 O-RAN Open Source. https://www.o-ran.org/software.

45 ONF SD-RAN. https://opennetworking.org/onf-sdn-projects/.

46 OAI OSA. http://www.openairinterface.org.

47 TIP openRAN. https://telecominfraproject.com/openran/.

48 GSMA Open Network. https://infocentre2.gsma.com/gp/pr/ONG/Pages/Default.aspx.

49 LF & O-RAN Software Community. http://o-ran-sc.org.

50 LF OPNFV. https://www.opnfv.org/.

51 LF Akraino. https://wiki.akraino.org/.

13

"One Layer to Rule Them All": Data Layer-oriented 6G Networks

Marius Corici and Thomas Magedanz

Frauhofer FOKUS, Berlin, Germany

13.1 Perspective

Up to 4G, the network system relied on optimizations of the communication protocols and on the offering of the same service for all subscribers [1]. With using protocols highly optimized for the radio and the core network and with optimizing the connectivity service offered to the subscribers, the networks were able to offer to a very large number of devices the specific features of reachability, mobility support, quality of service (QoS) support, and access control and security [2]. This type of optimization was pushed to the maximum by usage of binary protocols with complex to understand encoding schemas such as the Non-Access Stratum (NAS) protocol and by a minimal acceptance of differentiation between the connected devices requirements [3].

With 5G, a new level of efficiency was reached by the deployment of the network functions as software [4]. It made place for customization of the functionality itself and hence, of the communication protocols and of the adaptation of the policies toward the communication environments [5]. The flexibility in functionality provided the basis for a large number of algorithms and mechanisms to be implemented addressing the requirements of a high variation of use cases. This is done through gaining insight on how the networks are functioning and through using it for dynamically adapting the functionality [6].

For obtaining a new level of efficiency and flexibility, 6G will deploy even more network management and automation functionality, managing the unmanageable, also with the wide usage of machine learning and artificial intelligence techniques [7]. The main requirement of both these systems is the acquisition of information from the network itself and the generation of new information on the system either based on automatic behavior or through learning [8]. For this to be able to achieve its true potential, the data will have to be exchanged continuously, when and how needed, between the different network elements. Only in this way, the network will have a rapid enough reaction and adaptation to be immediately used as part of the devices' connectivity control.

Shaping Future 6G Networks: Needs, Impacts, and Technologies, First Edition.
Edited by Emmanuel Bertin, Noel Crespi, and Thomas Magedanz.

13.2 Motivation

With the development of the software system within 5G, a very large amount of data became available in the different network locations [9]. It includes an extensive subscriber profile with information from the mobile device, radio equipment, and core network. Also, an extensive network functioning and behavior profile can be generated using the infrastructure and the network function level monitoring information [10].

Additional to that, a very large amount of other data sources are steadily becoming available within the public domain such as extensive Earth Observation data using low-orbit satellites and unmanned automatic flying vehicles, sensor information from local deployments or large-scale networks, video surveillance information, etc. [11]. These sources of information can be used for further understanding of the network usage behavior and, as such, to more granularly customize it.

Furthermore, for the ultra-automation and ultra-flexibility expected from the 6G systems, there is a need that proper insights on the data will be generated in real-time [7]. For this, new promising machine learning techniques are developed [12]. However, these techniques could not function, they could not even be developed, without a very large quantity of data being extracted and made available at specific network locations where the algorithms are functioning. Furthermore, their effectiveness cannot be tested, and they cannot be further adapted without a continuous stream of data that enables the further insight generation and adaptation (Figure 13.1).

Additionally, with 5G an initial step toward ultra-customization of the services toward the subscriber needs was implemented [13]. This represented a first step toward an extensive emulation of the system toward the specific subscriber needs through the continuous adaptation of the configurations of the system. To be able to make this ultra-customization highly effective, it needs to have access to multiple sources, sometimes of heterogeneous data, as well as extensive development of algorithms converging the insight of this data.

However, at the current moment data does not flow [14]. Even though very large quantities of data are generated at every moment and even though a large amount of

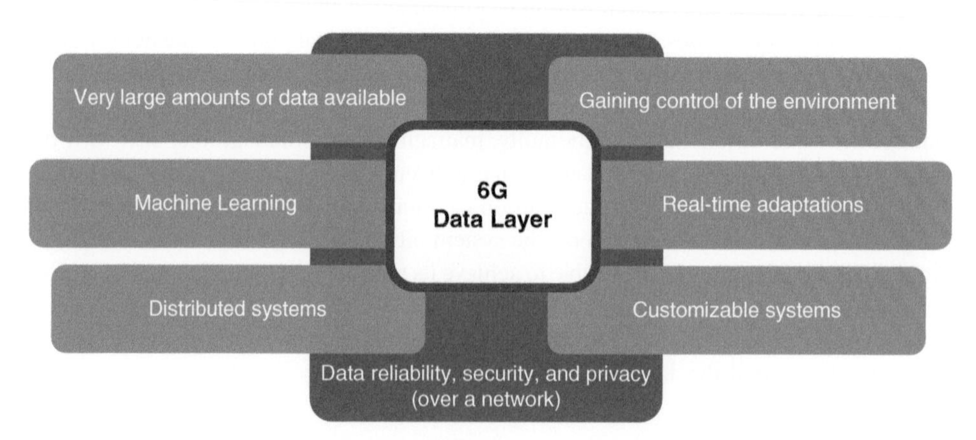

Figure 13.1 Technology evolution triggers enabling a new extensive data layer for 6G.

this data can be used for machine learning insight generation, this is done on a static, batch manner, most of the time. On a very punctual basis, an algorithm or a customization of configurations based on some specific source of data has proven that a specific optimization can be achieved. Currently, there are no graceful means to widely discover data and to exchange and to process it at a comprehensive network scale. This makes also the advancements in data processing, algorithms, and system adaptation based on insight rather limited.

With going in the direction of a data layer that regulates the way the data is exchanged from the discovery of the expected data sources, the end-to-end data exchange assurance, the expected privacy level, the curation of data in terms of quality and intelligent fixing of the specific incomplete data issues and especially in terms of being able to exchange across wide area networks and the Internet the expected quantities of data in the expected time intervals, this data blockage that we are seeing now can be overcome and a new level of network flexibility and adaptability can be achieved.

13.3 Requirements

For a data layer to prove its effectiveness in the different network domains, a subset of the following extensive set of requirements should be considered. The requirements presented here are ordered based on their importance to be fulfilled.

- Data privacy – one of the critical aspects for the enablement of a continuous exchange of the data within the same or between different systems remains the trust into the assurance of the data privacy. As the machine learning techniques evolve, the current assumed data privacy mechanisms, such as data anonymization, are highly challenged. Also, the usage of meshed up data can provide a large amount of information on a situation not only to improve the system but also to enable the malicious users. To be able to address this request, new mechanisms, always prepared to be adapted and exchanged, have to be provided. Furthermore, data exchange as considered in the next requirements has to be also protected.
- Data curation – to be useful within an automatic machine system, data has to be uniform. However, most of the data accumulated by a system has different formats, errors, or even data is missing. For this, a new data layer has to include a set of functions that are able to explore and profile the data received, to format it to make it consistent, as well as to improve the data quality. This can be achieved by the intelligent decisions on data errors, such as missing fields or measurement points, through the introduction of proper artificial data or through the elimination of the erroneous data element from further considerations. These operations are rather similar to the ones that a curator of a museum has to execute, with the large difference that in case of data it should be continuously executed on a continuous influx of data (like a museum on a fast forward mode where also a huge quantity of historical artifacts would be accepted).
- Graceful data exchange – to be able to use the data in an effective manner across the network system, data has to be effectively transmitted between the different network locations in the expected time interval and with the expected impact on the network.

Data has different shapes that have to be considered. This includes the size of the different data pieces as well as the frequency with which the data is batched and transmitted across the network. For example, data exchanges can be considered from the single bytes of sensor information up to terabytes of Earth Observation images. Furthermore, data may be exchanged at every second from the sensors; it may include one-time batches of historical data, or in most of the cases it may include regular short duration transmissions such as hourly or daily batches.

The transmission over the network should also consider not only when data is generated but also the expected receiving times. Most of the data has a rather high level of flexibility between the generation and the usage. This includes processing of the data even many days later after it was received. So, the impact on the network can be minimal, as the data exchange would represent mostly background data traffic.

- Data firewalling – there is a high level of opening of the network for being able to transmit data to other entities that would create a security breach unless properly managed. This includes the protection from unwanted external attacks and the possible disturbances in the data flows similar to any other communication over the network.
- Data discovery, identity, trust – to be able to use the data coming from other sources, the data has to be discovered. For this a proper presentation and marketplace should be considered with the specifics of the available data and how it can be accessed. For being able to trust the quality of the specific data, especially in critical decisions, the source of the data should be established on a non-repudiation basis. Especially in case of erroneous decisions taken based on errors coming from the data, a set of liabilities will have to be considered in such a manner that they both protect the source of the data as well as the user of the data. This becomes highly complex when the data is automatically exchanged and processed across long chains of data where it passes through multiple administrated domains.
- Data streams orchestration – to be able to make the most of the data it is utmost important to create data streams that pass through multiple processing and decision points. The data chains consist of multiple segments each with different communication characteristics. Furthermore, the data passes through multiple data firewalls and processing elements as well as through multiple curation elements. For the data to be exchanged without any issues, an orchestration mechanism has to be set in place that includes the different actors and enables to exchange the data across the network.

Most of these requirements are already responded in a minimal form by 5G networks. A future 6G network can build up on the existing functionality and bring it as the basis within a comprehensive data layer architecture. This would assure the smooth transition from the existing current network toward a data-oriented one.

Furthermore, the cohesion around a comprehensive data layer will provide the advantage of reducing the end-to-end decision cycle as well as the opportunity to enable the swapping of insight generation algorithms during testing phases, with this greatly accelerating the development of such algorithms through immediate feedback on their quality (Figure 13.2).

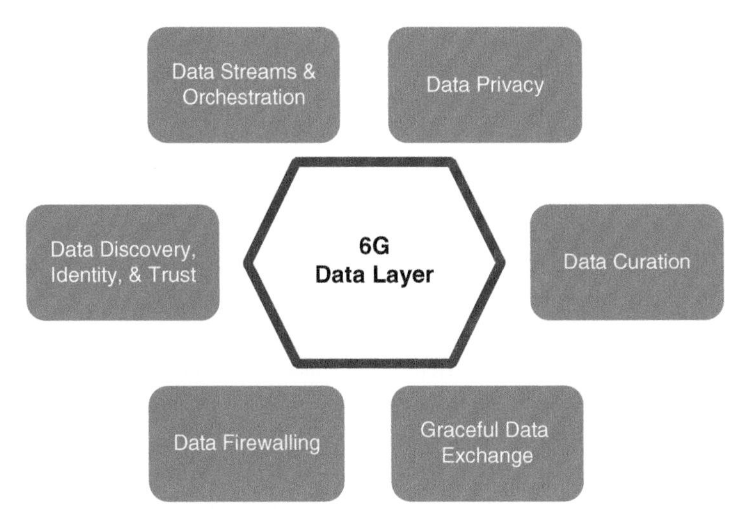

Figure 13.2 Major requirements of a 6G data layer.

13.4 Benefits/Opportunities

The data layer is acting as a middleware between the active system and the network management. It interacts with both through the exchange of data at different complexity levels: with different formats, quantities, or frequencies, batch and real-time. Furthermore, it can interact with the data layers or directly with other systems for exchanging the same type of data between different domains (Figure 13.3).

The active system has as main role to interact with the real-world sensors and devices and to provide the specific services be it radio and core network connectivity across a 6G system or be it an end-to-end service that transmits data between subscribers and the network.

Based on the information provided from the active systems on the real-world usage of the active system and the flexibility of its customization capabilities, the intelligence has the role to gain additional insight on how the system is used and to provide recommendations for a better usage.

The data layer may interact with the active system information or with external intelligence sources, providing more sophisticated mechanisms for system adaptation through the mesh-up of multiple data sources.

Being an intermediary layer between the existing network elements, the following major technology-oriented benefits can be immediately considered:

- With providing a flexible exchange of information between the active system and the intelligence layer – with the adding of any intermediary layers, a large flexibility is gained on how the information is exchanged between the already existing layers. In this situation, the data layer can route the information between different administrative domains and between different actors, as well as between different systems and insight generation algorithms.

- Enabling a continuous exchange of streams of information – one of the main missing elements in the current networks and needed for 6G is the ability of the system to sustain the machine learning and automatic decision elements with a continuous stream of data both for the continuous learning and for being able to influence the system. Lack of real-time information exchange limits the possibility of gaining highly customized insight to a given situation, thus decisions being limited mainly to new functioning polices of the overall system instead of immediate control decisions.
- Protecting the information – opening the information sources and actuation points needs trust into the network. With an additional data layer concentrating only on the data exchange, a new level of security-by-design can be achieved. This is especially important when the information is exchanged between different systems as a tight control is needed in order to be able to provide the appropriate privacy levels.
- Data convergence – as the data systems until now developed in a rather independent fashion for each active system be it radio or fixed access, core networks, optical transport of Internet backbone, the data formats, and their exchange mechanisms are highly different. In order to be able to use this data as part of separated intelligence functions or to transfer this to external entities, a data layer is required to provide a comprehensive convergence of the data formats and exchange mechanisms enabling translations and transcoding.
- Data curation – one of the aspects for the automation of the data exchange is to have a solution for all the exceptional cases such as the lack or the malformation of a specific form of data. Without this, the data would not be able to be processed by machines resulting into very limited possibilities for insight generation. Until now, the data curation was a data-specific process resulting into a fragmented environment. With the advent of the data layer, the data curation processes can also become automated and through this provide an additional automation into the end-to-end data chain processing.

In order to smoothen the adoption of such technologies within the future systems, the data layer builds on top of existing data exchange protocols. With considering the current deployments as part of the solution, a gradual adoption is possible giving the opportunities

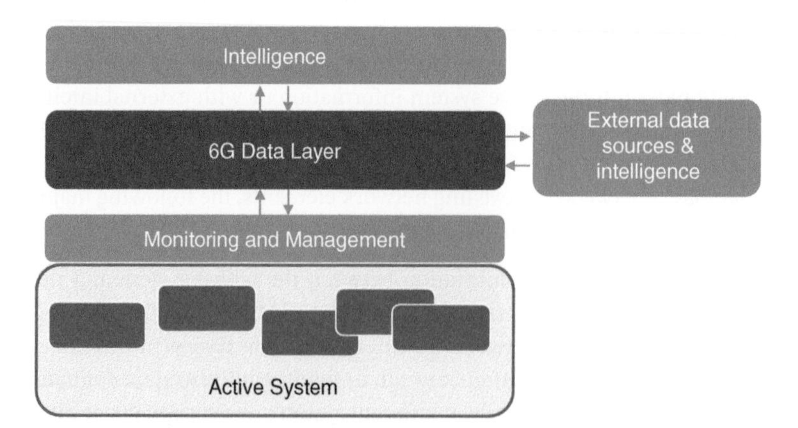

Figure 13.3 Data layer: a new middleware between the active systems and their management systems and the extended 6G intelligence.

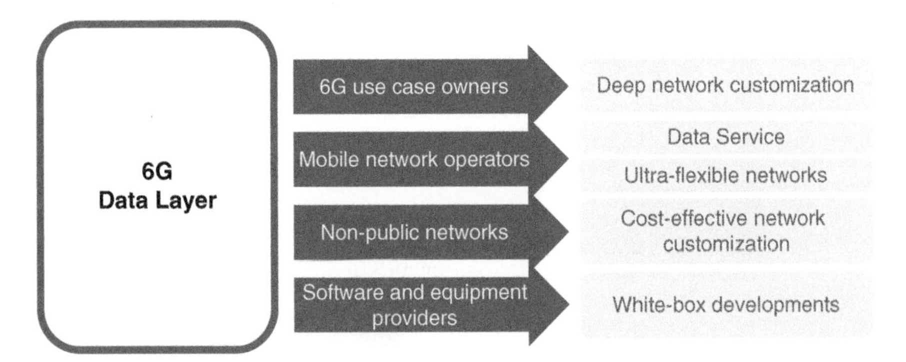

Figure 13.4 Major benefits of the 6G data layer.

for a snow-ball effect deployment with minimal initial invest and immediate feedback from the first deployments (Figure 13.4).

The adoption of a comprehensive 6G data layer brings a rather large number of benefits for the different stakeholders:

- 6G use case owners – a deep network customization can be achieved to address the limit of the different requirements. Especially important for this is the interaction with external sources or intelligence generation functions as to be able to mesh-up data from multiple domains.
- Mobile network operators – a data layer provides the opportunity to use in a privacy- and security-aware manner their large quantities of data, to mesh with data of other operators and from other sources such as Earth Observation, Internet of Things (IoT), local networks, etc. and through this to offer an external data service to be used as basis for third-party services and to provide an ultra-flexible network.
- Non-public network operators – the data layer is enabling a strong customization of the network without extra costs in new network functions. It facilitates the monetizing of own acquired data and the gaining of experience from other similar networks.
- Software and equipment providers – with being able to dynamically customize the network functionality according to local conditions, the providers can drastically reduce the cost of their specific use case deployment. Instead of custom software, a cost-effective white-box one can be offered, which gets customized later. Through moving of the focus from protocol development to situational optimization during runtime ultra-automation and ultra-flexibility levels are achieved.

13.5 Data Layer High-level Functionality

A new set of network functions is presented in this section at a high-level architecture level. They are included as part of a complete system where together with the existing functions are able to properly respond to the requirements of the specific developments as described earlier (Figure 13.5).

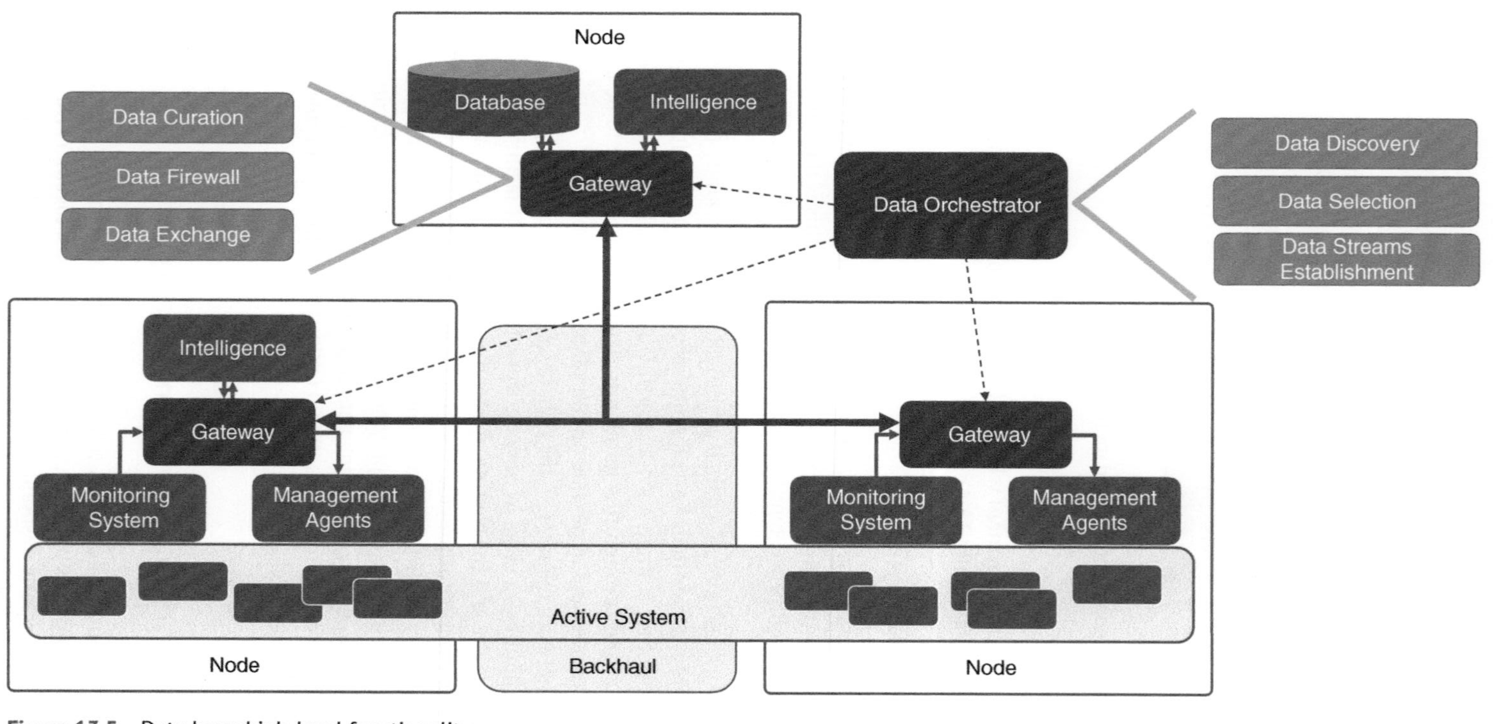

Figure 13.5 Data layer high-level functionality.

The architecture relies on a set of computing nodes, which may or may not have an active system, or a local intelligence generating specific insight. With this, the computing nodes include deployments of 6G active systems with local or with remote insight as well as third parties, which may pertain to other data acquisition and processing domains such as Earth Observation, recommendation engines, or IoT systems and apps.

In general, the active systems are the source of the data through their monitoring systems and the final recipients of the intelligence decision as adaptation indications toward the management agents. From the perspective of the active system, a data stream would be characterized by the size and frequency of the monitored data as well as of the insight indications received. It is expected that usually the monitored data will have a very high quantity and rather high frequency while the management decisions will be rather small and rare, however, requiring a very high transmission reliability.

Intelligence functionality is a generic representation of the insight generation functions. It can be represented by automatic adaptation engines such as the current policy-based decisions or machine learning models providing specific customized insight. From an intelligence perspective, the output data stream is smaller and directly proportional to the input data stream.

Also in this category enters the training of different new models, as a separate type of intelligence that requires to be installed only on specific nodes where enough compute resources are available and where a large amount of monitored information may be needed, usually in a batch form. For the training, the input data is significantly larger than the output data, which may be completely missed.

These nodes are interconnected with a generic backhaul. As to be able to cover most of the options for such a data layer, the backhaul spans from dedicated connectivity within a single network system, local or for longer distances up to large delay networks such as satellite interconnected systems or most common communication over a third-party connectivity provider with best-effort capacity and delay be it using a single network provider or multiple or across the Internet backhaul. Having such a large variation of the backhaul, the only assumption that could be made is that it supports the exchange of data between the different nodes.

The data layer middleware represents the functionality that brings the different active systems and intelligence nodes together. It is represented by two major functional blocks classified based on their features' types: the Gateway and the Orchestrator.

The Data Gateway represents the data middleware element within each of the nodes be it in charge mainly on the immediate interactions with the data sources and recipients. It represents the data plane of the data layer. As such, it includes specific communication functionality:

- Data exchange – two types of data exchange are considered: within the system and with other systems.

  The Gateway interacts within the node with the active system through its monitoring system and through its management system. Albeit the interfaces toward the monitoring systems are rather uniform due to the porting of the software networks within 4G and 5G, this is not the same for the management systems. In order to properly exchange data with the management systems, a new interface enabling the acquisition of external insight needs to be added. Furthermore, most of the current intelligence systems are functioning with batch history data and provide only generic insight, which cannot be immediately used in an active system. To progress toward a real-time intelligence insight, it is necessary that the Gateway can provide the monitored information in a stream manner to the insight elements.

The Gateway is also interacting with other external domains, specifically by communicating with equivalent Gateways that are deployed in other remote nodes connected with a backhaul. For the data exchange between the different domains, the Gateway needs to be able to include the following more granular functionality. First, the Gateway should be able to establish an exchange data path to the remote Gateway including the negotiation of the exchange protocols, data formats, and other communication-related policies specific to the data exchange and the backhaul. Second, it must be aware of the backhaul's momentary availability and capacity in order to be able to make appropriate decisions regarding when and with which parameters the data must be exchanged across the network. Third, it must be aware of the data that must be exchanged in a specific moment in time including possible policies in deferring the communication towards better time or capacity conditions. This feature is highly important for the deployment situations where the backhaul availability or capacity is highly varying such as in the case of mobile edge nodes or in the most common case where the backhaul is shared with other services especially with the active service. A large variation in opportunities and data exchange policies can be achieved when the domains are connected using multiple backhauls, giving the opportunities for load balancing across the different network connections.

- Data firewall – one of the critical elements for giving other parties access to own data is the assurance that the new interface opened will not create additional security issues. For the communication with external entities, the data Gateway should include a specific data firewall that is able not only to assure a secure connection at network level with the other Gateways but also to filter and select the data itself. With this a highly granular data-level firewall can be achieved assuring that the data that enters the system is checked against potential malicious intents. Also, the data firewall should be able to determine if a specific piece of data can be transmitted toward the corresponding Gateway through this assuring that the system is not leaking data toward unauthorized third parties. Especially, this functionality is important to be able to comply with the current privacy data regulation.
- Data curation – in most cases, the data is automatically acquired. To benefit from this, the data must be also processed in an automated manner. However, there are often situations where due to different side effects data was not acquired in a proper manner either by missing pieces of information or by missing completely. In these situations, data stream must be completed with appropriate elements depending on the information type and the way that it is going to be processed: generation of an empty data element or completion with synthetic data being the most common solutions. Considering that the data must be streamed in near-real-time to be able to make the system immediately benefit from the generated insight, it is needed that the data curation will be also fully automated based on predefined rules on how to handle such exceptions.

Furthermore, as the largest benefit will be obtained when the intelligence algorithms are supplied with data from multiple data source in different domains, a key aspect in the alignment of the data is the appropriate acquisition time timestamping. For this, during the initial connection establishment between Gateways, a system time should be determined as to be able to compute the time skew in the different nodes and to be able to systematically modify timestamps to address the same real-world moment, even when the time of the systems is different. Albeit this is a rather easy to execute operation, in practice, it requires a rather complex operation of clock synchronization, which may have to happen across a sometimes unreliable and long-distance backhaul.

For the Gateways to be able to perform the previously executed functionality, there is a need for a logically centralized control plane, named here Data Orchestrator, especially addressing the establishment of the data streams across the different domains. Specifically, the Data Orchestrator includes the following main functionality:

- Data discovery – the first step in any data exchanges is the discovery phase. To appropriately receive data from a specific source, a data discovery process has to be established. For this, first the source of the data will have to register its data to the data discovery function. The registration needs to include aspects related to the data itself (e.g. format, size, etc.), to the privacy and usage rights requirements (i.e. the restrictions in usage), and the Gateway through which this data could be accessed from.
- Data selection – after data is discovered by the user node, a selection process is executed. It presumes ultimately a binding between the source and the destination of the data, which would imply that all the specific security, privacy, and reliability aspects will be covered as well as the exceptional situations. Practically, at the end of the data selection phase, the two nodes have a contractual-like agreement on how and when the data will be exchanged with specific SLAs and procedures for exceptional situations. This will assure a certain level of end-to-end service reliability. For most of the situations, the data selection process can be fully automated by using a set of template SLAs.
- Data streams establishment – following the data selection process, a data path between the source and the destination Gateways is established. The data stream is regulated by a set of policies that bring the negotiated SLA to a set of real-system policies to be enforced both on the source and the destination of the data stream. This functionality is enforced on the Gateways while the specific decisions may be taken in a combination between Gateways and Orchestrator as a mediator between the two nodes.

Next to this basic functionality, a rather large additional functionality may be added to the system, such as procedures for data streams modification and termination or the basic secure connectivity between the Gateways with authentication and authorization in order to protect the data exchange. These procedures are not described here as they would push too many details to the reader.

13.6 Instead of Conclusions

The arguments presented within this chapter regarding a data layer within the networking ecosystem are based on a set of fundamental advancements within the networking world up to the first deployment of 5G and the advent of the 6G research. With these, we believe that a data layer will become part of the networking research in the next years, specifically as a new direction in which the network can be further optimized. As usual in these situations, there are two major alternatives to how this would happen: like a gold rush or like a railway network. The gold rush presumes that multiple systems will be developed in parallel, and each of the parties developing these systems will try to be the first to reach a large market share. As observed with other previous technologies where this happened, such as cloud virtualization platforms, a few major alternatives come to dominate the market at the final system deployment acting as de-facto standard and having rather complex interoperating elements. The railway network presumes the unification of the system around a single convergence standard in both the communication and the data presentation. This chapter advocates for the second option as the only way to highly optimize the functionality and to reach a higher market potential, same as with other technologies in the 5G environment and probably in the 6G environment.

To be successful within the convergence area, a rather strong initial momentum must be created in which the benefits across the whole stakeholders are made obvious. This can be achieved only with highly effective initial prototyping and demonstrations as well as through the awareness creation toward the community designing new intelligence algorithms on how important it is to have the data on time and at the given location.

Although the system proposed is mainly considering that the data sources will advertise the data while the data consumers will discover and select the data, this was done for technical reasons. From a market development perspective, the success of the data layer technology is more dependent on the intelligence stakeholders providing their capabilities toward the active system owners, especially when bringing together multiple data sources.

One step in this direction is provided by this chapter where an initial high-level architecture was presented. Following, to be part of such a data layer, one network owner must add to its network a Gateway that is bound to a logical Data Orchestrator. To fully benefit of the Gateway, at least one of the following additional connections have to be established: to the sources of data that could be shared in certain conditions, to the data actuation points that may use recommendations from the intelligence, or to the intelligence algorithms, specifically depending on the type of stakeholder.

As these interconnections are not necessarily related to the 6G radio or core network, albeit they are the most to benefit from its advancements, the same system can be easily deployed within a current 5G network, providing some immediate benefits. With this in mind, it is very easy to design a transition roadmap from the current data siloed systems toward a convergent system by starting with the deployment of the first Gateways and gradually binding them to source and destinations of data within the nodes. Following, new Gateways can be immediately added within the system, with minimal impact for the existing nodes, through this snowballing toward a comprehensive system.

References

1 Abed, G.A., Ismail, M., and Jumari, K. (2012). The evolution to 4G cellular systems: architecture and key features of LTE-advanced networks. *Spectrum*: 2.

2 Raaf, B., Faerber, M., Badic, B., and Frascolla, V. (2014). Key technology advancements driving mobile communications from generation to generation. *Intel Technology Journal* 18 (1): 12–28.

3 Dahlman, E., Parkvall, S., and Sköld, J. (2011). *4G: LTE/LTE Advanced for Mobile Broadband*. Academic Press. ISBN: ISBN: 012385489X.
4 Wang, F., Yang, L., Cheng, X. et al. (2016). Network softwarization and parallel networks: beyond software-defined networks. *IEEE Network* 30 (4): 60–65. https://doi.org/10.1109/MNET.2016.7513865.
5 Lu, J., Xiao, L., Tian, Z. et al. (2019). 5G enhanced servicebased core design. 2019 28th Wireless and Optical Communications Conference (WOCC), Beijing, China, 1–5.
6 Hernandez-Chulde, C. (2019). Intelligent optimization and machine learning for 5G network control and management. Highlights of practical applications of survivable agents and multi-agent systems. The PAAMS collection, 339–342. ISBN: 978-3-030-24299-2.
7 NTT DOCOMO (2020). White paper: 5G evolution and 6G. https://www.nttdocomo.co.jp/english/binary/pdf/corporate/technology/whitepaper_6g/DOCOMO_6G_White_PaperEN_v3.0.pdf.
8 Letaief, K.B., Chen, W., Shi, Y. et al. (2019). The roadmap to 6G: AI empowered wireless networks. *IEEE Communications Magazine* 57 (8): 84–90.
9 Hadi, M.S., Lawey, A.Q., El-Gorashi, T.E., and Elmirghani, J.M.H. (2018). Big data analytics for wireless and wired network design: a survey. *Computer Networks* 132: 180–189.
10 Gonçalves, L.C., Sebastião, P., Souto, N., and Correia, A. (2019). Extending 5G capacity planning through advanced subscriber behavior-centric clustering. *Electronics* 8: 1385. https://doi.org/10.3390/electronics8121385.
11 Japec, L., Kreuter, F., Berg, M. et al. (2015). Big data in survey research. *Public Opinion Quarterly* 79: 839–880. https://doi.org/10.1093/poq/nfv039.
12 Rohini, M., Selvakumar, N., Suganya, G., and Shanthi, D. (2020). Survey on machine learning in 5G. *International Journal of Engineering Research and Technology* 9: 569–576. https://doi.org/10.17577/IJERTV9IS010326.
13 Study on Enablers for Network Automation for 5G – Phase 2, document SP-190557, 3GPP, Sapporo, Japan (June 2019).
14 Mathur, N. and Purohit, R. (2017). Issues and challenges in convergence of big data, cloud and data science. *International Journal of Computer Applications* 160 (9): 7–12.

14

Long-term Perspectives: Machine Learning for Future Wireless Networks

Sławomir Stańczak[1], Alexander Keller[2], Renato L.G. Cavalcante[1], Nikolaus Binder[2], and Soma Velayutham[3]

[1] *TU Berlin/Heinrich Hertz Institute, Berlin, Germany*
[2] *NVIDIA, Berlin, Germany*
[3] *NVIDIA, Santa Clara, CA, USA*

14.1 Introduction

Wireless resources, especially spectrum and energy, are scarce. In addition, wireless links in the real world are distorted by noise and interference while their statistics can change rapidly due to mobility. This presents system designers of wireless networks with major challenges. Machine learning (ML) has great potential to address these challenges and significantly improve the performance of wireless networks. However, novel methods of ML must first be developed to meet the strict requirements of mobile applications on the one hand and to overcome limitations in wireless networks such as scarce resources on the other. This chapter presents some promising ML techniques for future radio access networks (RANs) and especially the physical layer.

We argue for hybrid methods that combine traditional data-driven ML approaches with model-based approaches including domain knowledge. In particular, we will consider kernel-based learning methods and deep neural networks in this chapter. The key to efficient ML solutions is the exploitation of structures inherent to wireless channels, transmission of signals, and derived and measured data. For example, it is of great importance to include a structure such as sparsity and compressibility right in the algorithms. Since data in mobile radio networks is not available at a single point but distributed across different locations, the novel ML methods must be suitable for a distributed implementation in order to efficiently use the scarce mobile radio resources. Finally, due to the highly dynamic nature of the mobile radio channel and constantly changing channel statistics, the methods must be capable of delivering robust results based on small datasets and under strict latency conditions. This requires a massively parallel and distributed implementation of future wireless networks. In fact, the massive parallelism offered by graphics processing unit (GPU) architectures allows for concurrent execution of certain RAN functions to

Shaping Future 6G Networks: Needs, Impacts, and Technologies, First Edition.
Edited by Emmanuel Bertin, Noel Crespi, and Thomas Magedanz.

achieve orders-of-magnitude acceleration. Based on open standards, a popular software ecosystem (Compute Unified Device Architecture [CUDA]), and speed of deployment on components-off-the-shelf (COTS), such truly software-defined radios (SDRs) driven by ML are highly attractive with respect to their total cost of ownership (TCO) and especially power efficiency.

Hence, telecommunication operators are welcoming the fusion of communication and ML. Besides the consumer business, such a technology also enables a completely new professional business for private networks, for example, self-hosted campus networks in factories with an ecosystem of specialized communication solutions.

14.2 Why Machine Learning in Communication?

There are at least three reasons for the use of ML in wireless networks and especially in RANs. First, the ever-increasing complexity of the communication infrastructure makes it increasingly difficult to develop precise mathematical models. As a result, the model error resulting from the discrepancy between models and real-world networks can become unacceptably large. This discrepancy can be bridged with ML. Second, since wireless networks operate in highly dynamic and resource-constrained environments, there is often no time to execute many iterations of asymptotically converging numerical algorithms. The truncation of such iterations leads to an additional source of error, which generally cannot be controlled very well. Again, ML can help to improve classical numerical algorithms to meet strict latency requirements in wireless networks. Third, data is collected in communication networks at different locations and has to be collected using limited communication resources. With ML, missing data can be predicted and reconstructed to save scarce resources.

The application of ML therefore offers several advantages: the complexity of communication networks will become more manageable as configurations may be learned and predicted. A reduced number of required iterations and measurements lead in general to higher efficiency in network operation. Similarly, decision-making may be faster based on online distributed learning, and robust learned predictions allow for anticipation rather than reaction possibly avoiding issues before they occur.

For ML to be a success in wireless communication, ML algorithms must meet some requirements and constraints. Often, only relatively small sets of training data are available and the statistics of the data can change over time. Therefore, there is a strong need for online ML algorithms with good tracking capabilities that provide robust results even with small uncertain datasets. The use of the so-called domain knowledge, for example, in the form of models or known correlations, is absolutely necessary to maintain important properties of the functions to be learned and thus significantly improve the performance of the ML algorithms. The limited capacity of wireless networks and the rapidly changing channel information pose additional challenges and call for distributed learning under communication constraints. Since distributed computation and data are inherent in communication networks, ML algorithms must be of low complexity and amenable to massive parallel implementation to meet the low-latency requirements of many wireless applications.

14.2.1 Machine Learning in a Nutshell

ML may help to overcome modeling and truncation errors. It may help to reconstruct and predict data. In a more abstract way, ML can be seen from the point of view of function approximation. Then, ML is the task of approximating an unknown function, which in some settings may be reduced to the problem of finding parameters in a parametric model representing a sought function. It should be emphasized that in many practical learning settings the task is to track time-varying functions. This case requires adaptive approximation methods that rapidly adjust to the changes. As already mentioned, fast adaptation is particularly important in highly dynamic RANs, which in turn sparks interest in online learning to take into account new measurements upon their arrival.

In principle, three kinds of learning may be distinguished: first, unsupervised learning has to learn from a given pool of data. Typical tasks are to identify clusters or structure of the data. The knowledge then may be used to classify new unseen data. Second, in supervised learning, the training process takes pairs of input arguments and output values. Once trained, output values are predicted for input arguments, where the inputs not necessarily have been part of the training. For instance, supervised learning is widely used for regression and classification tasks. Third, there is semi-supervised learning in the form of reinforcement learning, which aims to imitate the human way of learning: based on actions, an actor is provided reward. While in the beginning exploring the space of actions certainly is useful, over time the reward will be used to learn a policy that allows the actor to predict actions that promise the most cumulated rewards. This phase is called exploitation.

In the following, we briefly describe three promising methods for ML in RANs: kernel-based learning with projections, deep learning, and reinforcement learning.

14.2.1.1 Kernel-based Learning with Projections

Kernel-based learning methods have their origin in functional analysis, a branch of mathematics that deals with the analysis of function spaces, among other things. Of special interest in this context are the so-called reproducing kernel Hilbert spaces (RKHSs) of functions, which are a common tool for learning real or complex-valued functions on some predefined sets. Since an RKHS is a special Hilbert space, it is equipped with an inner product that induces a norm. In contrast to general Hilbert spaces, RKHSs have an important additional property: if two functions in an RKHS are close to each other in the induced norm, they are also close to each other pointwise. This is a crucial property, for example, to preserve the shape or other properties of the function to be learned. From a practical point of view, another key result in functional analysis implies that an arbitrary positive definite function gives rise to an RKHS. In fact, this statement can be even strengthened as follows: a function is a positive definite function if and only if it is a (reproducing) kernel for some RKHS that is unique. Therefore, in applications, the starting point for the development of kernel-based learning methods is the selection of a suitable positive definite function, which then has a reproducing property in the associated RKHS. This selection process usually requires some prior or domain knowledge about the learning problem at hand if good generalization performance is expected with few samples. A useful fact in this respect is that multiple kernels can be combined in an appropriate manner to define a new kernel function and the associated RKHS. This feature is often exploited to construct an RKHS containing functions able to mimic known characteristics of the estimand. For example, if the function to be learned is known to present linear trends overlapped with a "periodic" behavior, then we can combine linear kernels with kernels presenting some form of periodicity to obtain a new kernel and an associated RKHS for which existing learning tools are expected to show good generalization performance with few training samples.

The theory of finite- and infinite-dimensional Hilbert spaces is very rich, making it possible to systematically design algorithmic solutions able to provide performance or convergence guarantees. This is crucial when developing robust learning schemes for the dynamic setting of RANs, where latency should be low and training datasets may change rapidly. To address problems of this type, we can apply powerful algorithmic approaches based on projections onto closed convex sets in RKHSs. These algorithms are typically constructed by using computationally efficient projections onto simple sets, and they give rise to (massive) parallel implementations, thus offering significant latency reductions because the projections can be performed in parallel. They can also be used to incorporate domain knowledge or expert knowledge [1], thereby providing an elegant framework for the development of hybrid learning methods that use both model knowledge and datasets during training.

14.2.1.2 Deep Learning

With the advent of massively parallel processors, artificial neural networks and in particular deep neural networks have become practical in many applications. They traditionally belong to the class of data-driven (and model-free) ML methods and have recently attracted much attention in the context of wireless communication due to their successes in other areas such as image processing or speech recognition.

There are many types of neural networks, but it is conceptually reasonable to consider simple feedforward neural networks from which other types, such as convolutional and recurrent neural networks, can be derived. As the name already suggests, the information in feedforward neural networks moves from the input layer towards the output layer via one or multiple hidden layers. In particular, there are no cycles and loops in such networks. The function $y = F(x)$ relating the output $y \in Y$ of a neural network to its input $x \in X$ can be written in its canonical form as a concatenation of multivariate and vector-valued nonlinear functions that correspond to different layers:

$$y = F(x) = F_N\left(F_{N-1} \ldots F_2\left(F_1(x)\right)\ldots\right),$$

where $N \geq 1$ is used to denote the number of layers. Note that the function $F: X \rightarrow Y$ can be also seen as a mapping from a multidimensional input dataset X to a multidimensional output set Y. Each nonlinear function consists of a vector-valued affine function whose outputs are followed by a univariate nonlinear function. To be more precise, let $1 \leq n \leq N$ be arbitrary and, without loss of generality, assume that $F_n: R^k \rightarrow R^m$ where $k, m \geq 1$ correspond to the number of *nodes* in a layer. Then, the relation between the input and output of each layer n has the following structure:

$$F_n(u) = f\left(A^{(n)}u + b^{(n)}\right)$$

where $(A^{(n)}, b^{(n)}) \in R^{m \times k+1}$ is a matrix and $f(v) = (f_1(v_1), \ldots, f_m(v_m))$ is a nonlinear function composed of univariate functions $f_i, 1 \leq i \leq m$, that are also referred to as *activation functions*. The vector $b^{(n)}$ is often referred to as a bias term. While the activation functions are fixed, the entries of $(A^{(n)}, b^{(n)})$ – for simplicity we call them *weights* – are optimized subject to given constraints. Different choices of the matrix entries lead to different functions $F: X \rightarrow Y$. The set of all functions generated by a feedforward network is denoted by $S := \{F_N \circ F_{N-1} \circ \ldots \circ F_2 \circ F_1\}$. The main rationale for considering this function structure is that it is amenable to efficient implementations because different outputs of each layer n are coupled only through a linear function determined by the matrix $A^{(n)}$. An important question is, however, whether the imposed function structure entails an inherent loss in learning capabilities of feedforward neural networks.

The question is everything else but trivial since the task is to learn a multivariate function using compositions of univariate nonlinear functions and affine functions. To ensure learnability, it is necessary to guarantee that a function to be learned is approximable in some well-defined sense by a feedforward neural network. Therefore, since the sought function can be any continuous function,[1] the question is whether the imposed function structure allows us to design a feedforward neural network so that it can be trained to approximate any continuous function on a compact set. The approximability here means that any function in the Banach space $C(\Omega)$ of all continuous functions[2] defined on some compact set[3] $\Omega \subseteq X$ can be approximated by F defined above up to any error $\epsilon > 0$. In other words, S, which is the set of all functions generated by feedforward neural networks with N layers, is dense in $C(\Omega)$. In general, the approximation error depends on the univariate functions and the number of nodes, which might need to be adjusted as the error tends to zero.

The problem of representing (not approximating) continuous multivariate functions as a finite linear combination of univariate functions has a long tradition in mathematics and is closely related to the 13th problem of David Hilbert, who conjectured that not every continuous function has such a representation [2].[4] It took more than 50 years to resolve the problem. Indeed, by Kolmogorov's superposition theorem [3], any real-valued function $\phi: R^m \rightarrow R$ of m variables can be represented as a concatenation of two functions $\phi = g \circ h$ where the outer function $g: R \rightarrow R$ is univariate and the inner function h is a linear combination of m univariate functions. If the functions g and h are restricted to be continuous, then the basic statement is still true, but the number of linear combinations of univariate functions can be up to $2m+1$ terms and, instead of g, a set of univariate functions g_i is used.

We believe that the results of Kolmogorov and the later extensions [4–7] can be used in future research to stimulate the design of shallow feedforward networks with excellent learning capabilities that allow fast training based on relatively small datasets. The results are, however, not directly applicable to current feedforward neural networks since

1 For simplicity of presentation, we assume continuous functions here, but most of the statements made here also extend to noncontinuous sigmoidal functions.
2 Unless otherwise stated, we assume that the Banach space is equipped with the supremum norm. Note that every RKHS mentioned in the foregoing section is a Banach space but not vice versa. In particular, the Banach space considered here is not an RKHS.
3 In this chapter, a set is compact if and only if it is bounded and closed. So the assumption of compactness is not restrictive from a practical point of view.
4 More precisely, this would be a consequence of the conjecture of David Hilbert.

Kolmogorov's superposition representation involves different univariate functions that must be adjusted at least in part depending on the function to be represented. This implies that not only the weights should be optimized but also the univariate functions. The problem of optimizing univariate functions for recovery of sparse signals was for instance considered in [8].

Important results on the problem of universal approximation by feedforward neural networks can be found in [9]. In particular, assuming one hidden layer, it is shown (see also [10]) that an arbitrary continuous function on a compact set can be uniformly approximated by a finite linear combination of compositions of a fixed univariate function and affine functions, provided that the number of linear combinations is adjusted. This shows that the function set S generated by feedforward neural networks with only one hidden layer is indeed dense in $C(\Omega)$. The univariate activation function is assumed to be a continuous *discriminatory* function. The class of *discriminatory* functions consists of many important functions, including continuous sigmoidal functions defined to be $f(x) \to 1$ if $x \to +\infty$ and $f(x) \to 0$ if $x \to -\infty$.

The proofs in [9] are existence proofs and do not provide concrete constructions of neural networks. In particular, the work does not address the important question of how many linear combinations or, equivalently, how many nodes are needed to achieve the desired approximation accuracy. Also, the relationship between the number of nodes and the properties of the activation function are not considered. In general, the number of nodes can be exponential in the input dimension, which is not feasible for many practical applications. This problem was partly solved in [11] under the assumption of feedforward networks with one hidden layer and sigmoidal activation functions. The main result states that under some mild boundedness assumption on the input, the integrated squared error is of order $O(1/m)$, where m is the number of nodes. The crux is, however, that both the weights and the parameters of the sigmoidal activation function are adjusted to achieve the approximation scaling.

All these results show that feedforward networks with one hidden layer can be in principle used to learn any multivariate vector-valued function with an arbitrary precision, provided the number of nodes is large enough. Many nonlinear activation functions are, however, inadequate for practical implementations due to various hardware and cost constraints. The focus is therefore on a class of nonlinear functions that allow efficient implementation on commodity hardware (central processing units [CPUs] and GPUs). A particularly efficient implementation can be achieved with the ramp function defined to be $f(v) = max(v, 0)$ where the maximum is taken componentwise. The ramp function, which is referred to as the rectified linear unit activation (or simply ReLU) function, is a piecewise linear function. Consequently, we can conclude that in ReLU-based feedforward neural networks,

- $F_n(u)$ is a piecewise linear function for any given $(A^{(n)}, b^{(n)})$, and
- $F(x)$ is piecewise linear, since the concatenation preserves piecewise linearity.

It is clear that the ReLU function is not a sigmoidal function. However, reference [12] extended the previous works (including that of [9]) by showing that feedforward neural networks have the universal approximation property of [9] if and only if the activation function is not a polynomial almost everywhere. By this, we can conclude that any continuous function on a compact set is also approximable by ReLU-based feedforward neural networks. As in the case of sigmoidal functions, the number of nodes is extremely high for many approximation problems. As a result, the number of optimization variables grows, which renders the training of neural networks a difficult high-dimensional optimization problem. In fact, this problem is highly non-convex, so heuristic approaches are used to optimize the weights for a given neural architecture with respect to some predefined application-dependent metric.

Widely used training algorithms are based on stochastic gradient descent and backpropagation methods. They are in general very time-consuming and require a lot of computational power. In addition, neural networks are highly reliant on large amounts of data. It has indeed been found that the performance of deep neural networks is comparable to the performance of traditional ML methods for relatively small sets of training data, but their performance scales significantly better as the amount of training data increases ("big data"). This raises the question whether ML algorithms will have access to large amounts of data in RANs as well, so that the better scaling effects of neural networks can be exploited. Given the high dynamics and short stationarity intervals in RANs, this question is not easy to answer. For communication scenarios with high mobility, such as V2X communication, the stationarity interval is in the range of a few tens of milliseconds. This means that the distribution of data changes very quickly. In this case, retraining is usually necessary, as otherwise the performance of neural networks can deteriorate unacceptably. Due to the complexity of the underlying models and optimization problems, it is currently notoriously difficult to provide reliable estimates of how often and when retraining is necessary to ensure the required performance. However, if the amount of training data is insufficient or the available computing power is too limited, training and retraining becomes infeasible, especially under latency conditions.

One way to speed up the training is to choose a suitable network architecture. To this end, neural network architecture can for instance be designed by unfolding iterative algorithms or by incorporating a priori knowledge about the input signal such as sparsity.

In fact, the mathematical model underlying model-based and data-driven ML is not that far apart. There exist classic iteration schemes that resemble modern artificial neural network structures. Gradient descent is prominent in both model-based and data-driven ML. As an example, by combining the latter two principles, gradient descent may be used to find matrices that may approximate iteration schemes by only one step. Such approaches enable classic ML methods under latency constraints. The resulting algorithms very much resemble artificial neural networks. This indicates that hybrid ML methods are of central importance.

14.2.1.3 Reinforcement Learning

Reinforcement learning is motivated by how humans learn from their actions being rewarded.

The process of learning based on rewards rather than explicit training labels has been classified as semi-supervised learning. Implementing such schemes on computers allows for taking actions to reach goals with maximized reward, for example, to save energy by predicting when components of a network can be switched off, or for accelerating computation, for example, in radiation transport simulation [13]. The history of reinforcement learning and its principles are summarized in [14]. Next, we review the principles of the most recent progress in reinforcement learning [15].

Planning is at the core of modern reinforcement learning methods. Its goal is to predict the most rewarding action to take next. For that purpose three functions are required. First, the state of the environment, where actions are to be taken, needs to be mapped to a representation that the computer can operate on. Then, using this representation, the value of the current state is evaluated and a policy is predicted that assigns probabilities to the possible actions representing their expected rewards. Finally, the next state is computed by a selected action and the actual reward of the action is determined. The selection of the next state involves simulating a probability distribution using random elements, which is the reason for the method being named Monte Carlo tree search (MCTS). Sampling paths in the search tree according to the policy allows for exploring its likely more promising parts. Being able to predict a value of a node allows for pruning paths without fully exploring them. As compared to classic tree search, saving work on exploration by probabilistic selection and pruning by predicting values results in a much more efficient computation of a policy for the next action to be selected.

The aforementioned functions required for MCTS usually are of high dimension and rarely classic numerical representations are efficient. They are therefore approximated by neural networks. In order to train these neural networks, paths are generated by planning as described above. The neural network for the policy then is updated using the cross-entropy loss between the predicted policy in a node and the policy computed by MCTS. The neural network for the reward is trained on the difference between predicted and actual reward. Finally, the neural network for the value of a node is updated using the summed rewards predicted and discounted along a path, where in case the path has been pruned, the discounted prediction of the value in the pruned node is added.

Reinforcement learning as sketched above and detailed in [15] lends itself to planning and prediction tasks for autonomous wireless network operation and optimization.

14.2.2 Choosing the Right Tool for the Job

The performance of current artificial neural networks scales well with the amount of training data, and as a result they may outperform traditional ML algorithms such as kernel-based machines when huge datasets are available for training. However, when comparing deep learning to traditional ML methods in the context of mobile networks and, in particular, RANs, several inherent constraints and requirements associated with these networks need to be considered:

1) First of all, energy efficiency is becoming paramount. The energy costs represent a significant fraction of operators' operational expenditures and therefore mobile networks should not consume more energy than they need to meet the users' requirements.
2) While neural networks expose excellent performance on average, in RAN, the worst-case performance is often of interest as a certain quality of service (QoS) needs to be guaranteed. Moreover, many applications in RAN require good tracking capabilities.
3) The amount of data and time required to properly train a neural network may not be available since not only does the wireless channel change rapidly over time but also the distribution of the channel can change on the order of milliseconds.

4) Wireless communication resources are scarce so that training models and applying the models must be performed under strict communication constraints. This calls for distributed learning.

To meet the above requirements and constraints, the inherent limitations and characteristics of RAN (e.g. limited resources, dynamic environments) must be properly addressed. Key to this is the incorporation of domain knowledge, which must be considered in a systematic manner when designing ML algorithms. While traditional ML methods often provide a "natural" interface for incorporating domain knowledge (e.g. in the form of models), including domain knowledge in a systematic way in the design of neural networks is still a difficult problem. In conclusion, the choice of the ML tool to use depends on the setting it is being operated in.

Besides the choice of algorithms, selecting a platform for their implementation is at least as important. It is out of question that in order to be sustainable, algorithms need to be implemented on massively parallel processors. Considering the complete cycle of implementation and deployment, the choice of software ecosystem as well as the actual hardware platform play an important role with respect to cost and turnaround time.

14.3 Machine Learning in Future Wireless Networks

The RAN must guarantee reliable wireless connections to mobile devices, which, as mentioned earlier, is an extremely challenging task due to the inherent limitations and characteristics of the RAN. Selecting the right ML method for different tasks is therefore of great importance. When selecting the method, it is helpful to distinguish between the lower and upper layers of the communication stack. While the lower layers, such as the physical (PHY) and data link layers, have to deal with rapidly changing statistics and ephemeral events, at higher layers variations are much slower and more predictable. In particular, the PHY layer and media access control (MAC), which is at the heart of the data link layer, must adapt to changes in the wireless channel, which is prone to interference and exhibits high dynamics. Dynamically changing channel state information leaves only milliseconds to learn or track distributions and statistics. While online data-driven ML may be challenged by the lack of sufficient training data, online model-based ML appears to be more feasible.

In contrast, at higher layers, state changes slower and the statistics are stable over relatively long time. As a result, there is more time for training, and much more data is available. Moreover, since good models are usually not available, it seems that neural networks might be the preferred option for learning at higher layers. However, there are other challenges that need to be addressed carefully when training neural network models, including incomplete data (missing measurements over long periods of time), erroneous data (e.g. due to software errors), misaligned data (models are trained and used at different times), and dealing with time series because the assumption of independent, identically distributed samples is unrealistic in real systems.

In the examples that follow, data-driven, model-driven, and hybrid-driven ML algorithms for future wireless networks are explored along with reasoning about their domain of application.

14.3.1 Robust Traffic Prediction for Energy-saving Optimization

Switching off network components when there is little data traffic can save energy. However, switching components off and on involves significant costs and latency. In particular, just reacting to the capacity demand in a wireless network can lead to oscillations, which can cause link failures in the networks and must therefore be avoided.

A better alternative is to robustly predict traffic and with it the resulting capacity demand so that there is enough time to switch off and on network components and save energy. The capacity demand as a function of time and space is complex to model but can be learned efficiently. The learned spatial–temporal function can be used to predict low-capacity demand and, more importantly, to predict whether the demand will remain low for a long enough period of time so as to significantly reduce the energy consumption by switching off network components [16].

Learning time–space series of the capacity demand from environmental observations makes it possible to adapt the cellular network, e.g. to time of day, day of week, and holidays. The anticipated capacity demands even come with confidence intervals. As shown in [17, 18], a great part of densely deployed networks may be switched off without compromising QoS.

14.3.2 Fingerprinting-based Localization

The main idea of fingerprinting-based localization is to identify a signal feature that is relatively time invariant and has a strong relation to the location of the device that generated the signal. Once such a feature is identified, ML tools can be trained to map received signals to the location of transmitters. In particular, in massive multi-input multi-output (MIMO) systems, there is a strong correspondence between the channel covariance matrix and the transmitter location. This correspondence has been reported to change slowly over time. These two features of channel covariance matrices have enabled [19] to train neural networks to locate users with signals readily available at base stations. One potential limitation of the approach is that the availability of training data may be limited because the distribution of channel covariance matrices depends strongly on the operating frequency. Therefore, if there is a change in operating frequency, there is a corresponding change in the distribution of the signals, and it is widely known that the performance of ML tools, and, in particular, neural networks, may degrade severely with distribution changes.

To overcome the above limitation, [19] have also proposed to train and use the neural networks in a fixed reference frequency. In this technique, whenever there is a frequency change, the channel covariance matrix of the reference frequency is estimated from the channel covariance matrix of the target frequency. To this end, algorithms that estimate the angular power spectrum (APS) are typically used because the APS is known to be largely insensitive to changes in frequency, and it can be used to reconstruct the channel covariance matrix in a wide frequency range. Note that algorithms for APS estimation have been increasingly gaining attention in recent years because they find applications other than localization, and they can be based on pure model-driven methods [20–22], pure data-driven methods, or hybrid model- and data-driven methods [23].

14.3.3 Joint Power and Beam Optimization

For industrial communication, mmWave offers many advantages that make it amenable to establishing small production cells. As a consequence of the high frequency, there is less interference due to high attenuation and a higher immunity to jamming and eavesdropping.

Jointly optimizing power and beams allows for efficiently operating such a network. From a mathematical point of view, this is a mixed-integer problem, because power is a continuous variable, while beam width, beam direction, and base station assignment are discrete variables. As this problem is known to be non-deterministic polynomial-time hard, the tractable part of the computation is separated from the intractable one: given the discrete variables, the power control problem can be solved by iterative methods [24]. However, finding the received beam configuration given a power allocation and the transmitted beams requires a heuristic. This corresponds to a classification problem, which neural networks are very capable of learning. Other than in the previous section, the heuristic learned by the neural network is robust to changes in distribution. In addition, it is much more efficient as compared to solutions obtained by brute-force computation or simulated annealing.

14.3.4 Collaborative Compressive Classification

A primary application of neural networks is classification tasks. It is well known that classification results can generally be improved by combining outputs of multiple classifiers at a central point. We refer to this as a collaborative classification. In wireless networks, however, collaborative classification can be impractical if the multiple classifiers communicate with the central point (receiver) over the wireless channel (see Figure 14.1 for illustration). In particular, since wireless resources such as spectrum are scarce and must be used efficiently, transmitting high-dimensional outputs of multiple classifiers over the wireless channel might be too costly in terms of resource usage. A widely used approach to mitigate the problem is to compress the classifier outputs before transmission.

A key question is how to compress the outputs and how to reconstruct them at the receiver side without significantly degrading the performance of the overall classification. Since many signals in communication systems are sparse, we believe that exploiting sparsity is a key to improving the efficiency of collaborative methods in wireless networks.[5] We explain this using the example of collaborative compressive classification illustrated in Figure 14.1. Assuming a setting of multiple cameras targeted at a scene, each camera feeds its images into its locally running identical copy of an image recognition neural network. Each of these networks then computes a high-dimensional classification vector of its given image. Instead of now

5 Here a sparse signal is a finite-dimensional vector in which most entries are zero. This property is sometimes called hard sparsity.

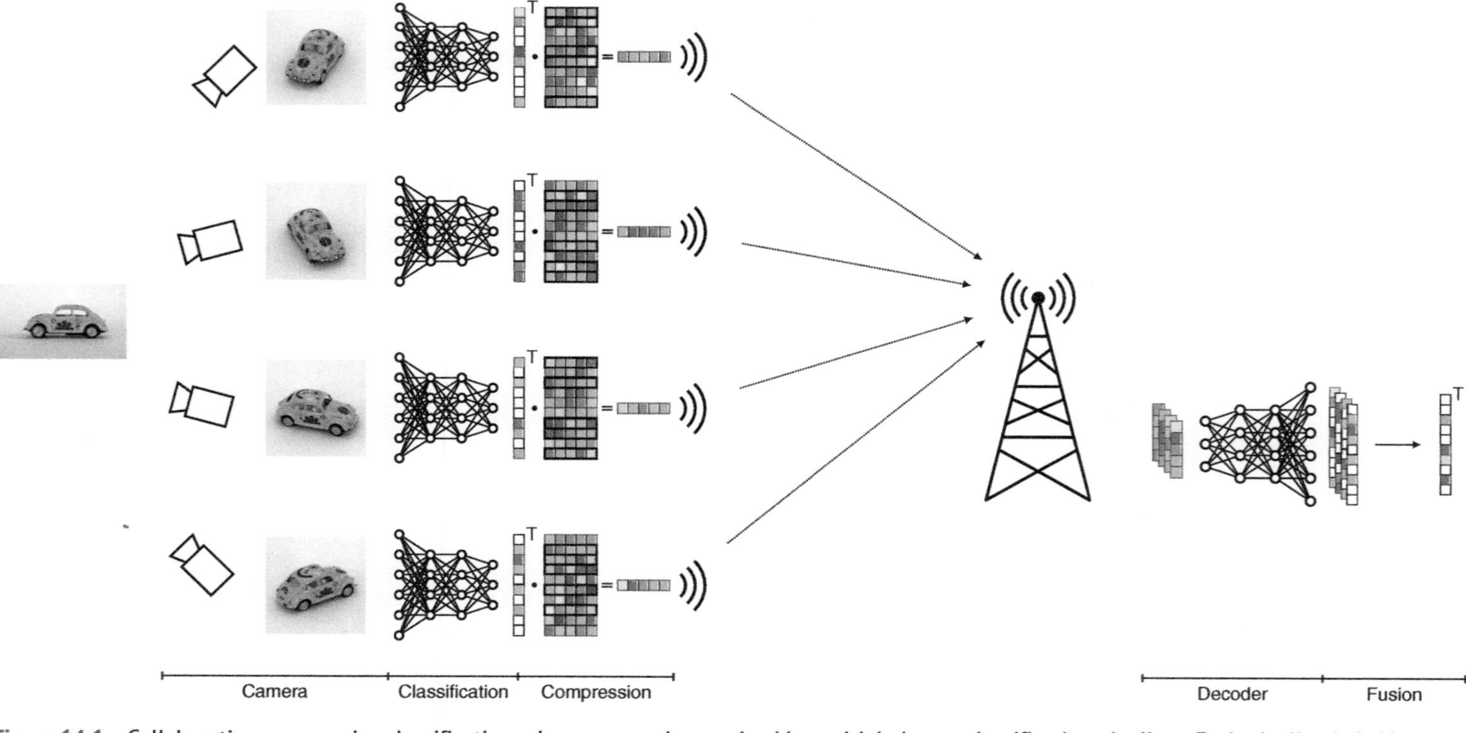

Figure 14.1 Collaborative compressive classification, where a scene is perceived by multiple image classification pipelines. Each pipeline is fed by images of a camera that are classified using an image recognition neural network. The resulting classification vector is sparse and compressed by multiplication with a randomly chosen matrix before transmission. At the receiver side, the sparse vectors are recovered or estimated by another neural network. The receiver-side fusion of the recovered vectors allows for a classification more reliable than the classifications right at the cameras.

transmitting the high-dimensional vectors to a receiver, bandwidth can be saved by compressing these vectors. This compression is achieved by multiplying the original vector with a fixed, randomly chosen matrix and transmitting the resulting low-dimensional vector. At the receiver side, these shorter vectors are collected from multiple cameras and processed by a neural network that recovers or estimates the original high-dimensional sparse vectors. The theoretical framework called compressive sensing provides conditions under which a full recovery of sparse vectors is possible. Now fusing the recovered vectors makes the classification results more robust when compared to any of the single classification at the transmitter sides.

Interestingly, the decoder neural network can be constructed from a fixed, randomly chosen matrix even without learning (see the next section). Based on a solid mathematical theory, combining neural networks and compressive sensing, the collaborative learning approach has been implemented on NVIDIA GPUs and demonstrated to the public at the GPU Technology Conference (GTC) in Munich, Germany, 2018.

14.3.5 Designing Neural Architectures for Sparse Estimation

As mentioned in the preceding section, many tasks in wireless communication are concerned with recovery or reliable estimation of a high-dimensional sparse vector from a low-dimensional (dense) vector obtained through linear compression. Such tasks basically boil down to estimating a vector $x \in R^N$ from an observation $y = Ax \in R^M$ with $M < N$, where $A \in R^{M \times N}$ is a fixed, known full-rank matrix called compression matrix. Since $M < N$, there are in general infinitely many solutions, and hence a reliable estimation of x, let alone recovery, is not possible unless some constraints can be imposed on the sought vector such as sparsity.[6]

As we already pointed out, compressive sensing provides conditions for a full recovery of sparse vectors as a solution of a convex optimization problem. We can therefore approach the solution arbitrarily close by using iterative methods such as the iterative shrinkage-thresholding algorithm (ISTA). However, those iterative methods are still too complex and too slow for important tasks at lower layers such as channel estimation, where reliable channel estimates have to be made available within a few microseconds. Interestingly, one iteration of ISTA resembles a hidden layer of a deep residual neural network. This becomes obvious as the soft-thresholding operator can be represented by two ReLU functions, which belong to the most popular activation functions in neural networks. Therefore, the ISTA can be efficiently implemented on neural architectures by unfolding its iteration into layer-wise structure corresponding to a neural network [25]. In [8], numerical evidence shows that unfolding only one step of the iteration and allowing both the nonlinearity and iteration matrix to be learned result in an algorithm that is feasible online while approximating the minimum mean squared error (MMSE) recovery sufficiently. This is a very interesting direction of research as the results not only are useful in practical applications of wireless communication but also indicate an option to reduce the depth of neural networks by optimizing activation functions. This indicates the potential of hybrid ML, i.e. the combination of classical algorithms and neural networks.

---

6 Note that we consider an estimation problem if signal recovery is not possible since for instance the observation is corrupted by noise. A reliable estimation means that some estimation error such as the mean squared error is sufficiently small (i.e. below a given application-dependent threshold).

However, many signals in wireless communication are not sparse in the strict sense but rather softly sparse. Soft sparsity is a weaker form of sparsity, where in a vector there are only a small number of relatively large values (magnitude-wise), while the majority of entries is close to zero. This kind of sparsity is for instance observed in the wireless channels so that soft sparsity is relevant for the task of channel estimation. Soft sparse signals can be means of a random vector that is uniformly distributed over generalized unit balls. Realizations of this random vector are softly sparse signals with the majority of entries being close to zero and a few large peaks [8]. Of particular interest is the l_1-ball, which is a convex set and lies at the core of compressive sensing. Since the l_1-norm is known to support sparsity, the l_1-ball can be used to approximate sparse signals and, in particular, sparse channel impulse responses. This approximation was used in [26] to design a neural architecture that is optimal in the sense of the nonlinear MMSE estimation of a vector that is uniformly distributed over the l_1-ball. In particular, it is shown how to map the optimal solution to this important estimation problem to a neural network architecture. Again this is an example of how to systematically incorporate domain knowledge (in this case soft sparsity) into the design of neural networks. A good choice of neural architecture reduces the time and the amount of data needed for training.

14.3.6 Online Loss Map Reconstruction

Being able to query the strength of a field in every point of a map comes in handy when predicting required rate and power [27]. Such loss map functions need to be learned from single samples of data submitted by the mobile user end devices. However, the distribution represented by the training data changes dynamically as the user end devices are moving. Hence, the function needs to be learned online, and there is not much time for training. In order to comply with real-time constraints, algorithms that run on massively parallel processors such as GPUs are key. A promising approach that can be used to derive iterative algorithms with those characteristics is the adaptive projected subgradient method (APSM), which can be seen as an extension of Polyak's subgradient method to the case where the cost function is time-varying, as required in online settings. A distinguishing feature that makes iterative APSM-based algorithms appropriate for implementation on GPUs is that they can be designed to use only simple projections onto closed convex sets in Hilbert spaces, and these projections can be computed in parallel. For example, to apply the APSM to radio map reconstruction, we can consider that a radio map is a function in an RKHS that maps geographical coordinates to the signal strength in that location. As users report their location and the corresponding signal strength, we construct closed convex sets that restrict the candidate functions (i.e. the possible estimates of the true radio map) to those consistent with the available measurements. In settings of this type, a wide variety of APSM-based algorithms can be easily derived, and these algorithms can, for example, take into account the number of projections that can be computed in parallel on a particular GPU architecture, as demonstrated in [27] and illustrated in Figure 14.2.

14.3.7 Learning Non-Orthogonal Multiple Access and Beamforming

In multiuser wireless systems, channel knowledge obtained from pilot symbols can be used to compute (often linear) receive filters that are able to separate data symbols of a desired

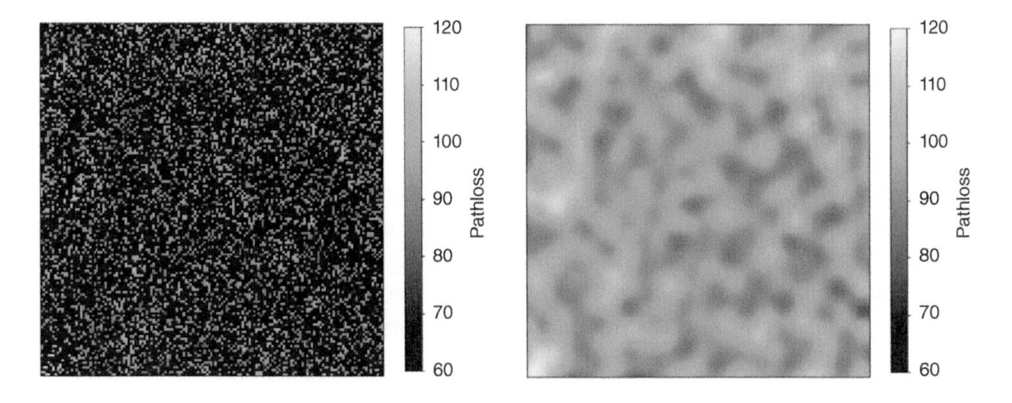

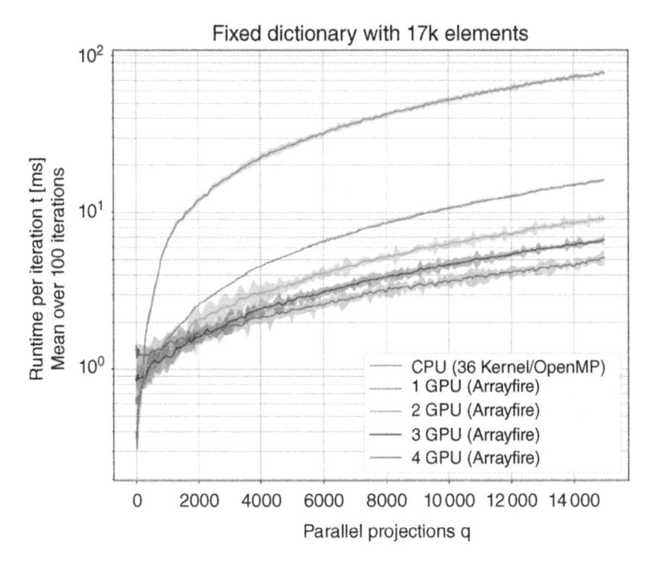

Figure 14.2 Top: Reconstruction of a continuous loss map (right) from a sparse set of data points (left) using the adaptive projected subgradient method (APSM). Bottom: Comparison of the runtimes of APSM on a CPU and several GPU configurations with respect to the number of projections executed simultaneously, i.e. in parallel. The GPU configurations are most efficient when executing the parallel numerical algorithm.

user from interference. Online ML methods based on the APSM (see the previous section) may be used to learn a receive filter directly from pilot symbols, without any channel estimates and without restricting the filters to be linear. The introduction of nonlinearities can be especially useful if the number of interfering users is larger than the number of receive antennas or if the users can be well separated in the power domain, as shown in [1] and, more recently, in [28, 29]. Similar to the approaches introduced in the previous sections, APSM-based algorithms for symbol detection assume that the nonlinear function to be estimated, which maps the received signal to the desired user's data symbol, belongs to an RKHS. As signals arrive at the receiver, knowledge of the desired pilots is used to construct closed convex sets that restrict the candidate functions. Iterative APSM-based methods using only simple projections onto closed convex sets are then derived by taking into account the computation power available to the receiver, and we note that the resulting adaptive nonlinear receivers can be implemented in real-time with existing GPUs.

14.3.8 Simulating Radiative Transfer

Across many industries, computer simulation has become an indispensable tool for both research and planning. Wireless networks are no exception. Especially, since with the advent of hardware accelerated ray tracing and ML algorithms from computer graphics, radio wave propagation can be simulated at high precision in real-time.

Given the geometry of an environment, like for example the streets and buildings of a town, tracing a ray from a point of origin into a direction determines the next point of intersection with the geometry. This basic operation is the foundation of high-precision radiation transport simulation [30].

As an example, the interactive simulation of a signal strength map using transport paths created by ray tracing and guided by reinforcement learning is principled. Given the location of a receiver in space, a first ray is traced into a random direction. Hitting a surface, a next ray is traced in the reflected direction. This process is repeated until the sphere around a transmitter is intersected, no more intersection is found, or a user-defined maximum path length is exceeded. In order to be able to intersect rays and transmitters, the transmitters are represented by small spheres instead of points. While this is an approximation already, finding paths that intersect antennas is hard for small spheres and large distances and becomes even harder with path length.

Reinforcement learning [31] can dramatically improve the efficiency of such transport simulations. For that, as shown in Figure 14.3 (left), the simulation volume is discretized. Each volume element (voxel) stores a probability distribution over the sphere. The direction of the first ray of a path is sampled proportionally to that distribution. Reinforcement learning starts with a uniform distribution, which is updated with the contribution of each path that connects to a transmitter. Hence, over the course of the simulation, directions

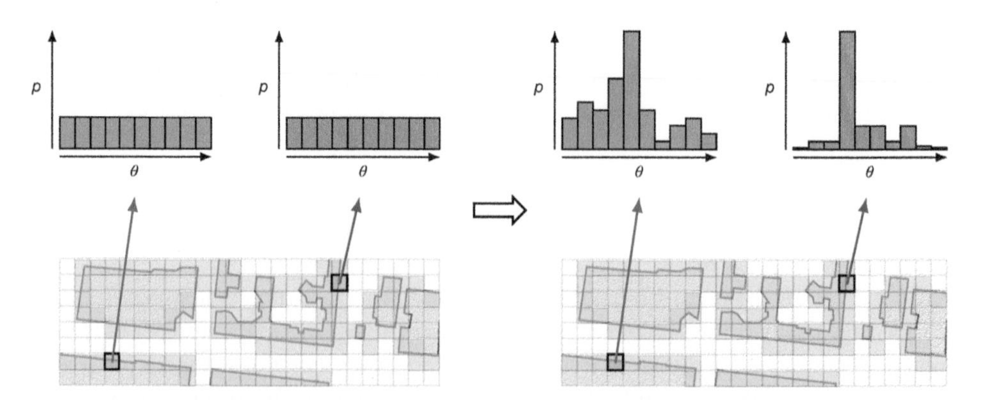

Figure 14.3 Quantizing three-dimensional positions of scattering events implicitly partitions them into clusters with a sparse voxel structure (bottom). Each cluster starts with a uniform probability distribution of finding a transmitter over directions θ (left), and over the course of the simulation each voxel learns a joint probability distribution (right).

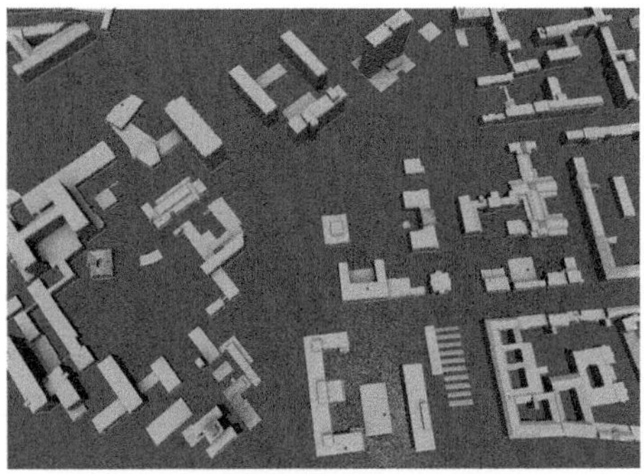

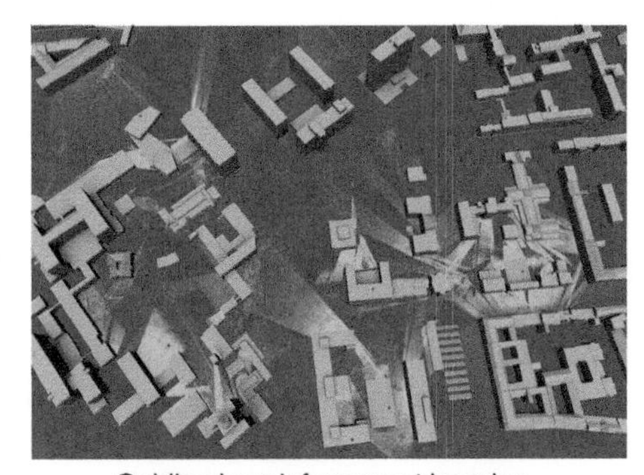

Figure 14.4 Strength of field for non-line-of-sight paths visualized on a user-defined measurement plane. Signal strength is calculated using a numerical simulation for each visible intersection of a primary ray with the visualization measurement plane and displayed in false colors, ranging from blue (low) to red (high). Guiding the simulation by reinforcement learning (right) dramatically accelerates the convergence as compared to a classic method, which simulates each path independently without using and sharing success probabilities over iterations of the simulation and across neighbors (left). Both images were computed in the same amount of time and include paths of lengths 2-5.

that connect to transmitters are selected with higher probability as illustrated in Figure 14.3 (right), which is key to efficiency. As only a small fraction of the voxels will be used in the simulation, the algorithm becomes very practical by hashing the voxels [32]. Figure 14.4 compares the benefits of reinforcement learning to the classic implementation.

It is straightforward to use the unidirectional reinforcement learning algorithm to improve the bidirectional path tracing approach explored by Taygur et al. [33].

14.4 The Soul of 6G will be Machine Learning

Telecommunication companies spend around 15% of their revenue on capital expenses (CAPEX), the bulk of it being network equipment and information technology. While most of the spendings have been on isolated workloads, more and more is transformed into telecommunication clouds. About 50% of revenue is invested in operational expenses (OPEX). Following the trend to migrate away from legacy networks, telecommunication companies have experienced a weak operational efficiency and a lack of competency. Assuming a year-over-year (YOY) revenue growth of about 5%, compute and network infrastructure actually would require five times as much. Therefore, costs have been kept low through procurement strategies. As a collateral, this has driven innovation out of the industry, has imploded the ecosystem, and consumer markets have saturated in penetration.

The way out is to pivot to enterprise solutions. The next generation of telecommunication business has to take a quantum leap in cost and focus on revenue growth. This requires to transform the companies into cloud-native centered around ML. It is to be foreseen that the immanent convergence of information and communication technologies will foster this innovation process based on software running on COTS including general-purpose accelerators and open interfaces between components.

ML will be instrumental in managing the complexity of future wireless communication in a sustainable and energy-efficient way. As an example, current cellular networks are based on a static partition of the wireless domain into cells. Future cell-less wireless networks may use unsupervised learning to assign user end devices to base stations dynamically. Such a dynamic assignment will allow for improved coverage guided by the actual physical conditions rather than an idealized cell structure. In such a scenario, semi-supervised learning, i.e. reinforcement learning, enables the tracking and prediction of both user bandwidth requirements and location. This in turn allows for load balancing by coarse grain scheduling across the network in combination with beamforming to efficiently link individual base stations and user end devices. As multiple base stations need to collaborate, learning algorithms will be implemented in a federated fashion.

In the same spirit, the interplay between networking elements will be optimized by ML. To enable such orchestration across equipment and software vendors, open interfaces will be paramount. A lot of potential still lies in cross-layer design, which aims at overcoming the classic separation of different layers such as PHY and MAC layers, being the enabler to improved federal scheduling involving a multitude of base stations.

14.5 Conclusion

As demonstrated by the variety of examples in this chapter, ML will help to cope with the increasing complexity of wireless networks. In fact, there will be a strong need for robust online ML methods that incorporate and exploit domain knowledge.

Key to a sustainable and energy-efficient realization of such ML algorithms will be massive parallelization in union with distributed computing. In order to cope with the expectedly large number of mobile devices, data needs to be processed right at the edge of the networks. Implementing algorithms and services on COTS will lower the TCO and add the value of easy deployment of services and core technologies.

Hence, the future of wireless networks natively will be based on a cloud infrastructure that runs the RAN. Besides, ML will support both network and customer operations as well as revenue generating services. This is in perfect analogy of how ML revolutionized and complemented computer graphics and high-performance computing by reducing brute-force computation. The approach becomes more profitable the more it is based on one open platform and one software ecosystem. It is the path forward for telecommunication providers and private networks, allowing them to focus on providing new services and revenues.

References

1 Theodoridis, S., Slavakis, K., and Yamada, I. (2010). Adaptive learning in a world of projections. *IEEE Signal Processing Magazine* 28 (1): 97–123.

2 Hilbert, D. (1902). Mathematical problems. *Bulletin of the American Mathematical Society* 8 (10): 437–479.

3 Kolmogorov, A.N. (1957). On the representation of continuous functions of several variables by superposition of continuous functions of one variable and addition. *Doklady Akademii Nauk SSSR* 114: 953–956.

4 Golitschek, M.V. (1980). Approximating bivariate functions and matrices by nomographic functions. In: *Quantitative Approximation* (eds. R. DeVore and K. Scherer), Academic Press: 143–151.
5 Golitschek, M.V. (1984). Shortest path algorithms for the approximation by nomographic functions. In: *Anniversary Volume on Approximation Theory and Functional Analysis*, Springer: 281–301.
6 Buck, R.C. (1982). Nomographic functions are nowhere dense. *Proceedings of the American Mathematical Society*: 195–199.
7 Sprecher, D.A. (2014). On computational algorithms for real-valued continuous functions of several variables. *Neural Networks* 59: 16–22.
8 Limmer, S. and Stańczak, S. (2016). Towards optimal nonlinearities for sparse recovery using higher-order statistics. IEEE International Workshop on Machine Learning For Signal Processing (MLSP).
9 Cybenko, G. (1989). Approximation by superpositions of a sigmoidal function. *Mathematics of Control, Signals, and Systems (MCSS)* 2 (4): 303–314.
10 Hornik, K., Stinchcombe, M., and White, H. (1989). Multilayer feedforward networks are universal approximators. *Neural Networks* 2 (5): 359–366.
11 Barron, A.R. (1993). Universal approximation bounds for superpositions of a sigmoidal function. *IEEE Transactions on Information Theory* 39 (3): 930–945.
12 Leshno, M., Lin, V.Y., Pinkus, A., and Schocken, S. (1993). Multilayer feedforward networks with a nonpolynomial activation function can approximate any function. *Neural Networks* 6: 861–867.
13 Keller, A. and Dahm, K. (2019). Integral equations and machine learning. *Mathematics and Computers in Simulation* 161: 2–12.
14 Sutton, R.S. and Bart, A.G. (2018). *Reinforcement Learning: An Introduction*, 2e, MIT Press.
15 Schrittwieser, J., Antonoglou, I., Hubert, T. et al. (2020). Mastering Atari, Go, chess and shogi by planning with a learned model. *Nature* 588: 604–609.
16 Cavalcante, R., Stańczak, S., Schubert, M. et al. (2014). Toward energy-efficient 5G wireless communications technologies: tools for decoupling the scaling of networks from the growth of operating power. *IEEE Signal Processing Magazine* 31 (6): 24–34.
17 Pollakis, E., Cavalcante, R.L., and Stańczak, S. (2012). Base station selection for energy efficient network operation with the majorization-minimization algorithm. 2012 IEEE 13th International Workshop on Signal Processing Advances in Wireless Communications (SPAWC): 219–233.
18 Pollakis, E., Cavalcante, R.L.G., and Stańczak, S. (2016). Traffic demand-aware topology control for enhanced energy-efficiency of cellular networks. *EURASIP Journal on Wireless Communications and Networking* (1): 1–17.
19 Decurninge, A., Ordóñez, L.G., Ferrand, P. et al. (2018). CSI-based outdoor localization for massive MIMO: experiments with a learning approach. 2018 15th IEEE International Symposium on Wireless Communication Systems (ISWCS): 1–6.
20 Cavalcante, R.L.G., Miretti, L., and Stańczak, S. (2018). Error bounds for FDD massive MIMO channel covariance conversion with set-theoretic methods. 2018 IEEE Global Communications Conference (GlobeCom): 1–7.
21 Miretti, L., Cavalcante, R.L.G., and Stańczak, S. (2018a). Downlink channel spatial covariance estimation in realistic FDD massive MIMO systems. 2018 IEEE Global Conference on Signal and Information Processing (GlobalSIP): 161–165.

22 Miretti, L., Cavalcante, R.L.G., and Stańczak, S. (2018b). FDD massive MIMO channel spatial covariance conversion using projection methods. 2018 IEEE International Conference on Acoustics, Speech and Signal Processing (ICASSP): 3609–3613.

23 Cavalcante, R.L.G. and Stańczak, S. (2020). Hybrid data and model driven algorithms for angular power spectrum estimation. 2020 IEEE Global Communications Conference (GlobeCom): 1–6.

24 Ismayilov, R., Holfeld, B., Cavalcante, R.L.G., and Kaneko, M. (2019). Power and beam optimization for uplink millimeter-wave hotspot communication systems. 2019 IEEE Wireless Communications and Networking Conference (WCNC): 1–8.

25 Hershey, J.R., Roux, J.L., and Weninger, F. (2014). Deep unfolding: model-based inspiration of novel deep architectures. arXiv preprint arXiv:1409, 2574.

26 Limmer, S. and Stańczak, S. (2018). A neural architecture for Bayesian compressive sensing over the simplex via Laplace techniques. *IEEE Transactions on Signal Processing* 66 (22): 6002–6015.

27 Kasparick, M., Cavalcante, R.L.G., Valentin, S. et al. (2015). Kernel-based adaptive online reconstruction of coverage maps with side information. *IEEE Transactions on Vehicular Technology* 65 (7): 5461–5473.

28 Awan, D.A., Cavalcante, R.L.G., Yukawa, M., and Stańczak, S. (2020). Adaptive learning for symbol detection: a reproducing kernel Hilbert space approach. In: *Machine Learning for Future Wireless Communications* (ed. F.-L. Luo), Wiley.

29 Mehlhose, M., Awan, D.A., Cavalcante, R.L.G. et al. (2020). Machine learning-based adaptive receive filtering: proof-of-concept on an SDR platform. *Proceedings of IEEE International Conference on Communications (ICC).*

30 Moser, S., Kargl, F., and Keller, A. (2007). Interactive realistic simulation of wireless networks. IEEE Symposium on Interactive Ray Tracing: 161–166.

31 Dahm, K. and Keller, A. (2018). Learning light transport the reinforced way. In: Monte Carlo and Quasi-Monte Carlo Methods 2016 (eds. A. Owen and P. Glynn) Mathematics & Statistics, Springer, Vol. 241: 181–195.

32 Binder, N., Fricke, S., and Keller, A. (2022). Massively parallel path space filtering. In: Monte Carlo and Quasi-Monte Carlo Methods 2020 (ed. A. Keller). Mathematics & Statistics, Springer.

33 Taygur, M.M., Sukharevsky, I.O., and Eibert, T.F. (2018). Computation of antenna transfer functions with a bidirectional ray-tracing algorithm utilizing antenna reciprocity. 2018 2nd URSI Atlantic Radio Science Meeting (AT-RASC): 1–4.

15

Managing the Unmanageable: How to Control Open and Distributed 6G Networks

Imen Grida Ben Yahia, Zwi Altman, Joanna Balcerzak, Yosra Ben Slimen, and Emmanuel Bertin

Orange Innovation Networks, France

15.1 Introduction

The tremendous increase in complexity of future networks finds its roots in the introduction of 5G families of use cases, namely, enhanced mobile broadband, massive Internet of Things (IoT), and ultrareliable low-latency communications, each of which represents an ensemble of services with specific technical requirements. In addition to technological advances, 5G introduces the softwarized, cloud native networks with centralized and edge clouds. The network cloudification provides incredible flexibility for dynamically evolving the network according to service demands. To manage complexity, the concept of network slicing has been introduced, providing a framework for the design, the management and orchestration, and the cohabitation of multiple logical networks sharing a common infrastructure. The network operator may need to support a large number of slices for different tenants, according to their service-level agreements (SLAs).

The deployment of 5G has recently started with still many remaining technological challenges. Certain 5G technologies and use cases will continue to evolve during the 6G era. Examples are the advances in virtualization, enhanced multi-vendor interoperability via standardized open interfaces (e.g. open radio access network [O-RAN] [1]), or efficient solutions for automated management and orchestration of network slicing. In parallel, the vision of 6G networks starts to build up, with the identification of use cases that will populate 6G. New use cases are associated with new technologies, such as quantum technology, metasurfaces for shaping propagation, electromagnetic-field-aware technologies, battery-free devices, or devices using new frequency bands. Artificial intelligence (AI) is considered as a central pillar in 6G network management, as a building block for new network technologies and as an enabler for new services.

New services aiming at improving quality of experience (QoE) are foreseen in 6G, such as tactile Internet, Internet of Everything (IoE), augmented reality, or high-precision 3D positioning. In terms of performance, 6G systems are expected to provide "simultaneous"

Shaping Future 6G Networks: Needs, Impacts, and Technologies, First Edition.
Edited by Emmanuel Bertin, Noel Crespi, and Thomas Magedanz.

wireless connectivity that is 1000 times faster than 5G. Standards are expected to support data rates up to 1 Terabytes. Certain new 6G services would require very large volumes of data as, for example, the exchange of holographic content and holographic type communications. The International Telecommunication Union (ITU) predicts that the trend of exponential growth will continue, and by 2030, the overall mobile data traffic will reach 5 zettabytes (ZB) per month.

From architectural advances, federating learning paradigms and the active research for massive communications, privacy, and distributions of actions fit well with the 6G requirements for diversified connectivities for massive IoT.

Furthermore, open virtual and software networks are adding complexity in the management of networks including in 6G: the variety of services with different quality of service (QoS) requirements, of technologies, and of possibly many coexisting slices and the geographical distribution of network nodes introduce a highly complex and distributed system to manage. A centralized Service Management and Orchestration (SMO) system will reach its limits. Local control and optimization capabilities are necessary as a means to efficiently perform dynamic configuration and resource allocation in the time scale of traffic dynamics. In this context, AI and machine learning (ML) can provide solutions for optimizing control and management tasks that cannot be performed efficiently in a centralized manner.

Several enablers can facilitate management complexity:

- *Data storage and parallel processing:* The capability to store and to process huge volumes of data in real-time by using big data technologies and federated learning (FL).
- *Automatic data preparation:* The capability to automatically clean, aggregate, and label the data for fully autonomous management systems.
- *Automatic orchestration decisions:* The capability to exploit data by means of data processing and ML techniques to extract useful information and to derive policies for managing the network. Policies can define conditions for modifying network configurations and for triggering optimization or control processes for given nodes, geographical area, or individual or a group of users.
- *Openness and support of network intelligence:* The capability to introduce policies, control, and AI/ML solutions via standardized open interfaces.

Standardization is a key enabler for introducing intelligence in the different segments and layers of the control protocol stack of the network, allowing it to adapt to varying conditions and to optimize users' QoS and QoE at different time scales. Openness in conjunction with network intelligence will stimulate the industrial players including third parties, vendors, and operators to develop control and AI/ML solutions that benefit from short deployment cycle.

15.2 Managing Open and Distributed Radio Access Networks

15.2.1 Radio Access Network

AI and ML (AI/ML) activities have been rich in the RAN segment, with some notable advances in the physical layer that could be part of the 6G technology. In the standardization arena, one can note the recent activities carried out in the O-RAN Alliance that give a significant momentum to AI/ML in the RAN segment [1]. The particularity of O-RAN is the standardization of open interfaces and intelligent controllers that support the operation of optimization, control, and AI/ML models.

1) *AI/ML in the control protocol stack:* Radio resource allocation is performed by different layers of the control protocol stack and can benefit from AI/ML algorithms, as is briefly described presently.

Radio Resource Control (RRC) layer: In the RRC layer, AI/ML techniques can benefit from different data sources to optimize access and mobility procedures. Typical example is the prediction of future users' locations and speed. Such predictions can feed intelligent mobility algorithms and can trigger a handover of a specific user or a group of users at the right time and to the best target cell, frequency, or technology. Example of data sources can be traffic maps that indicate the probability of choosing a given trajectory or Radio Environment Maps (REM) [2] that are constructed from geolocation measurements reported by mobile users. For each location the REM can provide radio quantities such as the received signal strength. The standard that is presently developed by O-RAN defines the necessary framework to embed AI/ML models and exploit various data sources.

Medium Access Control (MAC) layer: Different use cases involving the scheduler in conjunction with AI/ML have been recently studied and could be implemented if adequate interfaces and APIs (Application Programming Interface) are standardized. Different use cases involving intelligent schedulers can be envisaged, and two examples are briefly summarized. In the first, mobility prediction is used in conjunction with an REM. The scheduler can estimate the attainable rates along a portion of the future trajectory and selects users for scheduling in a manner to benefit from time and space diversity. This concept has been denoted as Forecast scheduling [3] or Anticipatory radio resource management [4] and can bring about considerable throughput gains. The second example considers coordinated scheduling of two neighboring cells with massive multi-input multi-output (M-MIMO) deployment [5]. We assume that both base stations are using a set of narrow beams for transmission, denoted as grid of beams. The narrow beams of the two neighboring cells can interfere users, mainly at cell edge. An AI/ML algorithm dynamically learns whether the transmissions over any couple of beams should be coordinated or not. Coordination means that the schedulers of the neighboring cells can transmit information on the two interfering beams on different (nonoverlapping) time intervals. The deployment of intelligent schedulers can be envisaged in 6G.

Physical (PHY) layer: AI/ML activity in the PHY layer is particularly innovative, and a few examples are described hereafter. A first research avenue that covers both radio and optical PHY layers is the use of AI techniques such as neural networks (NNs) and often Deep NN (DNN) to design and optimize distinct modules. Such modules include the encoder/decoder or the equalizer. An example where both the equalizer and the decoder are modeled as NNs is presented in [6]. Another challenging research axis deals with the end-to-end modeling of the PHY layer architecture as an autoencoder [7]. The transmitter, the channel, and the receiver of the communication system are implemented as a single NN, which reconstructs its input at its output. First contributions considered the autoencoder for specific channel models, whereas in a more recent work, it has been shown that the condition of explicit channel model has been removed at the expense of feedback sent from the receiver to the transmitter during the training [7]. AI/ML are used for optimizing M-MIMO beamforming, such as power allocation that optimizes energy efficiency by means of DNN [8], as well as cross-layer optimization involving M-MIMO beamforming (PHY layer) and the scheduling (MAC layer).

Recently, a new technology, the reconfigurable intelligent surfaces (RISs) or metasurfaces, has been proposed that allows the control of the electromagnetic environment [9]. The metasurface is equipped with integrated electronic circuits that can be programmed to modify the response of an incoming electromagnetic field, namely, its reflection and transmission in a customizable way. The RIS can be easily attached to walls or ceilings in indoor scenarios or embedded in facades of buildings in outdoor scenarios. Among the promising use cases is the enhancement of user-received signal in a situation of coverage outage. A mobile user can suffer from too severe attenuation due to high obstruction of the propagation waves. To avoid mobile outage, the RIS is programmed in a manner to steer reflections in desired directions of mobile users. The control of the RIS can be done at cell level, and more futuristic studies consider distributed control based on embedded intelligence at the RIS.

15.2.2 Innovation in the Standardization Arena

15.2.2.1 RAN

In the RAN segment, O-RAN Alliance is evolving the RAN standard with the aim of achieving more openness via open interfaces and supporting embedded intelligence [1]. A brief description of O-RAN activity related to the support of control loops and AI/ML models is presented. In O-RAN, control, optimization, and AI/ML algorithms can be deployed in two logical functions: the non-real-time (RT) Radio Intelligent Controller (RIC) and the near-RT RIC (see Figure 15.1).

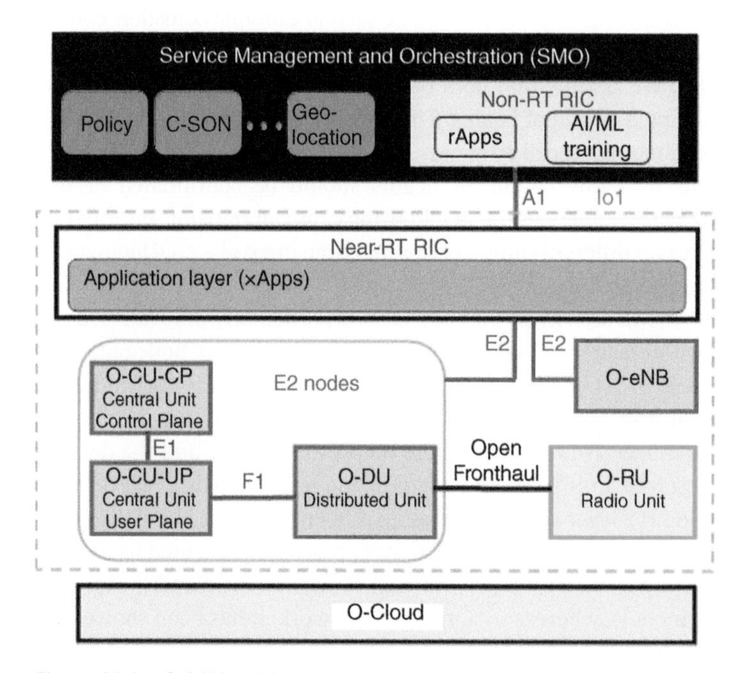

Figure 15.1 O-RAN architecture.

The **non-RT RIC** is a logical function internal to the SMO. It receives some intent from e.g. the Business Support System (BSS) and translates it to policies for the near-RT RIC to guide its operation. The non-RT RIC can host non-real-time control, optimization, and AI/ML models within applications denoted as rApps that operate in slow time scale varying from a second to hours and more. AI/ML models can be trained in the non-RT RIC, in the SMO, or in an external entity. The operation of rApps is supported by non-RT RIC or by SMO functions. A non-RT function can be an exposure function for rApps, and an SMO function can be data sharing or RAN inventory.

The **near-RT RIC** is a logical function that controls and optimizes E2 nodes' functions and resources. It can host control, optimization, and AI/ML models within applications denoted as xApps that operate at a time scale varying from 10 milliseconds to 1 second. The xApps are used to enforce policies from the non-RT RIC. The xApps control resource allocation for individual or group of users or parameters of radio resource management algorithms in E2 nodes. While both near- and non-RT RIC can serve as inference hosts for AI/ML models, initial training is recommended to be performed in the non-RT RIC (or SMO), namely, off-line.

Standardized interfaces and APIs: The non-RT RIC is connected to the near-RT RIC via the A1 interface. Three services are defined in the non-RT RIC and are supported by the A1 interface to guide the near-RT RIC: (i) A1-P – Policy Management Service. Example of policies can be QoS targets for a group or for all users in a cell, in a geographical zone or per slice instance. The near-RT RIC provides feedback on the policy execution to the non-RT RIC via A1. (ii) A1-EI – Enrichment Information (EI) Service. A1-EI aims at improving RAN optimization performance by providing information that is not available in the RAN. EI can be collected or derived in the SMO or in the non-RT RIC and sent to the near-RT RIC via the A1 interface. EI can also be provided by an external source through a direct secure connection with the near-RT RIC. Examples of EI can be information from an application server or from a geolocation server in the form of a portion of an REM that will be used to optimize resource allocation. (iii) A1-ML – ML Model Management Service that will support the use of ML models in the near-RT RIC. Full information on A1-ML will be available in a later stage.

The near-RT RIC is connected to E2 nodes via the E2 interface. This interface supports near-RT RIC services to allow its interworking with the E2 nodes, to subscribe to reports from E2 nodes and to control their operation. For example, following a triggering event, the near-RT RIC can trigger or suspend an E2 procedure, modify control commands, and apply policies. The condition(s) or the action(s) defined in a policy can be provided by an ML model via an xApp. The format and the possible values of metrics and actions (e.g. configuration values) in the policy are predefined.

The different blocks shown in Figure 15.1 are briefly described below:

- **O-CU-CP:** O-RAN Central Unit – Control Plane. This is a logical node hosting the RRC and the control plane part of the PDCP protocol.
- **O-CU-UP:** O-RAN Central Unit – User Plane. This is a logical node hosting the user plane part of the PDCP protocol and the SDAP (Service Data Adaptation *Protocol*) protocol.
- **O-DU:** O-RAN Distributed Unit. This is a logical node hosting the RLC (Radio Link Protocol)/MAC/PHY-high layers based on a lower layer functional split.

- **O-eNB:** An eNB (evolved Node B) that supports E2 interface.
- **O-RU:** O-RAN Radio Unit: a logical node hosting PHY-low layer and radio frequency (RF) processing based on a lower layer functional split.
- **O-Cloud:** A cloud computing platform comprising a collection of infrastructure nodes that are compliant to the O-RAN requirements to host O-RAN functions, the supporting software components, and the appropriate management and orchestration functions.

In the present standardization status of O-RAN architecture, the ML inference host, namely, the network function that hosts the ML model during inference operation can be located in both the non-RT RIC and in the near-RT RIC. A challenging evolution of O-RAN is supporting the deployment of control and ML algorithms in the O-CU and O-DU.

15.3 Core Network and End-to-End Network Management

The changing needs of users in terms of network utilization and the growing demand for new services need a change in terms of requirements for the next generation of networks. The most important requirement for 6G network will be the capability of handling massive volumes of data and very high-data-rate connectivity per device. Below requirements should be taken into account for 6G network:

- To perform smart resource management.
- To perform automatic network adjustment, provisioning, and orchestration.
- To use the discovered knowledge to optimize network performance.
- To exchange learned knowledge for smart connected devices.
- To use sensing as a service as possible new category for 6G applications.
- To use a smart device with embedded wireless communications as an access point.

Intelligence in the network will be a key aspect because of achievement of fully end-to-end automated network architecture. As described in [10, 11], the network will enable new verticals within the emerging holographic society, a future digital society empowered by holographic technologies through a wave of innovations to provide new communication services over federated, new infrastructures. Network will become user centric and will support new emerging use cases including new holographic media (holographic-type communication), haptic communication, etc. ITU-T [11] is already defining a Tactile Internet that is enabling haptic communication as the network that combines ultralow latency, ultralow packet loss, and ultrahigh bandwidth with extremely high availability, reliability, and security.

15.3.1 Network Architecture and Management

Beyond 5G will bring a convergence with non-3GPP (3rd Generation Partnership Project) access and also between mobile and fixed networks through a native Internet Protocol (IP) user plane, which should support services with ultralow latency and resiliency across the Internet and allow the development of service automation. The network slicing paradigm is a potential solution for a new perspective on network management since it contains essential features, such as automation, fast adaptation, and customization. Network slicing is key to realizing the potential of the 5G-6G network.

Beyond 5G would eventually replace the GPRS (General Packet Radio Service) Tunnel Protocol (GTP) offering flexibility in routing by considering the transport network characteristics [8]. Currently, 3GPP explores for this purpose the use of segment routing based on IPv6 (SRv6) [12] or locator/ID separation mechanisms [13]. SRv6 makes a QoS treatment within a node possible and also it enables a service instruction, e.g. directing a packet to a virtual machine [12, 14]. Locator/ID Separation Protocol (LISP) separates location addressing and node identity, offering IP-based dynamic encapsulation. It brings benefits in routing scalability, mobility, and network virtualization but suffers from limitations associated with the mapping process [13, 14].

To guarantee high QoE for interactive services, it is the primary need for deterministic networking (DetNet) that should provide end-to-end delivery of messages within a given time and to assure that every flow packet conforms to the desired delay bound and receives immediate and highly robust delivery or coordinated ultralow latency. Transport network alignment mechanisms are still open; however, taking advantage of DetNet, time-sensitive networking (TSN), and segment routing techniques is possible as described in [14–16].

Their intention is to create a network with features like time synchronization, zero congestion loss, reliability, and security for real-time applications, but TSN was designed for Ethernet and not mobile networks. DetNet operates on top of TSN, so integration for both with mobile networks need to be refined.

Software-defined network (SDN), AI, and network slicing technologies provide enabling tools for seamless integration of the heterogeneous network segments to maximize the benefits from their complementary characteristics to support various new use cases, dynamic adaptation to the change of service demand, and feedback about real-time user perception. Network management in 6G will be developed toward distribution, diversification, and intelligence. Network architecture will be adjustable, programmable, and automated. AI will be a driving force for development of next generations' networks. The design of the 6G architecture will allow the network to be smart, agile, and able to learn and adapt itself according to the changing network dynamics. It will evolve into slices allowing more efficient and flexible upgrades, and a new framework based on intelligent radio and algorithm-hardware separation to cope with the heterogeneous and upgradable hardware capabilities. Each slice should collect and analyze its local data and then exploit AI methods to upgrade itself locally and dynamically. 5G network extensions up to the next generation of the network will bring intelligent connectivity taking into account content-, relation-, context-, and behavior-aware networks. Network will need to meet the requirements about ultrahigh peak data rate, ultrahigh traffic capacity, ultralow latency, and connected intelligence with AI capability. 6G will provide the potential for many new applications, services, and solutions related benefits such as substantive improvements in the areas of sensing including sensors solving abnormal situations in the network, imaging, and location determination. 6G will require a new look at the service design. Fully AI empowered almost in all levels, from network orchestration and management to coding and signal processing in the physical layer, to manipulation of smart structures, and to data mining at the network and device level for service-based context-aware communications, can learn and envision the necessities of a user.

15.3.2 Changes in Architecture and Network Management from Standardization Perspective

In June 2020, the ITU-T has initiated an AI/ML in 5G challenge on the theme "How to apply ITU's ML architecture in 5G networks" to motivate researchers to identify and solve real-world problems based on standardized technologies developed for ML in 5G networks. It is envisioned that such efforts would be factored in later 5G development but will be materialized in a more concrete and pervasive manner in 6G.

Service-based architecture (SBA) already present in 5G will be further developed in standards, and it can be extended also to the user plane and not only to control and management planes as in 5G. SBA launches a new paradigm for organizing and operating network functions to allow interconnectivity, based on a micro-service architecture, instead of conventional point-to-point interfaces. With the assistance of SBA, it would be easier for beyond 5G networks to combine control and management plane services, enabling automation and value-added services at edge cloud. The SBA of 6G core network should be a significant cognitive function. This function should recognize target behavior and user characteristics but also adjusts the network services adaptively and dynamically through the unified service description method. The adoption of SBA in the RAN would ease the exposure of RAN analytics allowing the development of ML and QoS prediction. As described in [17], an integration of the SBA framework for the core network is possible into an end-to-end multi-plane (application, control, user, and management) and multi-segment (access, backhaul, and core), architected along similar lines as the zero-touch network and service management (ZSM) reference architecture described in ETSI ZSM 002 [17]. Zero-touch is about minimizing human intervention. The purpose is not only to eliminate errors but also to introduce agility in addressing service layer agreements (SLAs) and meeting the requirements of new services. The ZSM end-to-end architecture framework has been designed for closed-loop automation and optimized for data-driven ML and AI algorithms. In the core network, the SBA will lead in decomposing the network functions into more detailed independent services. In the RAN, 6G should separate the physical layer from the user plane.

In the end-to-end mandate-driven architecture (MDA), each segment can be managed by the separately specialized orchestrator. Mandate can be seen as a set of network services required from the underlying networks to achieve the required QoS. The selection of segments is based on context-aware understanding of the application or user environment [15]. In MDA monitoring is done per flow; in-band network, telemetry mechanism is used, where monitoring data is injected into data packets to learn about the end-to-end performance. Currently, packets using this mechanism are containing header fields that are interpreted as "telemetry instructions" by network devices. It enables network troubleshooting, advanced congestion control, advanced routing, and network data-plane verification. Network management in 6G will be supported by analytics to perform cognitive and autonomic functions. This can improve fault handling, security, SLAs, and accounting. Analytics can be used also for performance analysis and optimization with virtualization.

Standardization efforts in this network automation domain have led to the introduction of the network data analytics function (NWDAF) in the control plane and the management data analytics service (MDAS) in the management plane, for enhanced data collection and analytics functionalities within 3GPP.

15.3.3 Quality of Service and Experience

It is becoming clear that the required QoS pushed by some of the applications to extreme levels appear very challenging for 5G architectures. Current networks are not able to monitor QoS in the real-time and to answer whether QoS guarantees are met, which is a crucial piece of information still missing today. Currently in 5G, data flows with similar QoS requirements are transported on the same QoS level. Specific treatment for individual applications is not feasible due to the limited number of QoS levels, but in the new generation of network, it will rather require dynamic adaptation to changes taking place in the QoE level. The 5G QoS flow is the finest granularity for QoS forwarding treatment in the 5G system. All traffic mapped to the same 5G QoS flow receive the same forwarding treatment (e.g. scheduling policy, queue management policy, rate shaping policy, RLC configuration, etc.). Providing different QoS forwarding treatment requires separate 5G QoS flow. In 6G, QoS should contain more metrics relating to human perceptual-based factors impacting QoE such as gesture, face mimic, recognition, and physiology in addition to traditional metrics. Traditional key performance indicators (KPIs) usually cover latency, throughput, and packet loss rate. In the 6G, the computation capacity and storage also need to be considered, since intelligence will become the basic need. Future services should also be measured from the perspectives of situational awareness, learning ability, storage cost, computation rapidity, and capacity.

Edge AI is the combination of edge computing and AI. This has many advantages in case of network failure or saturated bandwidth by other services. As mentioned in [18], Edge AI may improve the scalability and can have a significant impact on QoE expressly for the applications used in the edge. It can also improve connectivity, data transmission, computation offloading, and capabilities for reactivity and proactivity. Intelligent real-time network management with quick adjusts and parameters' optimization approaches shall be employed to fulfil the different QoS demands and to provide high QoE. There is already a debate in many publications about the possibility of a transition from mobile edge computing to AI at the edge that can help in the progress and expansion of autonomous driving and in the implementations of intelligent devices. These intelligent/smart devices can control and manage their own data, improving the quality, processing data with small delay almost in the real-time, and probably taking into account changing behavior of users to adapt the QoE. The smart devices may provide opportunities for on-device data training, which can sense and learn from the local channel pattern, traffic pattern, moving trajectory, etc., to understand the features of user behavior and predict the network status. Smart decision-making devices can also help into a design of behavior that is relevant to our daily life and to offer services according to the needs of a specific user. Instant data that is generated by the user and relates only to the user can be computed and treated on the edge. This will result for example in a gain in latency because the data does not need to be sent across the network. FL is one of the most popular distributed learning algorithms, which enables data-driven AI on a large volume of decentralized data that reside on mobile devices. Thanks to the local treatment of data, the user data privacy is well protected and somehow it may affect the QoE. The most critical but at the same time very interesting aspect and something completely new in terms of quality is performance KPI related to learning time it takes to converge to a predefined accuracy level. Also, one of the new indicators may be

AI quotient for indexing and rating AI algorithms for comparing their capabilities in terms of providing various levels of QoS and QoE improvement.

The handover decision must consider the characteristics of users and infrastructures, for example, bandwidth, delay, signal strength of access links against the quality requirement of the on-going services, as well as load balance among network systems. During the handover decision, network coverage but also rapidity of the user should be taken into account. Autonomous network resource allocation with predicted needs about intelligent mobility and handover management of the user can come in handy for changing needs of users. With the prediction of user mobility, a proactive network resource adjustment can be designed to achieve a timely and smooth handoff. Also, due to the high network density, users may frequently move out the coverage of its associated network infrastructure, which results in a dynamic network topology. Autonomous resource allocation will provoke a dynamic network topology. AI techniques can be adopted to intelligently achieve mobility prediction and optimal handover solutions to guarantee communication connectivity. As mentioned in [19], AI and ML will be used to predict link loss events at high frequencies, to proactively decide on optimal handover instances and to determine optimal radio resource allocations for base stations and users. The dynamic network topology will change the service traffic distribution, previously optimal slice allocation, degrading network performance, and may even violate users' QoS requirements. When the network performance degrades to a predetermined threshold value, autonomous adjusting existing slices or creating new slices will be triggered, which incurs slice reconfiguration overhead. AI-based methods may come to provide accurate service-specific traffic prediction, which can effectively facilitate network resource allocation to accommodate service demands in the near future.

15.3.4 Standardization Effort in Data Analytics

It is certain that in 6G, there will be an even greater need to apply diagnostic, predictive, and prescriptive analytics. Velocity as the speed at which data accumulates but also the veracity that is the quality and origin of data and its conformity to facts and accuracy are important as the entry point into the data analysis process. Diagnostic analytics enable autonomous detection of faults and service impairments, identify the root causes of the anomalies, and ultimately improve the reliability and security of the network. Predictive analytics including ML the usage of several ML models to forecast future events such as traffic patterns, resource availability, user locations, or their behavior, their usual needs in terms of services but also resources but also can help with a need for specific personalized content. Prediction of QoS for one or more users in a given area will be a must in beyond 5G but also a richer user experience through localization services. Prescriptive analytics take advantage of the predictions to suggest decisions for resource allocation and network slicing. Such data may be provided for instance by NWDAF or management data analytics function (MDAF). The NWDAF [20] in the 5G core (5GC) is about to play a key role as a functional entity that collects statistics related to user mobility, load, communication patterns, QoS, and other information about different network domains and uses them to provide analytics-based statistics and predictive insights to 5GC network functions.

Analytics provided by MDAF [21] can help to perform root cause analysis and support a wide variety of services. MDAF allows an aggregation of performance and fault network statistics for cell(s). Together with NWDAF, MDAF enables complex analytics, e.g. predicting a user equipment (UE) QoS at a future location by combining mobility with performance perdition at specified cell(s) for beyond 5G. MDAF can expose one or multiple MDAS. A centralized MDAS can provide end-to-end or cross-domain analytics service, for instance resource usage or failure prediction in a network slice.

In 6G, we can also consider future evolutions of ONAP (Open Network Automation Platform) in terms of end-to-end network management, especially with DCAE (data collection, analytics, and events) subsystem. DCAE together with other ONAP components gathers performance, usage, and configuration data from the managed environment. This data is then fed to various analytic applications, and if anomalies or significant events are detected, the results trigger appropriate actions. In 3GPP there are some works about ONAP to define new requirements for MDAS.

15.4 Trends in Machine Learning Suitable to Network Data and 6G

15.4.1 Federated Learning

FL is a new iteration of ML, pioneered by Google in 2017. It is a collaborative approach to build ML models without any requirement on centralizing the train data. It is known as a step toward a larger paradigm that is "bringing the code to the data, instead of the data to the code." The main motivations of FL are privacy, ownership, energy consumption, and locality of data. FL enables devices (e.g. mobile phones in the Google pioneered blog) to learn in a collaborative way a shared prediction model without moving the data to a centralized and common data center [22, 23]. A simplified way of how it works can be described as follows: The device:

1) Receives the prediction model from the cloud,
2) The prediction model is then improved locally with the device data, and
3) The updates to the model are shared back to the cloud with encrypted communication.

FL to date is an active research area that is gaining more and more momentum. Several technical challenges are being tackled by the ML community. We cite here a few areas with higher relevance to the network and telecommunication context. FL research domains are numerous; the most active ones are about:

The optimal communication. ML model is using an optimization algorithm, e.g. stochastic gradient descent (SGD). The latter must run on large volumes of data as well as homogenous partitions across servers. It requires low latency and high bandwidth to get access and manipulate the train dataset. **FedSGD** denotes the baseline of FL. We notice several advances on this technique to overcome the latency inherited by the stochastic gradient technique [ref]. The basic approach is the following: a given device computes the gradient and sends the results to the server within the cloud, or the given device will start a loop of computing the gradient and ensuring the model updates before sending to the server. This type of loop is called the federate average, FedAVG. The rationale behind the

FedAVG is the following: FedSGD communicates the gradient for each iteration resulting in a heavy communication schema, while the FedAVG will only send the updated model after several iterations of gradient. The promise of FedAVG is of 10–100× less communication than the basic FedSGD. This approach is taking advantage of the device processors to rapidly compute updates. Table 15.1 summarizes a few research contributions in FL for optimal communication.

Secure aggregation was introduced by [27] as "a secure Multi-Party Computation protocol that uses encryption to make individual devices' updates uninspectable by a server, instead only revealing the sum after a sufficient number of updates have been received."

This is a must have for an FL-based approach. The main motivation as indicated is to avoid interpreting the client-device data. FL principles as cited above are to use only the element-wise weighted averages of the updates from each client-device vector. The secure aggregation protocol aims then to compute the weighted averages, thus allowing to ensure the privacy of the identity of the user and the corresponding updates and hence the client-device dataset [28].

Systemic view of FL. FL systemic view starts to emerge [29] in research. The major system components are the:

- Clients or the participants (e.g. user devices, network equipment, IoT devices, etc.).
- Cloud servers that manage the global model updates and synchronization.
- Communication framework including secure aggregation, communication protocol, privacy criteria, and all the needed computation to train the local ML models.

15.4.2 Auto-Labeling Techniques and Network Actuations

In previous mobile technologies, most of the existing solutions for the network management and orchestration are not fully automated since they still rely on human intervention. In fact, most of these models are either based on unsupervised learning where experts have to interpret the resulting clusters or based on supervised learning where data needs to be labeled. Labeling the data is particularly an expensive and a demanding task that is normally performed manually by domain experts via reporting tools. A most common automatic labeling approach uses predefined thresholds on some metrics or alarms, but practical experience proved that it is in most cases error-prone [30]. Thus, a

Table 15.1 Contribution to federated learning in optimal communication.

FedAVG	In FedAVG [22, 23] the local model parameters are element-wise averaged with weights proportional to sizes of the client-device datasets.
FedProx	FedProx [24] works on limiting the impact on the local updates.
Agnostic federated learning (AFL)	[25] AFL is a variation of the FedAVG, which works on optimizing a centralized distribution that is a combination of the client-device dataset distributions.
Probabilistic federated neural matching (PFNM)	[26] PFNM defines an approach to match the neurons of client NNs before proceeding to averaging them. Bayesian methods are also used to adapt to global model size and to face the data heterogeneity.

fully automated labeling mechanism seems inevitable in 6G networks, especially that we are expecting a fully automatic management. As a consequence, the huge amount of collected data should be automatically prepared and labeled for the training phase with no human intervention. It requires self-labeling approaches that reduce the burden on the AI service developer and the corresponding response bias. One solution is to use semi-supervised approaches in order to exploit any labels that already exist to predict the rest of the labels.

Essentially, semi-supervised methods use unlabeled samples to either modify or reprioritize the hypothesis obtained from labeled samples alone. This problem has been addressed by several approaches with different assumptions about the characteristics of the input data. Generative models [31] are used in the literature to learn a probability model by assuming that the data follows a determined parametric model. Transductive inference for support vector machines [32] can also be used. They assume that the classes are well separated and do not cut through dense unlabeled data. Semi-supervised approaches can also be viewed as a graph min-cut problem as explained in [33] assuming that if two instances are connected by a strong edge, their labels are likely to be the same. A successful methodology to tackle the semi-supervised classification problem is based on traditional supervised classification algorithms [34]. These techniques aim to obtain one (or several) enlarged labeled set(s), based on their most confident predictions, to classify unlabeled data. Self-labeled techniques are typically divided into self-training and co-training.

In the self-training process [35], a classifier is trained with an initial small number of labeled examples, aiming to classify unlabeled points. Then it is retrained with its own most confident predictions, enlarging its labeled training set. This model does not make any specific assumptions for the input data, but it accepts that its own predictions tend to be correct. The co-training process [36] assumes that the feature space can be split into two different conditionally independent views and that each view is able to predict the classes perfectly [37]. It trains one classifier in each specific view, and then the classifiers teach each other the most confidently predicted examples. These self-labeled techniques based on self-training and co-training follow an iterative procedure, aiming to obtain an enlarged labeled dataset, in which they assume that their own predictions tend to be correct [38]. Since this process of generating "correct" labels is too expensive, some researchers prefer to use a weak labeling approach that aims to generate less than perfect labels (i.e. weak labels), but in large quantities to compensate for the lower quality of the introduced labels.

While a fully automated labeling tool can reduce time and costs, the resulting labels may not be as accurate as those produced by a domain expert. The risk of using these labels is that ML models tend to learn the hidden behavior of the network and to generalize its training on the prediction data. In the context of mobile networks, both the assumption of "correctness" and the weak labeling approach seem dangerous since both imply a high risk on the model outputs if the data fed to the model are wrongly labeled. Serious consequences may result if the model outputs are inaccurate especially when an actuation decision should be triggered once some event is predicted by the model. Examples of actuation decisions range from equipment resets or configuration changes that can be ordered remotely, to hardware fixing or replacements with on-site intervention. This actuation mechanism is generally rule based with policies defined by domain experts.

In recent years, the automation of such decision-making has been addressed by researchers using deep learning and reinforcement learning. This includes connectivity problems, traffic management, fast Internet with strict latency requirements, etc. Reinforcement-learning-based approaches have several advantages with respect to the human treatment limits. Moreover, they allow to transfer and share information via the global environment, as they do not require a priori knowledge of the agent's behavior or environment to accomplish their tasks. However, such knowledge is usually acquired and learned repeatedly and autonomously by trial and error, which is a challenge in mobile networks since mobile operators are reluctant to let a reinforcement model to learn directly on their commercial network.

Besides, reinforcement learning is a challenging area in 6G networks due to a multitude of reasons such as large state space, complexity in the giving reward, difficulty in control actions, and difficulty in sharing and merging of the trained knowledge in a distributed memory node to be transferred over a communication network. Thus, while reinforcement learning offers some attractive properties mainly by autonomously learning the orchestration decisions without the need of labeled data, addressing large state and action spaces remains a challenge.

15.5 Conclusions

AI will play an important role in 6G networks at different levels. It will allow to support the introduction of new compelling services and to optimize management tasks in a complex environment by means of learning capabilities.

The distributed network architecture with centralized and edge clouds and the heterogeneity of network nodes and technologies will require different control processes to be performed at the edge. AI/ML can provide the framework to optimize control processes at different time scales in order to efficiently adapt the network to the traffic dynamics.

In the physical layer, recent research has demonstrated how AI algorithms such as NNs can be incorporated as technological building blocks of radio and optical access networks. 6G is expected to be the first network technology to offer full AI-as-a-Service (AIaaS) support, with the overall framework for delivering AI-oriented services such as data processing, model selection, training, and on-demand tuning.

The optimization of control processes can be achieved by embedding intelligence such as AI/ML algorithms in the different layers of the control protocol stack. To this end standardization will play a key role, as in the case of O-RAN. The use of AI/ML in order to optimize control tasks at the network edge impedes the operation of AI/ML models such as deep learning that are often computationally intensive. FL partly addresses this problem by allowing federated training to take place at distributed locations. Different AI/ML models used for control, management, and orchestration tasks in 6G networks need labeled data. The automation of such tasks requires self-labeling capability, which remains a challenging research avenue. Human intervention still seems mandatory to at least supervise and validate the outputs and the decisions taken by the automatic AI models.

References

1 Open RAN Alliance (2018). O-RAN: towards an open and smart RAN, White Paper.
2 Galindo-Serrano, A., Sayrac, B., and Ben Jemaa, S. (2014). Cellular coverage optimization: a radio environment map for minimization of drive tests. Cognitive Communication and Cooperative HetNet Coexistence, Springer, 211–236.
3 Zaaraoui, H., Altman, Z., Altman, E., and Jimenez, T. (2016). Forecast scheduling for mobile users. *Proceedings of IEEE PIMRC.*
4 Tsilimantos, D., Dimitrios, T., Nogales-Gómez, A., and Valentin, S. (2016). Anticipatory radio resource management for mobile video streaming with linear programming. *Proceedings of IEEE ICC.* Masson, M., Altman, Z., and Altman, E. Coordinated scheduling based on automatic neighbor beam relation, NETGCOOP 2020.
5 O'Shea, T. and Hoydis, J. (2017). An introduction to deep learning for the physical layer. *IEEE Transactions on Cognitive Communications and Networking* 3: 563–575.
6 Goutay, M., Aoudia, F.A., and Hoydis, J. (2019). Deep reinforcement learning auto-encoder with noisy feedback. *Proceedings of WiOpt.*
7 Zappone, A., Debbah, M., and Altman, Z. (2018). Online energy-efficient power control in wireless networks by deep neural networks. *Proceedings of IEEE SPAWC.*
8 Di Renzo, M. et al. (2019). Smart radio environments empowered by reconfigurable AI meta-surfaces: an idea whose time has come. *EURASIP Journal on Wireless Communications and Networking* 2019 (1): 1–20.
9 Open RAN Alliance (2020). O-RAN-WG1-O-RAN Architecture Description – v01.00.00. Technical Specification.
10 Network 2030. A blueprint of technology, applications and market drivers towards the year 2030 and beyond, ITU-T FG Network 2030.
11 New Services and Capabilities for Network 2030. Description, Technical Gap and Performance Target Analysis, ITU-T FG Network 2030.
12 RFC 8402 Segment Routing Architecture (July 2018).
13 RFC 6830 The Locator/ID Separation Protocol (LISP) (January 2013).
14 The road beyond 5G: a vision and insight of the key technologies, Konstantinos Samdanis and Tarik Taleb, IEEE Network Magazine (March 2020).
15 White paper on 6G networking, 6G Research Visions, No. 6 (June 2020).
16 Varga, B. (2020). In: *DetNet Data Plane: IP over IEEE 802.1 Time Sensitive Networking (TSN)* (ed. J. Farkas, A. Malis, and S. Bryant).
17 ETSI GS ZSM 002 V1.1.1 Zero-touch network and Service Management (ZSM), Reference Architecture (August 2019).
18 Lovén, L., Leppänen, T., Peltonen, E. et al. (2020). EdgeAI: a vision for distributed, edge-native artificial intelligence in future 6G networks.
19 Key drivers and research challenges for 6G ubiquitous wireless intelligence, 6G Research Visions (1 September 2019).
20 3GPP TS 23.503 3rd Generation Partnership Project. Technical Specification Group Services and System Aspects. Policy and charging control framework for the 5G System (5GS). Stage 2, V16.5.1 (August 2020).
21 3GPP TS 28.533 3rd Generation Partnership Project. Technical Specification Group Services and System Aspects. Management and orchestration, Architecture framework V16.4.0 (June 2020).

22 McMahan, B., Moore, E., Ramage, D., and Hampson, S. Blaise Agüera y Arcas: communication-efficient learning of deep networks from decentralized data. AISTATS 2017, 1273–1282

23 Federated Learning. Collaborative machine learning without centralized training data. Publication date: Thursday, April 6, 2017. Posted by Brendan McMahan and Daniel Ramage, Research Scientists.

24 Sahu, A.K., Li, T., Sanjabi, M. et al. (2018). Federated optimization for heterogeneous network. arXiv preprint arXiv:1812.06127.

25 Mohri, M., Sivek, G., and Suresh, A.. (2019). Agnostic Federated Learning.

26 Yurochkin, M., Agarwal, M., Ghosh, S. et al. (2019). Bayesian nonparametric federated learning of neural networks. arXiv preprint arXiv:1905.12022.

27 Bonawitz, K., Ivanov, V., Kreuter, B. et al. (2017). Practical secure aggregation for privacy preserving machine learning. *Proceedings of the 2017 ACM SIGSAC Conference on Computer and Communications Security*. ACM, 1175–1191.

28 Bonawitz, K., Salehi, F., Konečný, J. et al. (2019). Federated learning with autotuned communication-efficient secure aggregation. arXiv preprint arXiv:1912.00131.

29 Bonawitz, K., Eichner, H., Grieskamp, W. et al. (2019). Towards federated learning at scale: system design. arXiv preprint arXiv:1902.01046.

30 Li, M. and Zhou, Z.-H. (2007). Improve computer-aided diagnosis with machine learning techniques using undiagnosed samples. *IEEE Transactions on Systems, Man, and Cybernetics Part A: Systems and Humans* 37 (6): 1088–1098.

31 Fujino, A., Ueda, N., and Saito, K. (2008). Semisupervised learning for a hybrid generative/discriminative classifier based on the maximum entropy principle. *IEEE Transactions on Pattern Analysis and Machine Intelligence* 30 (3): 424–437.

32 Joachims, T. (1999). Transductive inference for text classification using support vector machines. *Proceedings of 16th International Conference on Machine Learning*. Morgan Kaufmann, 200–20910.

33 Blum, A. and Chawla, S. (2001). Learning from labeled and unlabeled data using graph mincuts. *Proceedingsof the Eighteenth International Conference on Machine Learning*, 19–26

34 Witten, I.H., Frank, E., and Hall, M.A. (2011). *Data Mining: Practical Machine Learning Tools and Techniques*, 3e. San Francisco: Morgan Kaufmann.

35 Li, M. and Zhou, Z.H. (2005). SETRED: self-training with editing. Lecture notes in computer science (including subseries lecture notes in artificial intelligence and lecture notes in bioinformatics), vol. 3518. LNAI, 611–621.

36 Blum, A. and Mitchell, T. (1998). Combining labeled and unlabeled data with co-training. *Proceedings of the Annual ACM Conference on Computational Learning Theory*, 92–100

37 Du, J., Ling, C.X., and Zhou, Z.H. (2010). When does co-training work in real data? *IEEE Transactions on Knowledge and Data Engineering* 23 (5): 788–799.

38 Triguero, I., García, S., and Herrera, F. (2015). Self-labeled techniques for semi-supervised learning: taxonomy, software and empirical study. *Knowledge and Information Systems* 42: 245–284.

16

6G and the Post-Shannon Theory

Juan A. Cabrera[1], Holger Boche[2], Christian Deppe[3], Rafael F. Schaefer[4], Christian Scheunert[5], and Frank H. P. Fitzek[1]

[1] *Deutsche Telekom Chair of Communication Networks, Technische Universität Dresden, Dresden, Germany*
[2] *Institute of Theoretical Information Technology, Technische Universität München, Munich, Germany*
[3] *Chair of Communication Engineering, Technische Universität München, Munich, Germany*
[4] *Chair of Communications Engineering and Security, University of Siegen, Siegen, Germany*
[5] *Chair of Communication Theory, Technische Universität Dresden, Dresden, Germany*

16.1 Introduction

A message is comprised of semantics, syntax, and statistics. The semantics focus on meaning, i.e. the relationship between abstract signifiers and what they stand for in the world (for instance, the set of abstract symbols and sounds "*t-r-e-e*," and the tree planted in the garden). The syntax refers to the rules that dictate the structure of the sentences in the language. Classical information theory was revolutionary because Shannon ignored the semantics and syntax of the messages and focused purely on the statistics, laying the foundations of information theory [1]. Semantics and syntax are qualitative and relative concepts, but the statistics of the message can be mathematically modeled and studied. Within the Shannon paradigm, there is a clear communication task, i.e. conveying quantities of a mathematically defined unit of measurement (the bit), which is determined purely by the statistics of the message, from one point to another. Syntax and semantics are determined a posteriori by the destination, outside the realm of information theory. This universality of the bit has allowed us to represent things within the same framework that appears to be different, such as images, audio, and video, and has moved the information age forward. Since our focus is on modern communication systems, in the rest of this chapter when we talk about a message, we are referring only to a bit stream, ignoring the semantics and syntax.

In addition to message transmission, there is also the transmission of the so-called gestalt information as shown in Figure 16.1. We use the German word "gestalt" as it is already used in other disciplines; the closest translation is "design" or "representation" (in Figure 16.1, we call it design and show it above the message). Here, the fundamental question is not what message the sender has transmitted, but whether it has transmitted at all,

Shaping Future 6G Networks: Needs, Impacts, and Technologies, First Edition.
Edited by Emmanuel Bertin, Noel Crespi, and Thomas Magedanz.

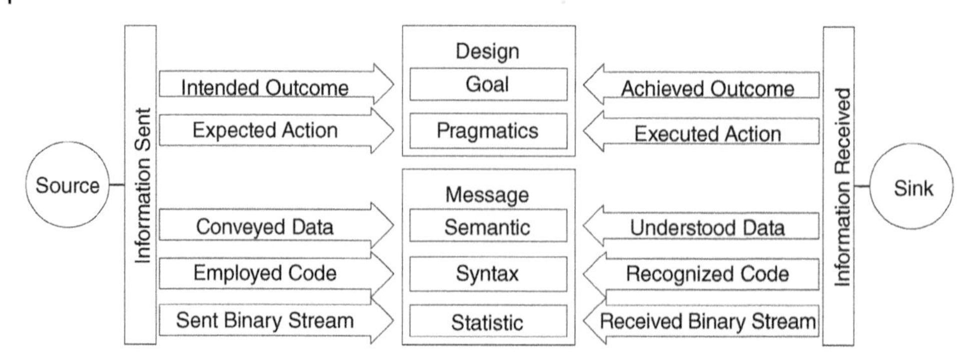

Figure 16.1 Hierarchical operating complex: message transmission and communication by means of transmitter–receiver interaction [2].

and with the purpose of achieving which goal. The study of the gestalt information transmission is what we refer to as the *post-Shannon theory of communication.*

Research on gestalt information is interdisciplinary, and other disciplines have addressed it. *"The medium is the message"* [3] was a provocative thesis of the Canadian media theorist Marshall McLuhan in the 1960s. He postulated his thesis in relation to the effect of political election campaigns, where it is no longer the content of a message that matters but only the medium (TV or radio) that implies the importance of the candidate to the audience. Although he referred to the emerging importance of television at that time, his thesis still has relevance to information theory 60 years later. The medium through which information is transported is sometimes more important than the message itself because it can produce the intended outcome. In 2006 the concept of the medium is the message was again introduced in [4] and used to describe how to exploit channel descriptor information in packet data communication networks to gain in communication efficiency.

Future communication networks such as 6G will impose strict requirements on reliability, latency, robustness, and security. Such networks will go beyond the classical use cases of voice and text transmission including event-driven communication and especially in the in-network control of cyber physical systems (e.g. tactile Internet, Industry 4.0, vehicular communication, and machine-to-machine communication). They will benefit from the post-Shannon theory, including new concepts such as physical layer service integration, resilience by design, physical layer security, and the convergence with quantum communication networks. For example, if a network node has the intended outcome of making a driverless vehicle brake, we know within the Shannon paradigm what to do. The network node must encode the message "brake" into a corresponding codeword and transmit it to the vehicle. The vehicle will brake once it decodes the received codeword into the order "brake" by asking the question "what was the transmitted message?" We must convey a message to produce a certain outcome, and we know that we can transmit up to 2^{nC} different messages, where n is the block length and C is the capacity of the channel over which the transmission takes place. The post-Shannon theory deals with extending Shannon's framework, which is limited purely to message transmission regardless of the intended action or outcome at the receiver. It aims to step back from this limited viewpoint by allowing different transmission frameworks that produce the intended outcomes at the destinations. This theory is not about replacing Shannon's framework but about extending it with the gestalt information. If we change the paradigm in our example, i.e. the vehicle asks instead "was the message "brake"?," then we are dealing with a message identification problem. The outcome would be the same, and the vehicle would brake, but within this identification framework, one can transmit $2^{2^{nC}}$ different identification messages [5, 6]. It shows a double exponential behavior! By changing the question from "what is the message?" to "what is the gestalt information?," we still produce the intended outcome for certain communication tasks, while at the same time increase the efficiency of our communication system exponentially when compared to the traditional Shannon approach. While Shannon solely considered the statistical behavior of a bit stream, post-Shannon theory exploits additional ancillary information. Therefore, there is the well-known bit stream of transported information and a virtual bit stream of the gestalt information.

Shannon stated in [1] "The fundamental problem of communication is that of reproducing at one point either exactly or approximately a message selected at another point." In this work, we aim to describe communication systems that consider problems beyond Shannon's concept and are capable of increasing the efficiency of communication for certain applications. We look at techniques that do not aim to reproduce the message at another point but instead, aim to produce an intended action or outcome at the receiver. Those techniques include but are not limited to the described message identification but also as the generation of common randomness (CR) between a pair of nodes. We will see that by considering the new post-Shannon paradigm, resources that were considered useless in terms of capacity and maximum transmission rates within the Shannon framework such as instantaneous feedback and the share of CR suddenly become relevant and capable of increasing the capacity of the system.

16.2 Message Identification for Post-Shannon Communication

The implementation of most existing communication, storage, and information processing systems is mostly based on Shannon's theory of communication [1]. In this communication model, it is always assumed that the receiver's design criterion is to reproduce the transmitted message. Shannon showed that one can transmit $\sim 2^{nC}$ different messages, where n denotes the block length and C the capacity of the channel. Due to this criterion, the coding scheme may become inefficient for some applications.

Especially if the goal of the sender or receiver differs from the Shannon scheme, it is prudent to think about alternatives. One of these alternatives was introduced by Ahlswede and Dueck [5] as the theory of identification via channels. In this theory, the goal of the receiver changes, as illustrated in Figure 16.2. In this scenario, the receiver has to determine if the sender transmitted a relevant message or not. The sender does not have any information about which messages are relevant for the receiver (in which case a single transmitted bit would be sufficient). This identification theory, therefore, belongs to the gestalt information in the post-Shannon framework. In the identification scheme, the transmitter encodes messages in such a way that the receiver can reliably and correctly decide whether or not the message in which it is interested was sent. The relevance of certain messages for the receiver can change during the application. In [5] the existence of randomized identification (RI) codes was shown such that one can transmit $\sim 2^{2^{nC}}$ different

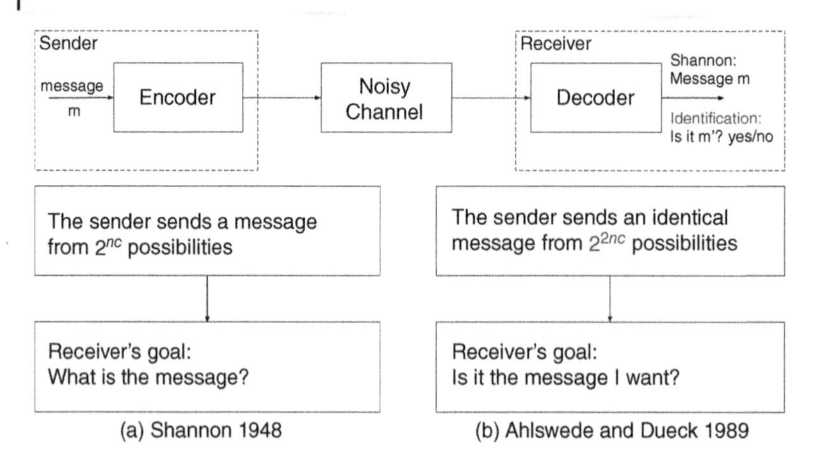

Figure 16.2 Communication model for (a) Shannon's scheme and (b) the identification scheme for a channel with capacity C.

identification messages over the channel. For RI codes we need randomized encoding. With Shannon's transmission method, we get the same capacity for codes with and without randomized encoding. We can identify a number of messages with RI codes that grow double exponentially with the block length of the codeword. This represents an exponential gain compared to Shannon's message transmission framework. Later, we will see that gains can be also made without randomized encoding for message identification. We consider the so-called deterministic identification (DI) codes in Section 16.2.3.

By changing the paradigm from a Shannon to a post-Shannon framework, it is possible to rethink some of the concepts that we take for granted. Some of the results within the old paradigm do not apply within the new one. In the case of message identification, there are some results that are surprising and can potentially increase the efficiency of future 6G communication systems. In particular:

- **The security of message identification with RI codes is free:** For message transmissions within Shannon's paradigm, it has been known since the 1970s that it is possible to achieve secrecy in message transmission in the presence of an eavesdropper at the cost of transmission capacity [7]. However, in the context of message identification, the capacity of the system is not reduced when secrecy of message identification is guaranteed.
- **Common randomness (CR):** It is known that the presence of CR (i.e. correlated observations of a random experiment) at the source and destination of a Shannon communication system does not increase the transmission capacity of the channel. However, CR increases the identification capacity of message identification systems [8].
- **Instantaneous feedback**: It is known that instantaneous feedback does not increase the traditional message transmission capacity of a channel [9]. This result is surprising in itself since feedback sounds like a valuable resource to have. However, this is no longer the case in message identification where instantaneous feedback, in fact, increases the identification capacity.

Let us provide an intuitive explanation of how identification codes operate and illustrate how they can achieve an exponential gain over Shannon's message transmission. To do this, first we need to remember how the encoding and decoding process works within Shannon's paradigm. Let us use Figure 16.3 to illustrate the problem of message transmission. In this scenario, a sender wants to transmit a message to a destination. There is a total of M possible messages ($u_1 \ldots u_4$ in Figure 16.3). The sender encodes the message, i.e. maps the message u_i to a codeword C_i of block length n. The goal of this encoding is to add redundancy to overcome possible errors in the transmission. Therefore, the encoding maps, one to one, the elements of a set of messages of size M to elements of a set of codewords of size 2^n, which is larger than M. In Figure 16.3, this is represented by the size of the ellipses. The codeword C_i is transmitted over a channel, and the channel can produce errors in the transmissions with a certain probability. This means that the channel can be understood as a function that maps an element C_i to a different element of the same set according to a probability distribution. The decoder, according to a decoding function, estimates the transmitted message. It maps the received element of size n to one of the possible M messages. In the figure, we visualize this by drawing some surfaces at the set representing the decoder. Each region D_i is mapped to an estimated message u_i. In the message transmission paradigm, the regions D_i should be designed to be disjoint, i.e. nonoverlapping. A decoding error occurs when an element C_i is mapped to an element in a region other than D_i, and it occurs with a probability λ. Shannon and Wolfowitz proved that if the size of the codeword n tends to infinity, it is possible to find a code that allows the encoding of 2^{nC} messages with a decoding error probability λ arbitrarily close to zero [1, 10]. In the model, C is the capacity of the channel, and it is a function of the probability distributions of the messages and the channel.

Message identification works in a different way. We illustrate its operation in Figure 16.4. In this scenario, the encoder also maps the N elements of possible identification messages to codewords. In this case, as illustrated in Figure 16.4 with a randomized encoding, the mapping of the identification message u_i lands on a region of the set of all possible codewords according to a probability distribution Q_i. This codeword is transmitted over the channel, which distorts it, in a similar fashion as in message transmission. The difference

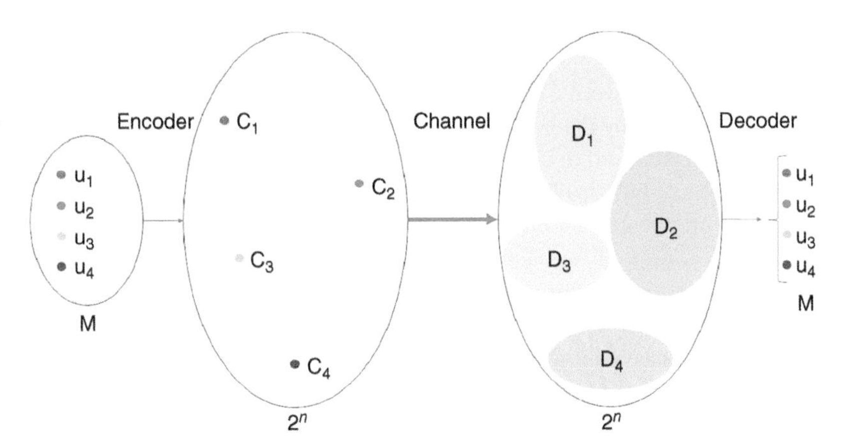

Figure 16.3 Geometric illustration of the classical message transmission. The encoder maps the messages to codewords. These codewords are distorted by the channel. And the decoder maps the received element to a possible transmitted message according to a certain decoding rule.

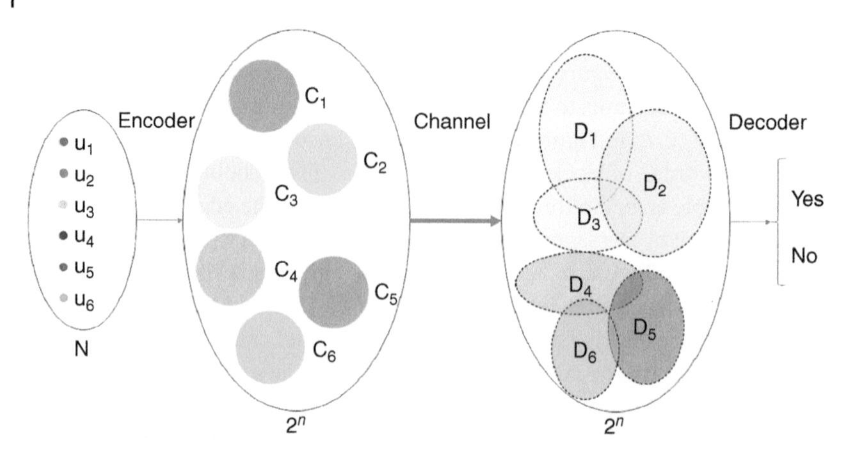

Figure 16.4 Geometric illustration of the message identification paradigm with a random encoding. The encoder maps a message to a codeword according to a probability distribution Q. The channel distorts the codeword, and a decoder maps the received element to a yes/no answer. Unlike Shannon's paradigm, the regions at the decoder decide whether or not the identification message of interest that was transmitted overlap.

in identification codes is that the regions considered by the decoder to identify a message overlap. The only condition considered when constructing an identification code is that of the probability of a *missed identification*, i.e. the receiver fails to identify that the message is λ_1, and the probability of a *false identification*, i.e. the event when the receiver accepts a false message is λ_2. These are referred to as type I errors and type II errors, respectively. By leveraging the possibility of overlapping regions at the decoder, it was proven in [5] that it is possible to identify $2^{2^{nC}}$ different messages with RI codes. For the gain with DI codes, we refer to Section 16.2.3. The overlapping of decoding regions allows exponentially more messages to be identified at the cost of the impossibility of exactly decoding the individual messages.

We illustrate in Figure 16.5 the difference between type I and type II errors in the identification decoding. The arrows indicate three scenarios for the channel output, given that the encoder transmitted the codeword C_1. If the channel output is outside D_1, then a type I error has occurred, as indicated by the bottom arrow. This kind of error is also considered in traditional message transmission. In identification, the decoding sets can overlap. If the channel output belongs to D_1 but also belongs to D_2, then a type II error has occurred, as indicated by the middle brown arrow. Correct identification occurs when the channel output belongs only in D_1.

In Table 16.1, we summarize the differences between message transmission and message identification.

The advantages of RI codes were shown in the pioneer papers by Ahlswede and Dueck [5, 6] for discrete channels. The first results for identification via the Gaussian channel have yet been established in [11, 12]. The Gaussian channel is known for its practical relevance. Furthermore, in [11] the authors also analyze multiple-input multiple-output (MIMO) Gaussian channels and provide an efficient signal-processing scheme. Furthermore, quantum channels are playing an increasingly important role. Here, too, the first result was

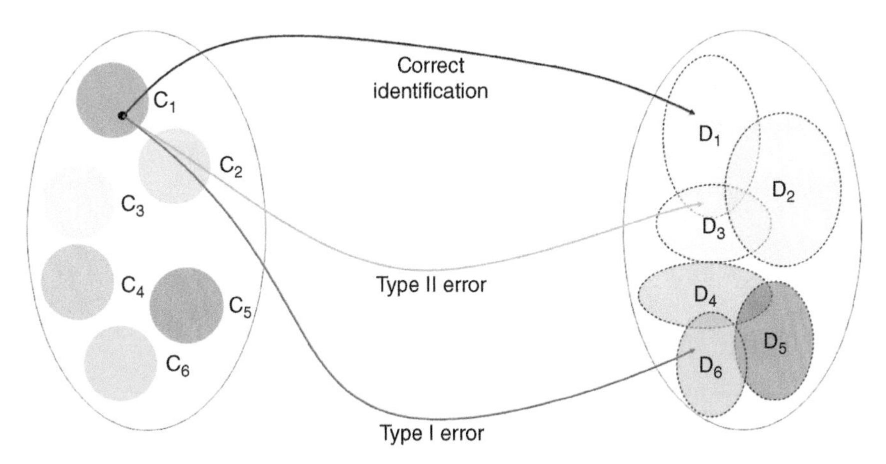

Figure 16.5 Geometric illustration of correct identification and type I and type II errors.

Table 16.1 Summary of message transmission and message identification.

	Message transmission	**Message identification**
Defined by	(n, M, λ)	$(n, N, \lambda_1, \lambda_2)$
Decoding regions	Disjoint	Overlapping
Errors when	• Channel output outside the decoding region (λ)	• Channel output outside the decoding region (λ_1) • Channel output inside the overlap of regions (λ_2)
Possible number of messages Using randomized encoding	2^{nC} (see [1] and [10])	$2^{2^{nC}}$ (see [5])

shown in 1999 in [13] for message identification via quantum channels. Also here, similar profits could be shown in the comparison to the message transmission. Work on robust and secure quantum channels appeared in [14].

16.2.1 Explicit Construction of RI Codes

We will provide an intuition on how RI codes are constructed. As we previously mentioned, the transmission and decoding of RI codes are prone to two types of errors, namely, transmission errors due to a noisy channel (type I errors) and false identification (type II errors) due to the code construction itself. It is possible to separate these kinds of errors and address them separately without losing identification capacity, i.e. it is possible to construct an identification code and use a channel transmission code to avoid type I errors. Then the identification decoder will experience a practically noiseless channel, and we can study identification codes in relation to type II errors independently of type I errors.

One way of creating an identification code is to make use of a traditional block code but use it differently. In Figure 16.6 we illustrate the operation of a block code where messages of k symbols in length are encoded into codewords of n symbols each. Each

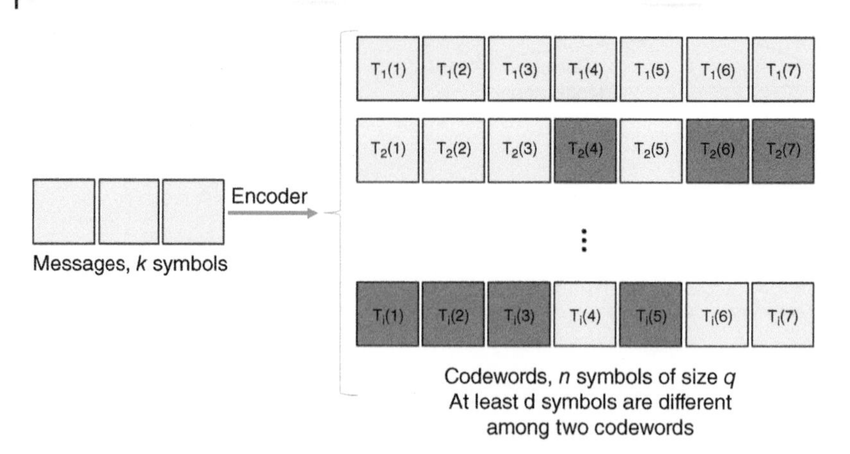

Figure 16.6 Illustration of a block code. Messages of k symbols are encoded into codewords of n symbols. Each symbol has a size of q bits. Between any two codewords, at least d symbols are different (Hamming distance).

symbol has a size of q bits. Between any two codewords, at least d out of n symbols are different. This is equivalent to stating that the hamming distance of the code is equal to d. In the figure, the symbol j of the codeword i is represented as $T_i(j)$. Within Shannon's message transmission paradigm, the n symbols of a codeword would be transmitted over a noisy channel. If a block code has a Hamming distance d=2t+1, it means that it can be used within Shannon's paradigm to correct up to t errors. This is the case because if a noisy channel changes t symbols in a codeword, then the decoder can confuse the channel output with a different valid codeword, and therefore, it will produce an erroneous decoding.

Each error correction code can be used as an identification code. For this purpose, the error correction code is not used as an input to the noisy channel. Rather, each codeword of the error-correction code is assigned to an identity, so there are as many identities as codewords in the error-correction code. Each codeword is then used as a mapping function by taking as input the number of a position in the codeword and giving as the output tag the corresponding symbol of the codeword at the input position. This implies a pre- and post-processing. The position and the symbol must then be channel coded via a transmission code if the channel is noisy. This is illustrated in Figure 16.7 where the sender wants to communicate that the event associated with identity 2 occurred. A decoder at the other end is interested in identifying if the event associated with identity 1 occurred. This is the event of interest. In the example, the encoder choses one symbol randomly. It is the 6th symbol in the figure, namely, $T_2(6)$. The encoder then transmits this symbol and the position, randomly chosen over a noiseless channel (remember that we assume a channel code is dealing with the channel errors).

Once the receiver gets the pair (6, $T_2(6)$), it computes the 6th symbol of the identity it is interested in. In this case, the decoder computes the symbol $T_1(6)$. Then it compares the symbol it computed with the received symbol. In the example of Figure 16.7, the different colors indicate that the two symbols are different. Therefore, the decoder can conclude that the transmitted identity was not the one it was interested in. The type II error in this RI code

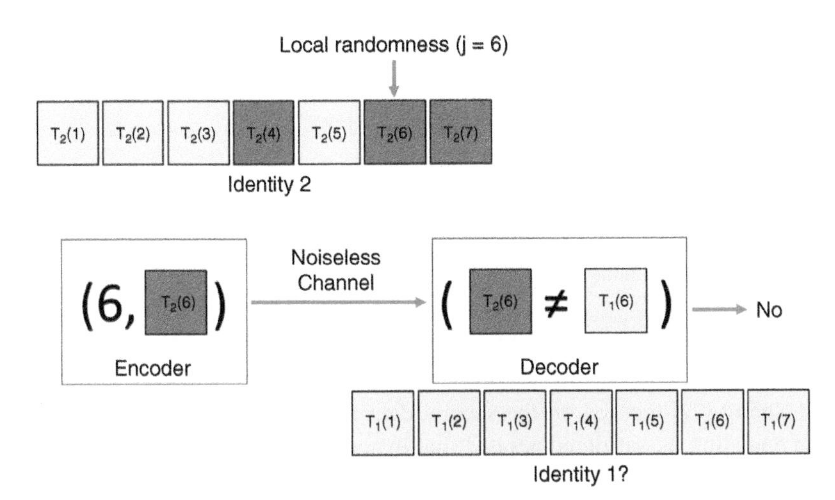

Figure 16.7 Transmission of an identity over a channel.

arises due to the fact that at least d out of n symbols between identity 1 and identity 2 are different (Hamming distance). Therefore, the error is bounded. More precisely, $\lambda_2 \leq 1 - \frac{d}{n}$.

With this particular code, we can see the benefits of message identification. In this example, the encoder needs to transmit q bits for the chosen symbol and n bits to indicate its position. With Shannon's message transmission, sending those bits would allow for the decoding of fewer messages than the number of identities that can be identified with a certain error. With the code of the example, the receiver can identify 2^{qk} identities, but it can only transmit 2^{q+n} messages.

This particular code in the example shows gains over Shannon's message transmission. However, to achieve the RI capacity, i.e. to allow the number of identities to grow double exponentially with the block length, other codes are needed. More details are given in [15, 16], but essentially two concatenated Reed–Solomon codes (a type of block code) can achieve the identification capacity. For more details about the construction of RI codes see [17–21].

16.2.2 Secrecy for Free

Future communication systems must support secure communication. A promising approach for implementing secure communication is the concept of physical layer security based on information-theoretic principles. This goes back to Wyner in the 1970s, who studied the so-called wiretap channel, providing the most basic communication scenario with a transmitter–receiver pair and a non-legitimate eavesdropper [7] as illustrated in Figure 16.8. To enable secure transmission over wiretap channels, two communication tasks must be solved within the Shannon approach: the legitimate receiver must be able to correctly recover the transmitted message with high probability, and the eavesdropper has to be kept uninformed about the transmitted message. It was shown that the secrecy capacity of the wiretap channel is reduced compared to the nonsecure capacity. There is a penalty in the secure transmission rate as a function of the quality of the eavesdropper channel.

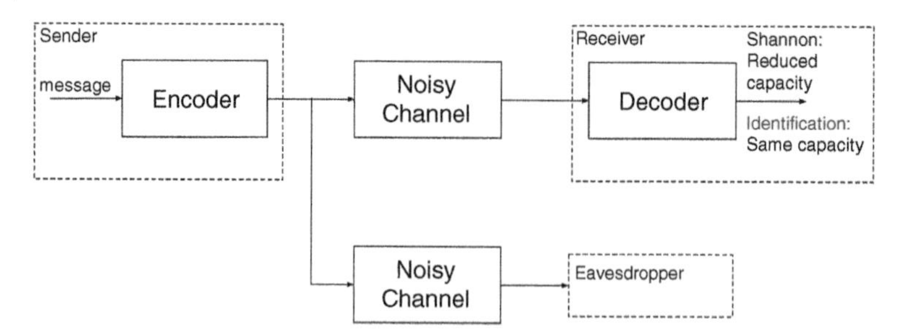

Figure 16.8 Wiretap channel model. An eavesdropper does not reduce the identification capacity.

Ahlswede and Zhang [22] focused on the secure identification over wiretap channels with randomized encoding. It was demonstrated that in contrast to secure message transmission, the secure RI capacity of the wiretap channel coincides with the capacity of the channel from the legitimate sender to the legitimate receiver. However, this is true only if the secrecy capacity is positive, or it is equivalent to say that as long as one bit can be identified with an arbitrarily small error probability, the presence of an eavesdropper does not reduce this capacity. This means that there is no penalty in the rate if we want to guarantee secrecy in the message identification task. These results are given for the most basic channels, but there already exist some generalizations for more broadly defined channels [23–25], and for continuous channels [11, 12].

16.2.3 Message Identification Without Randomness

For all the identification results presented so far, the sender needs a randomized source, and depending on this source, they encode the messages into the channel. In some applications it is difficult or not possible to implement such an RI code, because the encoder has no possibility to use local randomization.

Therefore, for applications such as molecular communication, it is also important to consider DI codes (earlier these were also called non-randomized identification [NRI] codes). In [26], the idea of identification was used for the first time to solve a problem in communication complexity. The paper [26] inspired Ahlswede and Dueck to do their research described above in [5]. In [5] and also in [27], it was stated that the DI capacity of a discrete memoryless channel (DMC) with a stochastic matrix W is given by the logarithm of the number of distinct row vectors of W. Instead of a proof, it referred to [28], which does not include identification and addresses a completely different model of an arbitrarily varying channel (AVC). The first rigorous proof of this statement was given in [29]. This result shows that in the deterministic setup, the number of messages scales exponentially in the block length as in the traditional setting of transmission. Nevertheless, the achievable identification rates are significantly higher than those of transmission. Furthermore, deterministic codes often have the advantage of simpler implementation and analysis. Furthermore, in [29, 30], the DMC and the Gaussian channel were analyzed with input constraints. Such a constraint is often associated with a limited power supply or regulation. It is noteworthy that the DI capacity with power constraint for the Gaussian channel is infinite. More precisely, the capacity is infinite in the exponential scale and zero in the double-exponential scale.

In the follow-up work [31], the authors then thought about whether there would be a different scaling so that the DI capacity of the Gaussian channel is finite. They found that for Gaussian channels, the number of messages scales as $n^{(nR)}$ and develops lower and upper bounds on the DI capacity in this scale. Furthermore, they consider deterministic identification for Gaussian channels with fast fading and slow fading, where channel side information (CSI) is available at the decoder. For slow fading, the DI capacity in the exponential scale is infinite, unless the fading gain can be zero or arbitrarily close to zero (with positive probability), in which case the DI capacity is zero. In comparison with the double exponential scale in RI coding, the scale here is significantly lower.

A very surprising result was achieved in [32]. Here the authors viewed DI via Gaussian channels with noiseless feedback. They have shown that if the noise variance is positive, any rate for identification via the Gaussian channel with noiseless feedback can be achieved. The remarkable result means that for any scaling selected, the corresponding DI capacity is infinite.

16.3 Resources Considered Useless Become Relevant

16.3.1 Common Randomness for Nonsecure Communication

The generation of CR between a sender–receiver pair as introduced by Ahlswede and Csiszár [8, 33] is another post-Shannon communication task that has benefits for communication systems. CR can be generated if the sender and receiver observe the outcome of the same random experiment. It is possible to use the knowledge provided by CR to implement correlated random protocols between the sender and receiver that could potentially lead to algorithms much faster than deterministic ones or those using independent randomness only. The maximum rate at which the sender and receiver can generate the CR is given as the CR capacity, and similar to Shannon's capacity, it tends to have an exponential behavior with the block length.

The model to generate CR between a pair of nodes is simple. Two nodes observe the result of a random experiment. Their observations are correlated but not necessarily the same. The pair can also communicate via a channel. The idea is that by communicating a few bits over the communication channel, together with their correlated observation of the random experiment, they can generate a large amount of CR. This randomness can later be used to greatly increase the identification capacity, to ensure secure message transmission and identification, and as we will see further in this chapter, to nullify the effect of jammers and attackers of the communication system.

CR has applications for secure communication within Shannon's framework since it functions as a secret key. However, surprisingly, the CR capacity also plays a major role in the identification capacity. If CR is available as a resource, it can boost the identification capacity of the communication, which contrasts with Shannon's message transmission where CR has no benefit for nonsecure communication. In Figure 16.9 we illustrate the idea behind CR. In general, if the sender and receiver have access to a random experiment, they can gather and store resources in the form of correlated random variables [8, 34], and the CR capacity is determined by the individual channels between the common random

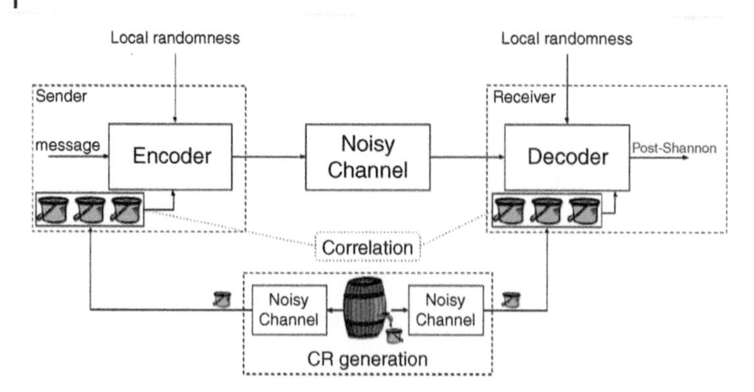

Figure 16.9 CR are resources generated by observing a random experiment. These resources are useful later, to increase the capacity of post-Shannon communication.

experiment and the sender and receiver. In Figure 16.9, the CR resource is illustrated with buckets of water that are generated by observing a random experiment and can be stored at both ends. When the sender communicates with the receiver, these resources can be used to increase the capacity of message identification. Within Shannon's paradigm, we achieve the same capacity with and without the additional CR, whereas in the case of identification, additional CR boosts the capacity.

In Figure 16.9 we show the channel model with possible random resources. While the local randomness and CR resources have no effect on the capacity in the Shannon picture, local randomness is necessary in the case of identification in order to increase the number of identities (i.e. identification messages). If we refer to the explicit construction of identification codes in Section 5.2.1, one can see how CR can increase the identification capacity. In the example given there, it was shown that a sender had to convey a coded symbol and the position of the symbol in a codeword to a destination. In the example, this position was a random number. If the sender–receiver pair has access to CR, then the bits to indicate the position of the chosen symbol do not have to be transmitted over the communication system. This in turn increases the identification capacity.

Later in this chapter, we will further address the fact that the post-Shannon communication task of generating CR has advantages within Shannon's paradigm of secure message transmission. But a possible implementation to leverage the CR resources is a system that slightly changes the protocol used for communication periodically. For example, the protocol can slightly change every minute, hour, or day. This approach resembles that used by the famous Enigma machine, where every day the protocol for encryption would change. At the time, radio operators carried sheets of paper describing the protocol to use for that particular date. In future communication systems, this implementation can be done easily with the help of software-defined networking (SDN) functionalities and CR generation. With SDN, the switches and routers in the network can be programmed to constantly change the protocol used within seconds, based on the observed CR, and without disrupting the communication.

16.3.2 Feedback in Identification and the Additivity of Bundled Channels

An interesting and counterintuitive result of Shannon's theory is that instantaneous and noiseless feedback does not increase the capacity of a DMC. This is not the case in post-Shannon communication. Ahlswede and Dueck proved that for message identification feedback increases the capacity of a DMC [6]. This result is particularly relevant for the case of multiple bundled channels. Intuition tells us that the overall capacity of a system with multiple channels is at least additive. Given two parallel channels, the overall capacity of the system should be at least the sum of the two capacities. This is the case for memoryless channels, but a general characterization for other types of channels (e.g. secure communication over varying channels) is an open problem to date. Moreover, there are examples that show that the capacity of bundled channels can be nonadditive, even in the presence of noiseless and instantaneous feedback. However, contrary to the context of Shannon's communication task, it was proven in [35] that the capacity of parallel channels is super-additive when doing message identification with feedback (IDF). This means that the overall capacity of the system is larger than the sum of the individual capacities. Furthermore, it is even possible that two parallel channels with zero IDF capacity can provide a capacity larger than zero when used jointly. This is known as super-activation, the strongest form of super-additivity.

16.4 Physical Layer Service Integration

16.4.1 Motivation and Requirements

To truly achieve the potential gains of post-Shannon communication, we have to break with the concept of a physical layer that is simply a transport channel. It is not enough to incorporate new communication tasks such as message identification (both secure and not), secure message transmission, and CR extraction with higher layer policies that assign different logical channels for the different services. Instead, the integration of these services has to be done at the physical layer. This becomes particularly relevant for future communication systems such as 6G where quantum communication will play an important role [36]. The authors of [37] discussed the importance of physical layer service integration for security. Instead of implementing the security with encryption at a higher layer while using the physical layer as a simple transport link, they discussed the potential benefits of integrating it as part of the physical layer unicast and multicast protocols. Physical layer security concepts [38] rely on information-theoretic principles that allow design security schemes with mathematically provable security. This is particularly relevant for national and governmental agencies for security that need to verify and guarantee certain security requirements. To this end, semantic security is desirable, which is the strongest requirement of security for which connections to the cryptographic notion of security have been established. Modular schemes for semantic security have been developed in [39] that work very well for all practically relevant channels. Semantic security also plays an important role for secret key generation, particularly in the privacy amplification step [40]. Securing future communication systems based on physical layer security show great potential and are an important and relevant research direction. It has been shown that in general, it is not possible to do this by an automatic approach, i.e. computer-aided algorithm design [41].

Nowadays, it is even more important to consider new network designs that achieve this integration, because novel capabilities introduced by post-Shannon communication, such as super-additivity and super-activation, can result in a more efficient multiuser medium access control but only if implemented with coding at the physical layer. A logical separation of the services at the physical layer results in less spectral efficiency and wasted capacity than a seamless integration of services exploiting coding and post-Shannon resources such as CR. A successful realization of physical layer service integration has been already demonstrated for classical communication systems [37, 42] and also for certain quantum communication tasks [36].

16.4.2 Detectability of Denial-of-Service Attacks

In the recent past, the number of cyberattacks against government units, organizations, research institutions, energy facilities, and many others has been steadily increasing, as for example reported in [43], and it is becoming one of the major threats of governments and agencies. See for examples [44–49]. Such attacks have the capability of damaging vital infrastructure significantly, and accordingly, are a significant concern. Most attacks are performed through (communication) networks by exploiting inherent weaknesses of communication protocols at higher layers or by exploiting erroneous user behavior based on insufficiently protected authentication mechanisms, weak passwords, and others. Usually, such vulnerabilities can easily be detected and fixed by using appropriate software solutions such as required software updates. Unfortunately, frequent updates for closing security vulnerabilities have become quite common nowadays, and it would be desirable to design software solutions that meet security requirements from the beginning, and that do not open the door for such security threats.

Furthermore, there are other types of attacks which are performed at the physical layer itself. Wireless communication systems are particularly vulnerable to these types of attacks given the nature of a shared medium. New technologies such as Software Defined Radio (SDR) allow malicious parties to jam expensive and complex wireless communication systems with just a small amount of money [50, 51]; all that is needed is a laptop connected to an SDR device and an antenna. Even though there are plenty of references that study the problem of communications under adversarial attacks, to date it is not clear whether or not such an attack can take place.

The problem of knowing if an adversarial attacker can, in principle, perform a denial-of-service attack (DoS), which in consequence disrupts the system and makes the communication impossible, was studied by the authors in [52]. In the paper, the problem of detecting such an attack was approached from a fundamental perspective. The authors were interested in understanding if it was possible to design an algorithm that could detect that a DoS was possible in the communication system so that the system could react. To study if a computer algorithm exists, the key is to make use of the concept of a Turing machine [53–55]. This is a mathematical model of an abstract machine that simulates any computer algorithm. It has no constraints in computing, memory, or storage capabilities. It is modeled as a machine that operates on an infinite tape in which there are symbols. The machine has a head that can read or write the symbols on the tape one at a time. It has a finite table of instructions and a state register that keeps track of the state of the machine. According to the current state and the symbol on the tape, the machine follows the instruction from the table of instructions. The instructions cause the head to write a symbol on the head, move the head to a different symbol or stay in place, and change to a new state (or stay in the same state).

This abstract machine provides a simple but very powerful model of computation that can compute any computer algorithm. Therefore, if it is shown that it is not possible to implement an algorithm for a Turing machine that detects a possible DoS attack, then it is not possible to design an algorithm that allows our classical (not quantum) digital computers to detect one.

The system model of communication in the presence of an attacker is simple and is shown in Figure 16.10. In the modeled system, a sender–receiver pair communicates over a noisy channel. The sender wants to convey a message to the destination. The sender encodes the message producing the codeword X in Figure 16.10. After the effects of the channel, the receiver receives the codeword Y, and if everything went as planned, it can decode the correct message. In the model, there is also a malicious jammer who aims to disrupt the communication. The jammer attempts to perform a DoS attack, i.e. to reduce the message transmission capacity to zero, thus preventing any reliable communication between the pair. We assume the jammer can influence the channel itself. We can model this by assuming the channel is an AVC [56].

An AVC is a type of channel widely studied in the literature. We can think of it in the following way. A non-varying wireless channel can be modeled as a transition matrix where the rows represent all possible codewords that the encoder can produce, the columns represent the different outputs of the channel, and the numbers of the elements of the matrix represent the probabilities that if the encoder produced a codeword X, the channel modified it to the codeword Y (what is received at the destination). So we can say that a non-varying channel is described by a single transition matrix that tells us the probability of receiving the codeword Y if we transmitted X. Since the channel is static over time, a single matrix is enough to describe it. On the other hand, if the channel varies over time (i.e. it is an AVC), we can still model it with transition matrices, but we need more than one. If the channel has multiple states, then we need one transition matrix per channel state to describe it. For example, if the sender transmits the codeword X, the probability of obtaining the codeword Y depends on the current state S of the channel. This model of an AVC is helpful because it lets us easily appreciate the effect of a jammer. A malicious entity who aims to disrupt the communications at the physical layer can make use of an antenna to produce different electromagnetic signals that can affect the probability that a receiver

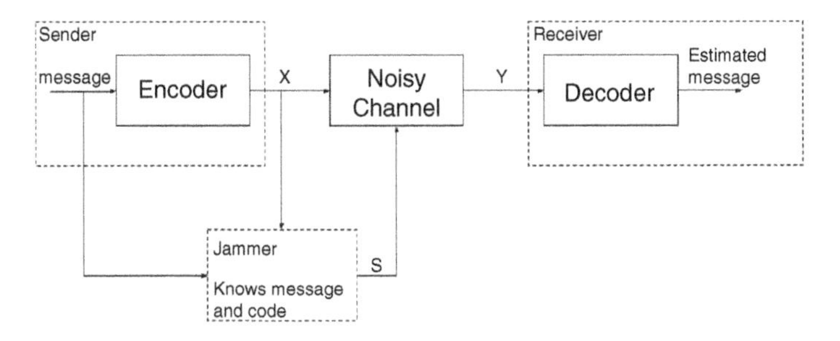

Figure 16.10 System model of a sender–receiver pair that communicates over a channel under the adversarial attack of a jammer. The jammer can disrupt the communications by having either partial knowledge (i.e. knowledge of the encoding and decoding functions) or total knowledge (i.e. partial knowledge plus knowledge of the actual message) of the message transmission.

obtains a particular codeword Y. In general, a jammer can produce a signal S that changes the state of the communication channel, as shown in Figure 16.10. AVCs provide a basic model for secure communication with passive eavesdroppers and active jammers. They account for the communication between the legitimate users; cf [21, 57, 58].

A non-varying channel is a function of the codewords' alphabets X and Y, while the AVC is a function of X, Y, and also the state S. There is a helpful definition for AVC that is the concept of symmetrizability [56]. An AVC is symmetrizable, roughly speaking, if it is possible to "simulate" a valid channel input, then it is impossible for the receiver to distinguish whether the channel output comes from a valid codeword sent by Alice or the jamming input of the jammer.

The results of [52] are interesting, because they prove that if an AVC is symmetrizable, then a jammer can perform a DoS attack. Furthermore, it answers a more important question for practical applications: Is it possible to design an algorithm that can tell if a DoS is possible in a given channel? The answer is no; the problem is semi-decidable. Given infinite compute and memory resources (a Turing machine), an algorithm would detect for which channels a DoS is not possible. However, for those channels where a DoS is possible, the algorithm would run nonstop. Therefore, if we run such an algorithm, and after some time it has not stopped, then we cannot know if it is because the channel can suffer a DoS, or because the channel cannot suffer such an attack, and the algorithm will stop at the next iteration. Subsequently, it has also been shown that feedback from the receiver to the transmitter does not enable the algorithmic detection of such attacks [59].

With these results we can ask two important questions. First, what are the implications for post-Shannon communication? And second, what are the implications for 6G technologies?

To answer the first question, we also refer to further results in [52]. The authors prove that the inclusion of the post-Shannon communication task of CR can help. It is proven that if both the sender and receiver have access to CR, a DoS attack is not possible. This solves the issue of semi-decidability and asks for the introduction of post-Shannon communication tasks, as illustrated in Figure 16.11. CR, by design, makes the communication system more resilient in the presence of attackers [52, 59, 60]. A smart attacker can

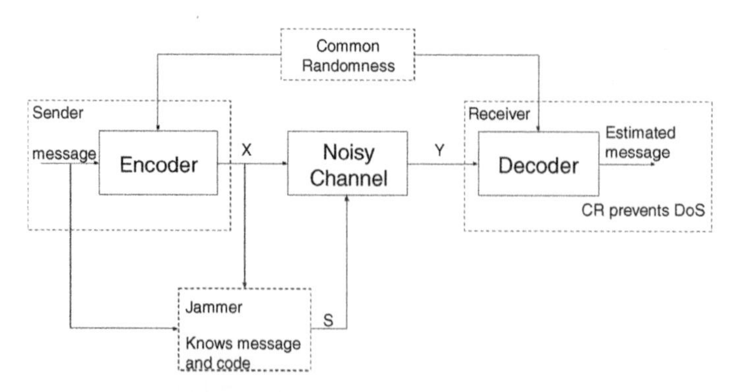

Figure 16.11 In the presence of the post-Shannon resource of CR, a jammer cannot produce a DoS attack [23, 52].

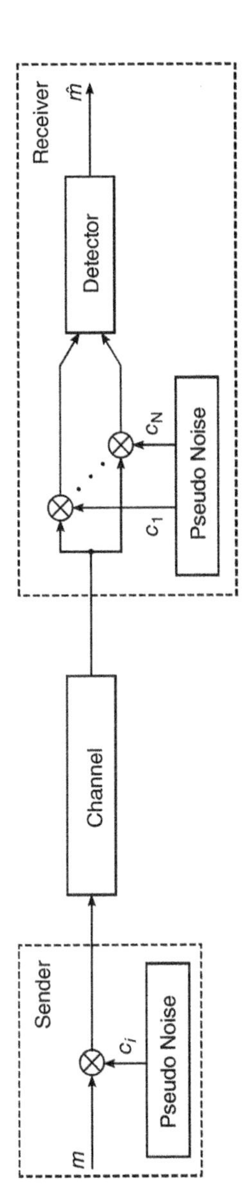

Figure 16.12 Transmission concept for the medium is the message as introduced in [4].

negatively affect the communication because it does not simply introduce noise to the system. There is not much that we can do within Shannon's paradigm to protect the system if this attacker can do an overwrite of the transmitted codewords, becoming undetectable at the receiver. However, if the sender and the receiver have access to CR, as shown in Figure 16.12, this resource allows them to overcome the jamming and prevent possible DoS attacks. Interestingly enough, it has been proven that the computational costs for a jammer increase exponentially with the block length of the messages, while the CR resources that are needed increase only linearly [23]. This means that just a few bits of CR are sufficient to prevent a strong jamming attack. CR has also been investigated within Shannon's paradigm. Already the initial results for AVCs show the importance of CR, as it boosts the capacity of a symmetrizable AVC from zero to a positive value. Although these results were known from the 1960s and 1970s, CR has been ignored completely in the design of modern communication systems. Therefore, the inclusion of CR generation as a post-Shannon task integrated into the physical layer designs becomes important. We see that CR not only boosts the capacity of message transmission from zero to nonzero in some cases, but it also provides a solution to the DoS detection problem. It has been shown that this approach of making the system resilient by exploiting CR is also working for quantum communication [61, 62].

To answer the second question, these results are particularly relevant as there is a trend in 5G/6G communication systems toward SDN and network function virtualization (NFV) by breaking the vertical integration, by introducing the ability to program the network, and by virtualizing classes of network functions into building blocks, which are then accordingly connected to create certain communication services. The results in [52, 59] demonstrate the limitations of the trend toward software-focused solutions, and they ask for the introduction of the integration of some of the services at the physical layer for 6G communication systems as post-Shannon communication tasks.

In addition to this, these observations raise the question of whether this comes from inherent limitations of Turing machines and if there are more powerful computing models that are able to overcome this issue. Accordingly, it is a very interesting research direction to study new hardware platforms for the network infrastructure such as quantum computing and neuromorphic computing. This becomes apparent as so-called Blum–Shub–Smale machines are capable of detecting DoS attacks [63]. The same problems naturally arise for secure communication [52, 59] and also for jamming attacks in quantum communication networks [36]. Similarly, in the quantum case, Blum–Shub–Smale machines are powerful enough to detect DoS attacks [64].

16.4.3 Further Limits for Computer-Aided Approaches

It is interesting to observe that computer-aided system design also has fundamental limits in other areas of engineering. In [65] the task of remote state estimation and stabilization via noisy communication channels of disturbed systems in robotics and control is considered. In [66], a surprising link between this problem and Shannon's theory of zero-error communication was established. By applying results from [67, 68], it is analyzed in [65] if, on the set of linear time-invariant systems paired with a noisy communication channel, it is uniformly decidable by means of a Turing machine whether remote state estimation and stabilization is possible. The analysis incorporates scenarios both with and without channel feedback, as well as a weakened form of state estimation and stabilization. In the broadest sense, these results yield a fundamental limit to the capabilities of computer-aided design and autonomous systems, assuming they are based on real-world digital computers.

16.5 Other Implementations of Post-Shannon Communication

In this section we present other post-Shannon communication schemes that make use of the medium used to transmit information in order to encode gestalt information.

16.5.1 Post-Shannon in Multi-Code CDMA

The authors in [4] introduced the concept "the medium is the message," which describes how to exploit channel descriptor information in packet data communication networks to gain in communication efficiency. The idea is to use the medium, in this case the channel descriptor, to convey additional information. In their scenario of a wireless system that employs pseudo-noise (PN) sequences to achieve spread spectrum, the authors obtained gains in capacity of around one order of magnitude higher than the standard systems, using spread spectrum. Similar to multi-code code-division multiple access (MC-CDMA), each pair, sender and receiver, has a set of PN sequences it uses to communicate. Each of these sequences is a channel descriptor.

In Figure 16.12 we show the novel spread-spectrum transmission approach in [4]. If a sender wants to communicate S bits to the receiver and has a set of H possible PN sequences, then it can encode information by its choice of one particular PN sequence out of the available H. In practice, there are $N_d = H$ bits that can be encoded in this way. Therefore, the sender can remove the first N_d bits out of S, use them to choose one channel descriptor out of the available H, and transmit the remaining bits over the air interface. Note that the interference in the system is reduced because the N_d bits are not transmitted over the air. The receiver will detect that a transmission using one particular PN sequence occurred. They will decode the transmitted bits as they would normally do in a spread spectrum transmission but will also decode and append the extra N_d bits depending on the PN sequence involved in the transmission. Concerning Figure 16.1, the N_d bits are modulated as the gestalt information, and the rest of the bits are classically transmitted following Shannon's framework. This approach can be used in any other technology where the total number of available channel descriptors is larger than the number of channel descriptors that can be used simultaneously. Because of the interference-limited nature of the system, there are more PN sequences available in MC-CDMA than are simultaneously usable.

As an example, assume we have eight PN sequences preassigned for a sender–receiver pair. Then we can encode 3 bits as gestalt information by the choice of the PN sequence. Furthermore, assume that only 2 bits are transmitted over the air at each time slot. Assume the sender wants to transmit the binary sequence $S = (001111010100100)$, it would take the spreading sequence c_{001} to transmit the bit value 11 in the first time slot (the first 3 bits identify the spreading sequence and the following 2 bits are transmitted over the air with the chosen sequence), the spreading sequence c_{101} to transmit the bit value 01 in the second slot, and the spreading sequence c_{001} to transmit the bit value 00 in the last time slot.

16.5.2 Waveform Coding in MIMO Systems

In [69], the idea of [4] is extended to multiple antenna systems where the capacity and security are improved. The information to be conveyed is coded onto the communication medium used, and as a consequence, the link security increases. With a single data stream, the response at the sensor array at the receiver's side is the communication medium. And in the case of transmitting multiple data streams, the medium refers to all the orthogonal channels that result from the singular value decomposition of the channel transfer matrix. The main idea behind this technique is to code phase information on the orthogonal channels. This is done by rotating the constellation of the received signal. The destination can still detect the differential-phased coded signals, but an eavesdropper located at a different location will experience a different rotation, even if it is as close as one-tenth of a wavelength from the intended receiver. This enhances communication security.

The principles in this section are not limited to systems with (i) MC-CDMA and (ii) MIMO with spatial modulation. They can also be extended, for example, to (iii) orthogonal frequency-division multiplexing (OFDM) with index modulation. In terms of gestalt information, all these systems follow a common basic concept: given a resource set S with n elements, one can choose all different subsets of k elements from S. The set of all k-combinations forms a combinatorial number system of degree k. To transmit a message from the sender to the receiver, the sender maps the message to one of the subsets of k elements and uses those resources for transmission. The resource set S equals a set of different codes in (i), a set of different transmit antennas in (ii), and a set of different subcarriers in (iii).

16.6 Conclusions: A Call to Academia and Standardization Bodies

Future communication systems will have requirements and capabilities that were not foreseen by Claude Shannon in his seminal paper, "A Mathematical Theory of Communication." At the time, the engineers at Bell Labs were interested in the transmission of voice and telegraph text messages, instead of event-driven communication or the control of cyber physical systems. The use cases of 6G go beyond voice and text. Therefore, we need to consider disruptive and more efficient mechanisms and systems that can deal more efficiently with the new requirements of future systems. Furthermore, the trend of 5G communication systems has been the softwarization at higher layers, by means of SDN and NFV, of multiple tasks such as resource allocation. However, we have seen in this chapter that higher layer softwarization has its limits. It has been proven that it is not even possible to implement an algorithm that detects a DoS attack in the simplest of communication systems. Therefore, the academia has to focus research on new post-Shannon communication schemes such as identification and CR extraction, among others presented in this chapter. The benefits of such techniques can potentially increase the efficiency of the communications, e.g. we saw that from 2^{nC} possible messages within Shannon's paradigm, techniques such as message identification can transmit $2^{2^{nC}}$ possible messages with randomized encoding. This new framework tells us that some resources not useful within Shannon's paradigm are now not only useful but also actually very powerful. We see for example how CR, noiseless feedback, and multi-descriptor channels play an important role in post-Shannon communication. Therefore, we want to motivate the standardization bodies to address these new network requirements that were not needed before, by making use of novel technologies that allow us to exploit post-Shannon communication techniques, as well as perform the integration of multiple services at the physical layers to overcome the limitations of algorithms implemented by softwarization at higher layers.

Acknowledgments

This work was supported in part by the German Research Foundation (DFG, Deutsche Forschungsgemeinschaft) as part of Germany's Excellence Strategy – EXC 2050/1 – Project ID 390696704 – Cluster of Excellence "Centre for Tactile Internet with Human-in-the-Loop" (CeTI) of Technische Universität Dresden, in part by the German Federal Ministry of Education and Research (BMBF) within the national initiative for "Post-Shannon Communication (NewCom)" under Grants 16KIS1003K, 16KIS1004, and 16KIS1005; in part by the German Research Foundation (DFG) within the Gottfried Wilhelm Leibniz Prize under Grant BO 1734/20-1 and within Germany's Excellence Strategy EXC-2092 – 390781972 and EXC-2111 – 390814868; and in part by the German Federal Ministry of Education and Research (BMBF) within the National Initiative for "Molecular Communication (MAMOKO)" under Grant 16KIS0914.

Holger Boche would like to thank Vince Poor for interesting discussions on post-Shannon communication and finite block length performance, Argyaswami Paulraj for interesting discussions on physical layer design of 6G communication systems, and Christan Arendt from the BMW Group for valuable discussions on the compensation of DoS attacks for vehicle-to-everything (V2X) applications. He further would like to thank Robert Schober for the suggestion of the application of identification for molecular communication and interesting discussions on the deterministic encoding of biological data. He further would like to thank Ning Cai for interesting discussions on deterministic encoding in the theory of identification.

References

1. Shannon, C.E. (1948). A mathematical theory of communication. *Bell Syst. Tech. J.* 27 (3): 379–423.
2. Edwin, K.W. (1994). *Mensch und Technik: Methoden Systemtechnischer Planung.* Trans-Aix-Press.
3. McLuhan, M. and Fiore, Q. (1967). The medium is the message. *N. Y.* 123: 126–128.
4. Fitzek, F.H.P. (2006). The medium is the message. *Proceedings of IEEE International Conference on Communications*, Istanbul, Turkey, 5016–5021.
5. Ahlswede, R. and Dueck, G. (1989). Identification via channels. *IEEE Trans. Inf. Theory* 35 (1): 15–29.
6. Ahlswede, R. and Dueck, G. (1989). Identification in the presence of feedback – a discovery of new capacity formulas. *IEEE Trans. Inf. Theory* 35 (1): 30–36.
7. Wyner, A.D. (1975). The wire-tap channel. *Bell Syst. Tech. J.* 54 (8): 1355–1387.
8. Ahlswede, R. and Csiszár, I. (1998). Common randomness in information theory and cryptography. II. CR capacity. *IEEE Trans. Inf. Theory* 44 (1): 225–240.
9. Shannon, C. (1956). The zero error capacity of a noisy channel. *IRE Transactions on Information Theory* 2 (3): 8–19.
10. Wolfowitz, J. (1978). *Coding Theorems of Information Theory*, 3e. New York: Springer Verlag.
11. Labidi, W., Deppe, C., and Boche, H. (2020). Secure identification for Gaussian channels and identification for multi-antenna gaussian channels. arXiv:2011.06443.
12. Labidi, W., Deppe, C., and Boche, H. (2020). Secure identification for Gaussian channels. *Proceedigs of IEEE International Conference on Acoustics, Speech, and Signal Processing*, Barcelona, Spain, 2872–2876.
13. Löber, P. (1999). Quantum channels and simultaneous ID coding. PhD thesis. Universität Bielefeld, Germany.
14. Boche, H., Deppe, C., and Winter, A. (2019). Secure and robust identification via classical-quantum channels. *IEEE Trans. Inf. Theory* 65 (10): 6734–6749.
15. Moulin, P. and Koetter, R.. (2006). A framework for the design of good watermark identification codes. Security, Steganography, and Watermarking of Multimedia Contents VIII; International Society for Optics and Photonics, SPIE: Bellingham, WA, USA. vol. 6072, 565–574.
16. Verdú, S. and Wei, V.K. (1993). Explicit construction of optimal constant-weight codes for identification via channels. *IEEE Trans. Inf. Theory* 39 (1): 30–36.
17. Derebeyoğlu, S., Deppe, C., and Ferrara, R. (2020). Performance analysis of identification codes. *Entropy* 22 (10): 1067.

18 O. Günlü, J. Kliewer, R. F. Schaefer, and V. Sidorenko, "Doubly-Exponential Identification via Channels: Code Constructions and Bounds," in Proc. IEEE Int. Symp. Inf. Theory, Melbourne, Australia, July 2021.

19 Mattia Spandri, Roberto Ferrara, Christian Deppe, Reed-Muller Identification, arXiv:2107.07649, 2021.

20 Verdú, S. and Wei, V.K. (1993). Explicit construction of optimal constant-weight codes for identification via channels. *IEEE Trans. Inf. Theory* 39 (1): 30–36.

21 Kurosawa, K. and Yoshida, T. (1999). Strongly universal hashing and identification codes via channels. *IEEE Trans. Inf. Theory* 45 (6): 2091–2095.

22 Ahlswede, R. and Zhang, Z. (1995). New directions in the theory of identification via channels. *IEEE Trans. Inf. Theory* 41 (4): 1040–1050.

23 Boche, H. and Deppe, C. (2019). Secure identification under passive eavesdroppers and active jamming attacks. *IEEE Trans. Inf. Forensics Security* 14 (2): 472–485.

24 Boche, H., Schaefer, R.F., and Poor, H.V. (2020). Secure communication and identification systems – effective performance evaluation on Turing machines. *IEEE Trans. Inf. Forensics Secur.* 15: 1013–1025.

25 Boche, H. and Deppe, C. (2018). Secure identification for wiretap channels; robustness, super-additivity and continuity. *IEEE Trans. Inf. Forensics Security* 13 (7): 1641–1655.

26 JaJa, J. (1985). Identification is easier than decoding. Annual Symposium on Foundations of Computer Science (SFCS), 43–50.

27 Ahlswede, R. and Cai, N. (1999). Identification without randomization. *IEEE Trans. Inf. Theory* 45 (7): 2636–2642.

28 Ahlswede, R. (1980). A method of coding and its application to arbitrarily varying channels. *J. Comb. Inf. Syst. Sci.* 5, 1: 47–54.

29 Salariseddigh, M.J., Pereg, U., Boche, H., and Deppe, C. (2020) Deterministic identification over channels with power constraints. arXiv preprint arXiv:2010.04239.

30 Salariseddigh, M.J., Pereg, U., Boche, H., and Deppe, C. (2021). Deterministic identification over channels with power constraints. IEEE International Conference on Communications, IEEE.

31 Salariseddigh, M.J., Pereg, U., Boche, H., and Deppe, C. (2020). Deterministic identification over fading channels. IEEE Inf. Theory Workshop, Riva del Garda, Italy, extended version available at https://arxiv.org/abs/2010.10010.

32 Labidi, W., Boche, H., Deppe, C., and Wiese, M. (2021). Identification over the Gaussian channel in the presence of feedback, Proceedings of IEEE International Symposium on Information Theory, Melbourne, Australia, extended version available at https://arxiv.org/abs/2102.01198.

33 Ahlswede, R. and Csiszár, I. (1993). Common randomness in information theory and cryptography. I. Secret sharing. *IEEE Trans. Inf. Theory* 39 (4): 1121–1132.

34 Boche, H., Schaefer, R.F., and Poor, H.V. (2019). Identification capacity of correlation-assisted discrete memoryless channels: analytical properties and representations. *Proceedings of IEEE International Symposium on Information Theory*, Paris, France, 470–474.

35 Boche, H., Schaefer, R.F., and Poor, H.V. (2018). Identification over channels with feedback: discontinuity behavior and super-activation. *Proceedings of IEEE International Symposium on Information Theory,* Vail, CO, USA, 256–260.

36 Bassoli, R., Fitzek, F.H.P., Boche, H. et al. (2021). *Quantum Communication Networks.* Springer International Publishing.

37 Schaefer, R.F. and Boche, H. (2014). Physical layer service integration in wireless networks: signal processing challenges. *IEEE Signal Process. Mag.* 31 (3): 147–156.

38 Poor, H.V. and Schaefer, R.F. (2017). Wireless physical layer security. *Proc. Natl. Academy Sci. U.S.A.* 114 (1): 16–26.

39 Wiese, M. and Boche, H. (2020). Semantic security via seeded modular coding schemes and Ramanujan graphs. *IEEE Trans. Inf. Theory* 67 (1): 52–80.

40 Wiese, M. and Boche, H.. Mosaics of combinatorial designs for information-theoretic security. preprint, arXiv:2102.00983.

41 Boche, H., Schaefer, R.F., and Poor, H.V. (2020). Secure communication and identification systems – effective performance evaluation on turing machines. *IEEE Trans. Inf. Forensics Security* 15: 1013–1025.

42 Wyrembelski, R.F. and Boche, H. (2012). Physical layer integration of private, common, and confidential messages in bidirectional relay networks. *IEEE Trans. Wirel. Commun.* 11 (9): 3170–3179.

43 Kreutz, D., Ramos, F.M.V., Verissimo, P.E. et al. (2015). Software-defined networking: a comprehensive survey. *Proc. IEEE* 103 (1): 14–76.

44 Marchetti, M., Colajanni, M., Messori, M. et al. (2012). Cyber attacks on financial critical infrastructures. In: *Collaborative Financial Infrastructure Protection* (eds. R. Baldoni and G. Chockler), 53–82. Berlin, Germany: Springer-Verlag.

45 Amin, S. and Giacomoni, A. (2012). Smart grid, safe grid. *IEEE Power Energy Mag.* 10 (1): 33–40.

46 Nicholson, A., Webber, S., Dyer, S. et al. (2012). SCADA security in the light of cyber-warfare. *Comput. Secur.* 31 (4): 418–436.

47 Choo, K.-K.R. (2011). The cyber threat landscape: challenges and future research directions. *Comput. Secur.* 30 (8): 719–731.

48 Kushner, D. (2013). The real story of stuxnet. *IEEE Spectr.* 50 (3): 48–53.

49 Perez-Pena, R. (2013). Universities face a rising barrage of cyberattacks. [Online]. http://www.nytimes.com/2013/07/17/education/barrage-of-cyberattacks-challenges-campus-culture.html.

50 Pärlin, K., Alam, M.M., and Le Moullec, Y. (2018). Jamming of UAV remote control systems using software defined radio. *Proceedigs of 2018 International Conference on. Military Communications and Information Systems*, Warsaw, Poland, 1–6.

51 Lichtman, M., Poston, J.D., Amuru, S.D. et al. (2016). A communications jamming taxonomy. *IEEE Secur. Priv.* 14 (1): 47–54.

52 Boche, H., Schaefer, R.F., and Poor, H.V. (2020). Denial-of-service attacks on communication systems: detectability and jammer knowledge. *IEEE Trans. Signal Process.* 68: 3754–3768.

53 Turing, A.M. (1936). On computable numbers, with an application to the Entscheidungsproblem. *Proc. Lond. Math. Soc.* 2 (42): 230–265.

54 Turing, A.M. (1937). On computable numbers, with an application to the Entscheidungsproblem. A correction. *Proc. Lond. Math. Soc.* 2 (43): 544–546.

55 Weihrauch, K. (2000). *Computable Analysis – An Introduction.* Berlin, Germany: Springer-Verlag.

56 Csiszár, I. and Narayan, P. (1988). The capacity of the arbitrarily varying channel revisited: positivity, constraints. *IEEE Trans. Inf. Theory* 34 (2): 181–193.

57 Wiese, M., Nötzel, J., and Boche, H. (2016). A channel under simultaneous jamming and eavesdropping attack – correlated random coding capacities under strong secrecy criteria. *IEEE Trans. Inf. Theory* 62 (7): 3844–3862.

58 Nötzel, J., Wiese, M., and Boche, H. (2016). The arbitrarily varying wiretap channel – secret randomness, stability, and super-activation. *IEEE Trans. Inf. Theory* 62 (6): 3504–3531.

59 Boche, H., Schaefer, R.F., and Poor, H.V. (2021). On the algorithmic solvability of channel dependent classification problems in communication systems. IEEE/ACM Transactions on Networking.

60 Boche, H. and Arendt, C. (2017). Communication method, mobile unit, interface unit and communication system. German Patent DE102 017 207 185A1.

61 Boche, H., Cai, M., and Cai, N. (2019). Message transmission over classical quantum channels with a jammer with side information: message transmission capacity and resources. *IEEE Trans. Inf. Theory* 65 (5): 2922–2943.

62 Boche, H., Cai, M., and Cai, N. (2020). Message transmission over classical quantum channels with a jammer with side information; correlation as resource and common randomness generating. *J. Math. Phys.* 61 (6): 062201.

63 Boche, H., Schaefer, R.F., and Poor, H.V. (2021). Real number signal processing can detect denial-of-service attacks. *Proceedings of International Conference on Acoustics, Speech, & Signal Processing*, Toronto, Canada.

64 Boche, H., Cai, M., Poor, H.V., and Schaefer, R.F. (2021). Algorithmic detectability of denial-of-service attacks on classical-quantum channels, in preparation.

65 Boche, H., Böck, Y., and Deppe, C. (2021). On the semi-decidability of remote state estimation and stabilization via noisy communication channels, arxiv, 2021.

66 Matveev, A. and Savkin, A. (2007). Shannon zero error capacity in the problems of state estimation and stabilization via noisy communication channels. *Int. J. Control.* 80: 241–255.

67 Boche, H. and Deppe, C. (2021). Computability of the channel reliability function and related bounds. arxiv:2101.09754.

68 Boche, H. and Deppe, C. (2020). Computability of the zero-error capacity of noisy channels. arxiv:2010.06873.

69 Zhou, X., Kyritsi, P., Eggers, P.C.F., and Fitzek, F.H.P. (2007). The medium is the message: secure communication via waveform coding in MIMO systems. *Proceedings of IEEE Vehicular Technology Conference*, Dublin, Ireland, 491–495.

Index

1-bit A/D converter usage 95
2T CO_2/year/person objective 64
3rd Generation Partnership Project (3GPP) 24–26, 28, 29, 31–34, 36, 101, 261
  Service and Systems Aspect (SA) specification group 102
4G cellular standards 187
5G core (5GC), NWDAF in 264
5G network 9, 25, 27, 29, 32–33, 188
  service-based architecture (SBA) 262
5G New Radio (NR) broadband standard for satellites 102
5G-non-terrestrial network (5G-NTN) 101–103
5G QoS flow 263
5G to 6G transition 2–3
6G data layer 221–232
6G early development 1–2
6G future service 2
6G major trend 8
6G network architecture 6–7
6G-non-terrestrial network (6G-NTN) 105–114
6G requirement 223–225
6G technology evolution 222–223
6G use case owner 227

a

Active networking 134–135
Active phased array power consumption *vs.* RIS 74, 75
Adaptive projected subgradient method (APSM) 248
Advanced network optimization and decision 205
Advanced self-organizing network (SON) 189
AI-aided method 44–46
AI-as-a-Service (AIaaS) support 268
AI-embedded RAN architecture 208–212
AI model inference service for xApp 212
Airborne network platform 107, 108
Air interface protocol processing function 213
Analogue beamforming (ABF) approach 92, 93
Angular power spectrum (APS) estimation, algorithm for 244–245
Anomaly detection (AD) 206
Anticipatory radio resource management 257
Application programming interface (API) 32–33, 137, 259–260
Application-Specific Integrated Circuit (ASIC) 103
Arbitrarily varying channel (AVC) 280, 285–287
Artificial Intelligence (AI) 17, 47–50, 95, 109, 112, 156–157, 168, 199, 255, 256, 261
Artificial neural network 238, 241, 242
Augmented reality (AR) 12, 17, 154, 173
Auto-labeling technique 266–268

Shaping Future 6G Networks: Needs, Impacts, and Technologies, First Edition.
Edited by Emmanuel Bertin, Noel Crespi, and Thomas Magedanz.

Automated driving (AD) 15
Automated multi-domain service production
 see also Multi-domain service
 frameworks and assumptions 170–173
 research challenges 183–185
Automatic anomaly analysis 206–207
Automatic data preparation 256
Automatic orchestration decision 256
Autonomic functions for networking 262
Autonomous network resource allocation 264

b

Base station (BS) 69, 110, 187, 244, 245, 252, 257, 264
 deployment, with directional transmissions 42–43
 energy consumption 40, 41
Beamforming 7, 42, 70, 72, 74, 79, 82, 104, 109, 111, 112, 252
 analogue 92
 digital 92
 hybrid 92
 M-MIMO 257, 258
 non-orthogonal multiple access and 248–250
Beam hopping, in satellite 104
B5G network architecture 69
Bilateral agreement, for multi-domain service delivery 174–175
Blockchain 142
Blum–Shub–Smale machine 287–288
Border Gateway Protocol (BGP) 177, 178
Broadband access, demand for 69
Broadband baseband technology access 95
Business and Operation Support System (BOSS) 200, 201
Business-to-business (B2B) market, use cases for 12–13
 digital twin 15
 financial world 20
 health and well-being 17–19
 industry and manufacturing 11, 13
 public safety 16–17
 smart transportation 15–16
 Smart-X IoT 19
 teleportation 13–14

c

CDMA2000 25
Cellular network 1, 5, 11, 13, 17, 55, 56, 65, 102, 104, 112, 127, 244, 252
Centralized edge training 48
Centralized Service Management and Orchestration (SMO) system 256
Centralized unit data analysis (CUDA) function 200–202
Channel estimation, of RIS-enabled networks 83
Channel modeling and simulation, for NTN 109
China Mobile Communications Corporation (CMCC) 207
Citizen Broadband Radio System (CBRS) 111
Cloud-based computing 136
Cloud-based wireless networking system 194
Cloud computing and virtualization 197–198
CO_2 emission reduction 56–57, 59–61, 64–65
Coexistence technique 111–112
Collaborative compressive classification, in wireless network 245–247
Collective perception message (CPM) 15
Common Public Radio Interface (CPRI) 191–194
Common randomness (CR) 273, 274, 281–282
Compute-first networking (CFN) 135–136
Computer-aided system design, limits for 288
Computing and networking 133–134
Computing-in-the-network system, 133–134
 see also Compute-first networking (CFN)
Consensus management 151–152
Consistency management 151–152
Content-centric networking (CCNx) 135

Convolutional neural network (CNN) 153–154
Cooperative Intelligent Transport Systems (C-ITS) 15
Core-edge continuum 143–148
Core network management 260–265
Cost, size, weight and power consumption (C-SWaP) design 76
Co-training process 267
COVID-19 pandemic
  and media evolution 118
  video analytics, need for 154
COVID-related economic crisis of 2020 57
C-RAN 190–195
CubeSat satellite 105
Customized data collection and control, of nRT-RIC 212–213
Cyberattack security 119
Cyclic prefix orthogonal frequency division multiplex (CP-OFDM) 112–113

d

Data and flow aggregation, for on-path computing 150
Data Center (DC) 63
  computing in 150–152
  and edge network 133
Data consistency, in multi-domain service 184–185
Data convergence 226
Data curation 223, 226, 230–231
Data discovery 231
Data discovery, identity, trust 224
Data-driven service deployment 138
Data firewall 230
Data layer high-level functionality 227–231
Data orchestrator 231
Data privacy 223
  challenges, in RAN 200
  and security preserving algorithms 141–142
Data processing, in RAN 200
Data selection 231
Data storage and parallel processing 256
Data stream establishment 231
Data stream orchestration 224
Decoupled User/CP 204
Deep learning, 44, 46, 238–241 *see also* Machine learning (ML)
Deep neural network 235
Deep reinforcement learning (DRL) 44
Denial-of-Service (DoS) attack 284–288
Deterministic identification (DI) code 274, 280
Deterministic networking (DetNet) 261
DiffServ Code Points (DSCP) 180–181
Digital beamforming (DBF) approach 92, 93
Digital minimalism 62
Digital Sobriety 5
Digital sovereignty 170, 180
Digital Twin 14, 169
Disaggregation principle 7
Distributed denial-of-service (DDoS) attack 120
Distributed file system 139
Distributed trust system 139–140
Distributed unit data analysis (DUDA) function 200–202
Dynamic 6G network 107
Dynamic multi-domain service parameter negotiation 176

e

Eco-conscious networking, power savings for 141
Ecological crisis 56–57
Edge AI 263
Edge co-inference 49–50
Edge computing 39, 41
  evolution of 119
Edge inference 47, 49
Edge learning, energy consumption models in 41–42
Edge training 47
  energy-efficient 48–49
Electromagnetic (EM) metamaterial 70
Electromagnetic wave propagation at THz 94
Embedded RAN intelligence 199–202
End-to-end network management 260–265

End-to-end principle 137–138
Energy consumption models 41–42
Energy consumption reduction incentive 59–60, 64–65
Energy crisis 57
Energy efficiency 56, 65–66
Energy-efficient edge training 48–49
Energy-efficient model aggregation policy, accuracy of 49
Energy-efficient network planning 42–43
Energy-efficient THz semiconductor 94
Energy-efficient waveform 112–113
Energy overconsumption reduction incentive 60–61
Energy spreadsheet, of 6G network 40–42
Enhanced mobile broadband (eMBB) service 27
Environmental crises 56
Event-triggered decision consistency, in multi-domain service 184
Evolved License Shared Access (eLSA) 111
Extended Community BGP attribute 177
eXtended Reality (XR) 16, 149
  and multimedia 154–155

f
Federate average (FedAVG) 265–266
Federated Learning (FL), 48, 263, 265–266 *see also* Machine Learning (ML)
FedSGD 265
FG-30 (Focus Group on Technologies for Network 2030) 24
Field-programmable gate array (FPGA) technology 103
Financial world 19
Fingerprinting-based localization 244–245
Firewall 224, 230
Fixed antenna array 92
Flexible addressing system 124, 127–129
FlexNGIA (Flexible Next-Generation Internet Architecture) 148
Fog computing 194, 195
Fog-RAN 194, 195
Forecast scheduling 257

g
GAIA-X 126
Gateway 142, 144, 145, 152, 153, 229–230
General Data Protection Regulation (GDPR) 120
Geometric scattering 70
GEO satellite 103, 105
gNB (next-generation NodeB) 102
GPP-based hardware 215
Graceful data exchange 223–224
Gradient compression technique 48
Green Network 188, 189
Ground-based network platform 107, 108
GSMA Open Networking Initiative 216

h
Health and well-being 16–18
Heterogeneous 6G network 107, 108
Heterogeneous network interconnection 129
Heterogeneous non-terrestrial 6G network 109
High-altitude platform (HAP) 101
High data transmission cost, in RAN 200
High-gain antenna 92, 93, 95
High-precision and deterministic service 122–124
High-risk vendor (HRV) 28, 29
High throughput protocol 126–127
High throughput satellite (HTS) 103
Holographic communication 118
Homogenizable RIS 72, 73
Hybrid beamforming approach 92, 93

i
ICE-AR project for advanced augmented reality (AR) 135
Industrial automation 153–154
Industrial Ethernet technology 119–120
Industry and manufacturing 12
Inexpensive antenna, arrays of 78–80
Information-centric networking (ICN) 134, 135
Information, Communication, and Data Technology (ICDT) deep convergence embedded 6G networks 198–214

Information protection 226
Infotainment 14–15
Inhomogenizable RIS 72, 73
In-network caching 151
In-network computing 135, 136, 139–141, 143, 148, 149, 151, 159
Instantaneous feedback, and message transmission 274
Intelligence, 199 *see also* Embedded RAN intelligence
Intelligence functionality 229
Intelligent reflectors, applications of 70
Intelligent transport service 169
Inter-domain virtual private network (VPN) service 167, 172–173
Interface of business owner's computation logic 179, 180
International Mobile Telecommunications (IMT) 24
  and generations of mobile network technology 25, 26
International Telecommunication Union (ITU) 24, 256
Internet of Intelligent Things 152–153
Internet of Things (IoT) 1, 6, 10, 19, 23, 47, 82, 105, 111, 112, 114, 120, 127, 128, 133, 135, 139, 142, 144, 145, 152, 154–157, 227, 255, 256
Internet Protocol (IP) 117, 129
  challenges 120, 121
  next-generation 122–127
IPv4 Address-Specific Extended Community attribute 177
IPv6 Address-Specific Extended Community attribute 177
Iridium Next 103
Iterative APSM-based algorithm 248, 249
Iterative shrinkage thresholding algorithm (ISTA) 247
ITU-T Focus Group Technologies for Network 2030 106

j

Joint Communication Sensor System (JCSS) 91
Joint computation and communication resource management 49
Joint power and beam optimization 245

k

Kernel-based learning method 235, 237–238
Key performance indicators (KPIs) 21, 22
Key-value (K-V) storage 151

l

LEO constellation 103, 105
Locator/ID Separation Protocol (LISP) 261
Long-term evolution (LTE) 25
Low phase noise and THz system 95
LTE Cat-M 105
LTE Cat-NB (NB-IoT) 105
Lyapunov-based device scheduling policy 49

m

Machine Learning (ML) 112, 199, 205, 235, 251–252, 256
  methods in RAN 237–242
  model 47
  for network management 156–157
  in wireless network 236, 243–251
Machine-to-machine communication 119
Management data analytics function (MDAF) 264–265
Management data analytics service (MDAS) 262
Mandate-driven architecture (MDA) 262
ManyNets support 125–126
Market representation partners 26
Massive IoT (MIoT) 27
Massive machine type communication (mMTC) 10, 105, 106
Medium Access Control (MAC) layer, AI/ML technique in 257
Message identification
  feedback 282–283
  for post-Shannon communication 273–281
  with random encoding 275, 276
  without randomness 280–281
Message transmission and communication 271, 272

Message transmission *vs.* identification 276, 277
Metamaterial-based RIS implementation 80
Metasurface-based RIS 72, 74
Metawave-implemented solution 71
Middleware 225, 229
MIMO beamformer *vs.* RIS and phased array 74, 75
MIMO systems, waveform coding in 289
Minimalism in feature 60
Minimalism in scale 60
Minimalism in use 60
Minimalist phone 63
Mobile edge computing (MEC) 47, 195–197
Mobile load balancing *see* Traffic steering
Mobile WiMAX 25
MobilityFirst 128
Monte Carlo tree search (MCTS) 242
Multi-codecode-division multiple access (MC-CDMA), post-Shannon communication in 288–289
Multi-domain 6G network, need for 168–169
Multi-domain service 171
 delivery automation 175–181
 delivery framework 174–175
 roles of and interactions between tenants of 173, 174
 subscription framework 178–179
Multimedia 118
 computing support for 154–155
Multiple high-gain antenna 92
Multi-RAT cooperation energy-saving system (MCES) 206

n

Named data networking (NDN) 135
Nanosatellite, 105 *see also* CubeSat satellite
National and international regulation, sustainable 6G 65
Near-real-time QoE optimization 207
Near-real-time RAN intelligent controller (nRT-RIC) 208–209, 211–212, 259
 "condition + action" policy 213
 customized data collection and control 212–213
 pure control instruction 213
NetCache system 151
Network actuation 267
Network architecture 6–7
Network data analytics (NWDA) 199, 201, 202
Network data analytics function (NWDAF) 262, 264, 265
Network function virtualization (NFV) 46, 197
Network Index Address (NIA) 129
Network infrastructure technology 5–6
Network intelligence 260
 openness and support of 256
Network Slice-as-a-Service (NSaaS) service 173
Network slicing paradigm 260–261
Network virtualization (NFV) 137
Next-generation distributed intelligent systems 138
Next-generation fronthaul interface (NGFI, xhaul) 191–194
Next-generation Internet Protocol 122–127
Next-generation IoT and Intelligence 152–154
Next-generation NodeB (gNB) 192, 194, 199
N23 interface 202
Non-orthogonal multiple access (NOMA) 111
Non-orthogonal multiple access and beamforming 248–250
Non-public network operator 227
Non-real-time RAN intelligent controller (NRT-RIC) 208–210
Non-real-time RAN intelligent controller, of O-RAN 112
Non-RRM xApp 212
Non-RT RAN intelligent controller (NRT-RIC) 259
Nonsecure communication, common randomness for 281–282
Non-terrestrial network (NTN)
 in 5G 101–103
 in 6G 105–114

NRT-RIC framework 209, 210
NTN *see* Non-terrestrial network (NTN)

o
O-Cloud 260
On-board processor (OBP) 103–104
Online loss map reconstruction 248, 249
On-service orchestrator 172
Open and distributed radio access network management 256–260
Open API for rApp 209
Open API for xApp 214
OpenFlow 137
Open Generalized Address system 128, 129
Open Interface 204
Openness, defined 199
Open Network Automation Platform (ONAP) 265
Open Networking Initiative 216
open radio access network Alliance (O-RAN Alliance) 31
Open radio access network (O-RAN) standard 112
OpenRAN 202–205
Open RAN Policy Coalition 31
OpenRAN project 215
Open reference design (white box) hardware 203
Open-source communities 216–217
Open-source projects, for edge and radio access network 216, 217
Open-source Software 203, 205
Open systems interconnection (OSI) model 120
Open Testing Framework (OTF) 216–217
Open *vs.* traditional RAN interface 202
Optical transport network (OTN) technology 124
Optimal communication, Federated Learning in 265–266
O-RAN Alliance 214–215, 256–258
O-RAN Alliance architecture 204
O-RAN Central Unit–Control Plane (O-CU-CP) 259
O-RAN Central Unit–User Plane (O-CU-UP) 259
O-RAN Distributed Unit (O-DU) 259
O-RAN evolved Node B (O-eNB) 260
O-RAN Radio Unit (O-RU) 260
O-RAN Software Community (OSC) 216, 217

p
P4 approach 127
Paxos 151–152
Peak-to-average power ratio (PAPR), of transmission waveform 112
Performance and scalability, in multi-domain service 185
Phased array *vs.* MIMO beamformers and RIS 74, 75
Photonic microwave oscillators 95
Physical (PHY) layer, AI/ML activity in 257–258
Physical layer channel optimization modeling 199
Physical layer service integration 283–288
Policy control function (PCF) 199
Post-Shannon perspective 8
Post-Shannon theory of communication 272
Prediction-based radio resource management optimization 199–200
Privacy-preserving protocol 126
Proactive network resource adjustment 264
Programmable data plane 145–147
Programming Protocol-Independent Packet Processors (P4) language 146–147
Protocol-independent switch architecture (PISA) 146
Pseudonymization technique 126
public safety communication 15–16

q
Quality of Experience (QoE) 154, 159, 189, 255, 263–264
Quality of Service (QoS) 6, 16, 44, 74, 111, 122, 141, 159, 196, 242, 256, 261, 263–264
Quantum networking 148
Queueing theory 42
QUIC (Quick UDP Internet Connections) protocol 133

r

Radiative transfer simulation 250–251
Radio access network (RAN) 42–47, 102, 196
  6G-NTN 109–110
  machine learning methods in 237–242
  management, open and distributed 256–260
  revolutionary evolution of 187, 188
Radio access network big data analysis network architecture (RANDA) 200–202
Radio access network energy consumption model 40–41
Radio access technology (RAT) 8, 12, 40, 107
Radio application (rApp) 209
Radio Environment Map (REM) 257
Radio fingerprint-based traffic steering 207
Radio Resource Control (RRC) layer, AI/ML technique in 257
Radio resource management (RRM) 213
Radio resource management, energy-efficient 44–46
Railway network 232
Randomized identification (RI) code 273–274, 277–279
Random linear network coding (RLNC) 157–158
RAN intelligence
  embedded 199–202
  use cases 205–208
RAN Intelligent Controller (RIC) 208–212
RAN Virtualization, Modularity, and Open-source Component 2041
Real-time (RT) RAN intelligent controller, of O-RAN 112
Rebound effect 56–59
Reconfigurable Intelligent Surface (RIS) 43, 70–71, 258
  vs. active phased array power consumption 74, 75
  advantages and limitations 74–78
  vs. phased array and MIMO beamformer 74, 75
  research areas and challenges 82–83
  size of 76, 77
  types 72–74
Reconfigurable payload 110
Regenerative on-board processor 104
Regenerative payload 110
Regulatory aspects, in spectrum sharing 111
Reinforcement learning 241–242, 250–252, 268
Reinforcement learning-based node clustering method 44
Release 17 5G-NTN technology 102, 103
ReLU-based feedforward neural network 240, 241
Renewable energy utilization 47
Reproducing kernel Hilbert space (RKHS) 237–238
Requirements of 6G data layer
  data curation 223
  data discovery, identity, trust 224
  data firewalling 224
  data privacy 223
  data streams orchestration 224
  graceful data exchange 223–224
Research area and challenges, of RIS implementation 82–83
Research challenges
  automated multi-domain service delivery 183–185
  6G-NTN 107–114
Resource discovery, computing and networking 140
Resource optimization problem, in RIS 82–83
RESTful API 33
RFocus 78–79
RIS *see* Reconfigurable Intelligent Surface (RIS)
Robotic telesurgery 17
Robust traffic prediction, for energy-saving optimization 244
RRM xApp 212

s

Satellite-based network 101
Satellite network 5

Satellite parameter, for system-level calibration 113, 114
Scalable radio frequency carrier bandwidth 113–114
ScatterMIMO prototype 79–80
SDN-enabled domains 171–172
Sector selection with high-gain antenna 92, 93
Secure aggregation, Federated Learning in 266
Secure communication 279–280
Security and privacy, IP-based network 126
Security of message identification with RI code 274
Security of operation, in multi-domain service 184
Segment routing based on IPv6 (SRv6) 261
Self-labeled technique 267
Self-Organizing Network (SON) 3, 189
Self-training process 267
Semantic addressing 124–125
Semantics and syntax 271
Semi-supervised learning 237
Semi-supervised method 267
Service-based architecture (SBA) 33, 262
Service-based interface (SBI) 33
Service function chaining (SFC)-based multi-domain service traffic forwarding 181–183
Service function chaining (SFC) migration 46, 47
Service-level agreement (SLA) 173, 255
Service Management and Orchestration (SMO) framework 209
Service provisioning with NFV and SFC 46–47
SFC computation logic 181–182
Shift Project 63
Single-domain service 171
Smart radio environment 70
Smart transportation 14–15
Smart-X IoT 18
Soft Network 188–189
Software and equipment provider 227
Software-defined CP 204
Software-defined networking (SDN) 127, 136–137, 144, 145, 261, 282
Software defined radio (SDR) 284
Software-defined radio-based (SDR-based) software 215
Softwarization of networking 137
Spaceborne network platform 107, 108
Sparse estimation, neural architecture design for 247–248
Spectrum sharing technique 110–112
Standard development organizations (SDOs) 26
Standardization 23, 24, 34, 35
  and politics 28–30
Standardized interface 259–260
Standards developing organization (SDO) 122
Standards, 3GPP 31, 32, 34
Starlink constellation 103
Stochastic gradient descent (SGD) 48
Stochastic gradient descent and backpropagation method 241
Structure pruning 49
Subscription framework, multi-domain service 178–179
Supervised learning 237
Sustainable 6G
  CO_2 emission reduction 56–57, 59–61, 64–65
  digital minimalism 62
  ecological crisis 56–57
  energy consumption reduction incentive 59–60, 64–65
  energy crisis 57
  energy efficiency 56, 65–66
  energy overconsumption reduction incentive 60–61
  environmental crisis 56
  minimalist phone 63
  national and international regulation 65
  opportunities and risk 61–62
  rebound effect 56–59
  Shift Project 63
  status quo 62–63
  2T CO_2/year/person objective 64

Sustainable 6G (*cont'd*)
technological innovation 57–59
value creation 63–64
Syntax and semantics 271
Systemic view, of Federated Learning 266

t
Technological innovation 57–59
Telecom Infrastructure Project (TIP) 215
Telecom satellite innovation 103–105
Telemedicine service 169
Telemetry 155–156, 262
Teleporting 12–13
Tenant 171
multi-domain service, roles of and interactions between 173, 174
Terahertz (THz) access network 5
Terahertz (THz)-band communication 41
Terahertz (THz) communication 5
Terahertz (THz) system challenges 91
Terahertz (THz) wireless communication 89
Time-sensitive networking (TSN) 124, 261
Traffic steering, radio fingerprint-based 207
Transmit distance, of RIS 76–78
Transparent payload 110
Transport layer protocol 141
Turbocharged edge 195–197
Turing machine model 284–285
Type I and type II errors, in identification decoding 276, 277

u
Ultrareliable low-latency communication (URLLC) 10, 12, 27, 39, 106
Universal Mobile Telecommunications System (UMTS) 25
Unlicensed spectrum band 111
Unsupervised learning 237
User-defined network operation 127

v
Very high throughput satellite (VHTS) 103
Virtualization and cloud computing 197–198
Virtual network function (VNF) consolidation 46, 47
Virtual reality (VR) technology 12, 17, 154, 173

w
Waveform coding, in MIMO systems 289
White box hardware 205
Wireless big data (WBD) 199–202
Wireless network, machine learning in 236, 243–251
Wiretap channel model 279, 280

x
x-application (xApp) 211–212
XR *see* eXtended Reality

z
Zero-touch network and service management (ZSM) end-to-end architecture framework 262

Printed and bound by CPI Group (UK) Ltd, Croydon, CR0 4YY